SCANNING ELECTRON MICROSCOPE OPTICS AND SPECTROMETERS

SCANNING ELECTRON MICROSCOPE OPTICS AND SPECTROMETERS

Anjam Khursheed

National University of Singapore, Singapore

World Scientific

NEW JERSEY • LONDON • SINGAPORE • BEIJING • SHANGHAI • HONG KONG • TAIPEI • CHENNAI

Published by

World Scientific Publishing Co. Pte. Ltd.

5 Toh Tuck Link, Singapore 596224

USA office: 27 Warren Street, Suite 401-402, Hackensack, NJ 07601

UK office: 57 Shelton Street, Covent Garden, London WC2H 9HE

British Library Cataloguing-in-Publication Data
A catalogue record for this book is available from the British Library.

SCANNING ELECTRON MICROSCOPE OPTICS AND SPECTROMETERS

ISBN-13 978-981-283-667-0
ISBN-10 981-283-667-5

Printed in Singapore.

Dedication

I would like to dedicate this book to three people.

The first dedication goes to Alan Dinnis, who supervised my Ph.D between the years 1979 to 1983 at the University of Edinburgh in Scotland. It was Alan who first introduced me to the challenge of how best to combine electron energy spectrometers together with scanning electron microscopes, and much of the content in this book is the fruition of work we started together over 30 years ago.

The second dedication goes to my wife Francesca, thank you for your support, patience and understanding.

The third dedication goes to my three year old daughter Elettra, this just to wish you well, and thank you for being a part of my life.

Anjam Khursheed

Preface

The subject matter of this book deals with the question of how to combine electron energy spectrometers with scanning electron microscopes (SEMs). For approximately three and half decades, starting from around 1968, this question attracted considerable research interest, fuelled primarily by the application of testing integrated circuit devices. After this period, due to changes in the way integrated circuits were manufactured, direct electron beam testing was no longer viable, and general interest in the subject dwindled. However, over the last 15 years or so, there is evidence of renewed interest. This is because, more applications, beyond the testing of integrated circuit devices are coming to light. Combining electron energy spectrometers with SEMs provides a way of obtaining depth information for multi-layer samples (Rao and Robinson, 1996), mapping the dopant concentration across p-n junctions (Kazemian *et al.*, 2006), monitoring negative charging of insulator samples (Belhaj *et al.*, 2008), and acquiring elemental identification on the nano-scale (Hoang and Khursheed, 2009).

Spectrometers are used in a wide range of different charged particle beam instruments, ranging from electron microscopes, mass analyzers and particle accelerators. Although this book focuses on their use in the SEM, ideas and methods are borrowed from their use elsewhere. There is in particular, a natural synergy with Scanning Auger Microscopy (SAM), which also combines a scanning focused electron beam with electron energy spectrometers. Both techniques share much in common. This book aims to present spectrometer developments in the context of the SEM to a wide audience of research scientists, with the intention of not only improving SEM design, but also of enriching the design of other focused electron/ion beam instrumentation, as well as contributing to the design of spectrometers in general. The following pages are aimed at scientists, engineers and graduate students whose research area or study in some way involves the use of focused electron/ion beam instrumentation and/or charged particle spectrometers. It can be used both as an introduction to these subjects, as well as a guide to more advanced topics about SEM redesign.

The book is mainly a book of ideas. It emphasizes the concepts underlying design of SEMs and spectrometers, as opposed to providing practical engineering details. Wherever possible, spectrometers are discussed under one common framework, so that their relative strengths and weaknesses can be more readily appreciated. This will be done for the most part, through simulations and derivations carried out by the author himself. If not explicitly stated, the reader can assume that the programs used for simulation were written by the author. A detailed description of the numerical methods underlying the author's simulation programs can be found elsewhere (Khursheed, 1999).

Lastly, I would like to acknowledge help from various members of my research group. My postdoctoral fellow, Mans Osterberg, carefully proof-read most the manuscript and gave many detailed helpful suggestions. He carried out simulations of a quadruple-octupole aberration corrector and provided core information from which Chapter 4 (Aberration Correction) and Appendix 2 (Introduction to Multipoles) were written, a big thank you for all your help Mans. Help also came from my Ph.D student, Hung Hoang, who performed 3D simulations of a SE detector in a conventional SEM, and carried out simulations of various spectrometer designs, including the second version of my second-order focusing toroidal analyzer, a novel modified fountain analyzer design of his own, the cylindrical mirror analyzer, and the spherical deflector analyzer. He also carried out important experimental work. Particularly appreciated are the secondary electron/backscattered energy spectra from the second-order focusing toroidal analyzer prototype reported in Chapter 6. A Masters student, Saravanan

Musuwathi Kesavan, helped with the analytical derivations of field distribution expansions, paraxial ray equation, and on-axis aberration coefficients presented in Appendices 1.0–1.3. An undergraduate student Zhang Teng helped in the simulation of the parallel-plate/fountain and hemispherical analyzers, and my long standing research engineer, Karuppiah Nelliyan, provided many experimental SEM images.

Contents

Chapter 1

Conventional SEM Design

The Scanning Electron Microscope (SEM) is undoubtedly one of the most important electron beam technology inventions of the twentieth century. Ever since the first commercial instrument appeared in the early 1960s, the design of the scanning electron microscope has been constantly evolving and improving.

Recognition that the SEM was much more than a powerful microscope came early. In addition to providing high resolution topographical images of a specimen's surface, it was realized that the SEM could also provide a variety of different contrast modes, where information relating to things like surface voltage is embedded in its output signals. In many cases, in order to quantify these contrast mechanisms, electron energy spectrometers are required. The purpose of this book will be to summarize electron spectrometer designs that have in one form or another been proposed or tried out in the SEM over the past 50 years, and point the way for future development.

Before understanding how spectrometers relate to SEMs, it is first important to describe how the SEM functions. In this chapter, the basic layout of a conventional SEM and its underlying principles will be summarized. Special attention will be given to basic concepts in electron optics.

1.1 Introduction to the SEM

Figure 1-1 shows a schematic drawing of a typical SEM. The SEM column consists of an electron gun, two condenser lenses, an objective lens, an electron detection system, and a set of deflectors, all operating in a vacuum. The electron gun provides a source of electrons and accelerates them to energies in the range of 1 to 30 keV. The smallest beam cross-section at the source, called the "crossover", is subsequently demagnified by the three-stage electron lens system, so that an electron probe of diameter 1 to 10 nm carrying a current of 1 to 100 pA is formed at the specimen surface. For higher currents, 1 to 10 nA, the electron-probe diameter increases to 0.1 to 1 μm. In most SEMs, the electron beam emerges from the final lens into the specimen chamber, where it interacts with the near-surface region of the specimen to a depth of approximately 1 μm and generates signal electrons used to form an image.

A deflection system in front of the final lens scans the electron probe in a raster-like fashion across the specimen and operates in synchronism with a computer display monitor, or cathode-ray tube (CRT). Two pairs of magnetic deflection coils are usually used. The image magnification M_I of the specimen is defined to be the ratio of the linear size of the viewing screen to the linear size of the raster on the specimen. The magnification can be increased simply by decreasing the scan-coil current, keeping the image size on the CRT constant.

Electron detectors are used to collect various kinds of signals generated from the primary beam/specimen interaction. Figure 1-2(a) depicts a schematic representation of this process. As the primary beam penetrates the specimen surface, it scatters electrons from a range of different depths, some of which escape from the surface. Scattered electrons that escape from close to the specimen surface are known as secondary electrons, and are created by inelastic collisions. Electrons that are scattered back from deeper levels are known as backscattered electrons, and they are generated by multiple elastic collisions. Secondary and

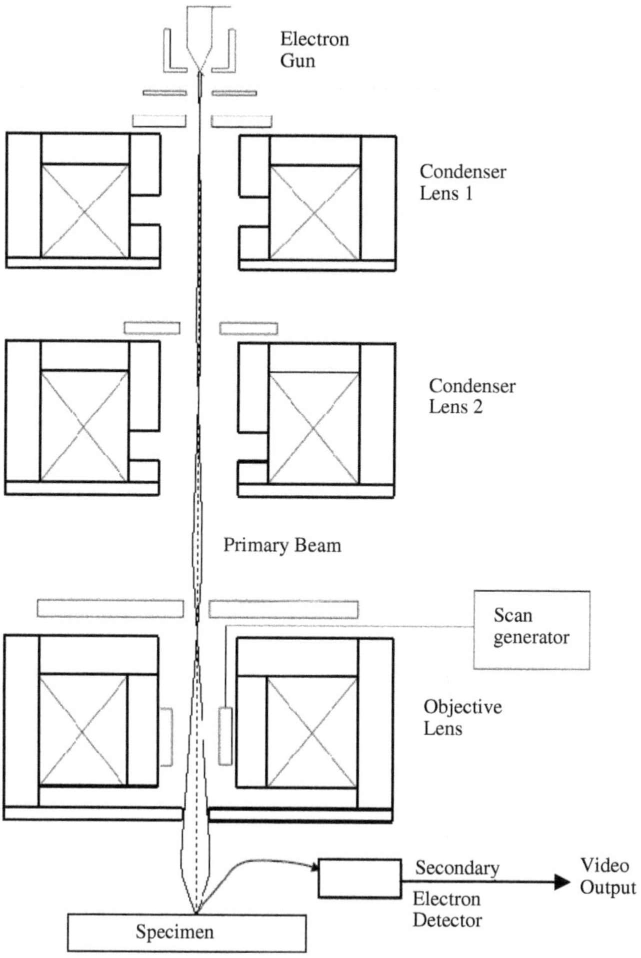

Fig. 1-1. Schematic drawing of a SEM.

backscattered electrons make up the two most common signals that are used to form the SEM image.

Figure 1-2(b) depicts the form of the energy distribution of the electrons that are scattered back from the sample. Secondary electrons have low energies, and by convention are defined to be those electrons that have

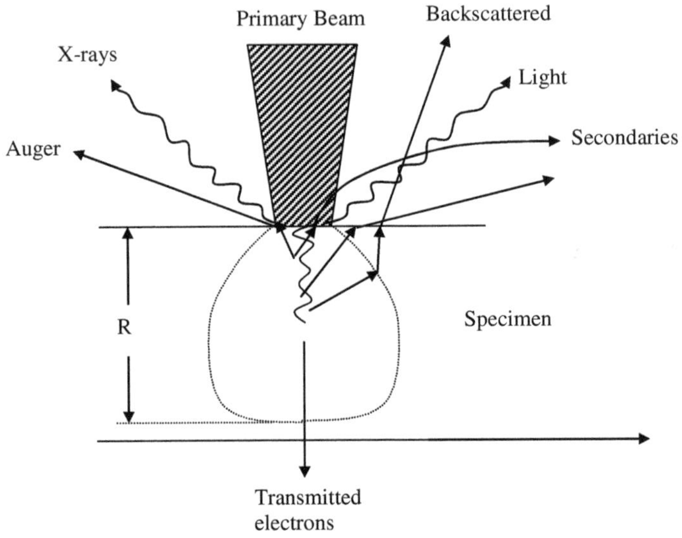

(a)

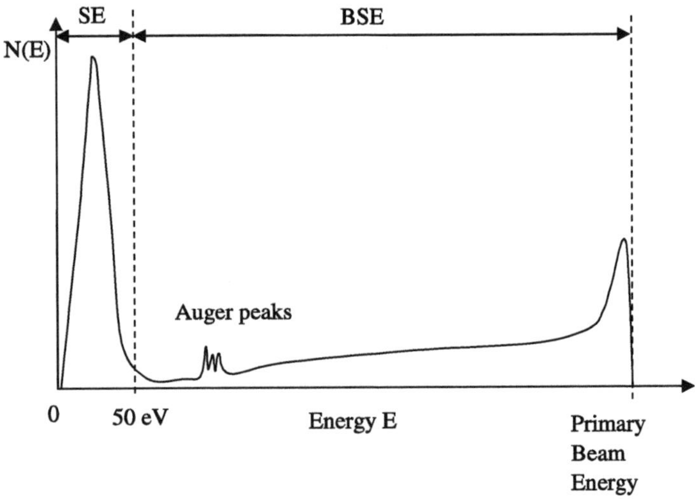

(b)

Fig. 1-2. Beam/Specimen interaction: (a) Interaction volume and emitted signals, (b) Energy spectrum of electrons that leave the specimen.

energies less than 50 eV. Most of them however, lie in the 0.5 to 5 eV energy range. Backscattered electrons are defined to lie in the broad energy range of 50 eV up to the primary beam energy. Auger electrons emanate from inner atomic shells, and their energies show up as characteristic peaks in the energy spectrum.

It is important to understand the way the primary beam is scattered inside the specimen, since it affects the final image resolution and signal yield. Monte Carlo simulation programs have been developed for this purpose. Figures 1-3(a) and 1-3(b) show Monte Carlo program plots of primary beam scattering in silicon. The primary beam energies of 1, 5, 10 and 20 keV are used with a specimen thickness of 50 nm and 1 μm. These results were generated from a program called McforVB written by Joy (1995). They show that both the number of scattered electrons that escape from the surface and their area of emission is critically dependent on the specimen thickness and primary beam energy. For thin specimens (less than 50 nm) at high primary beam energies, the area of emission for backscattered electrons is much smaller than it is for bulk specimens (> 1 μm). This is because a substantial number of primary beam electrons are transmitted through the specimen. Also apparent from these results is the reduction of the emission area as the primary energy is decreased. At 1 keV, the width of the emission area is as small as several nano-meters, while for 20 keV, it is close to a micron. This has considerable importance for high resolution images. The SEM resolution at high beam energies on bulk specimens will be limited by the interaction volume, where the depth of the interaction volume, R, can easily exceed the diameter of the electron probe. Nano-meter resolution at high energies can only be obtained on relatively thin specimens (less than 50 nm thickness). On the other hand, at low primary beam energies, the size of the interaction volume is considerably reduced, and is generally comparable to the minimum probe size of the primary beam. Also important but not shown in the results presented in Figs. 1-3(a) and 1.3(b), is the material type. In general, the higher the atomic number, the shallower the interaction volume.

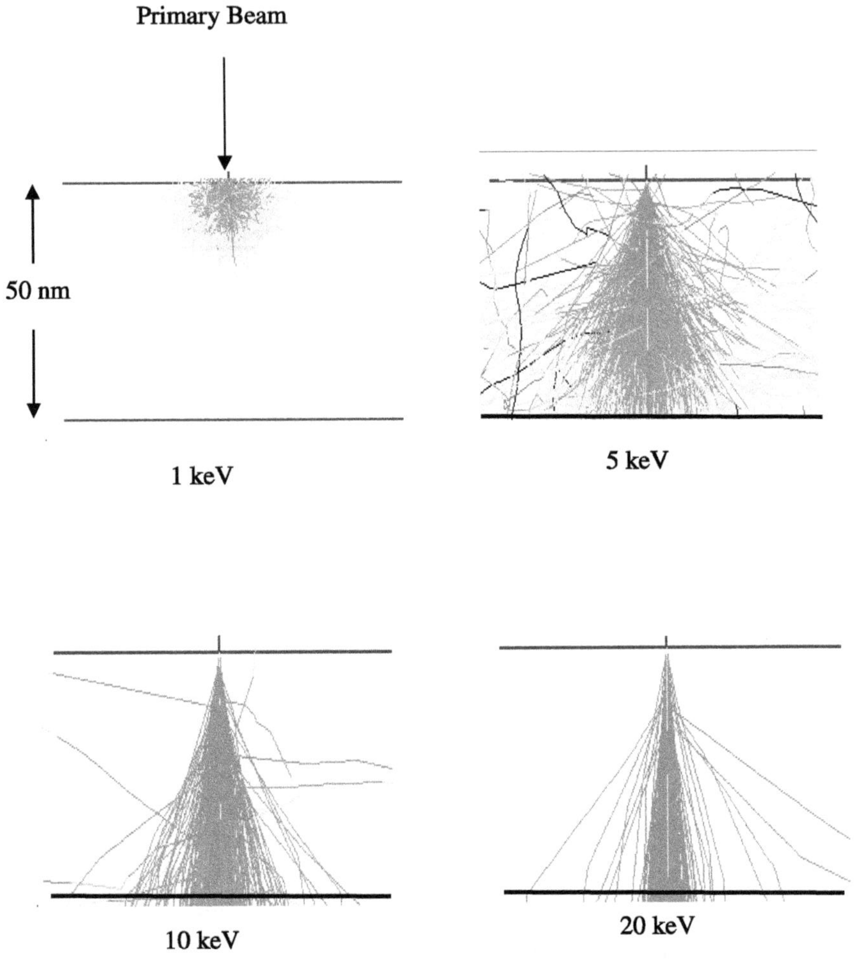

(a)

Fig. 1-3. Monte Carlo simulation of primary electrons striking a silicon sample at energies 1, 5, 10 and 20 keV: (a) 50 nm thick sample, (b) 1 μm thick sample.

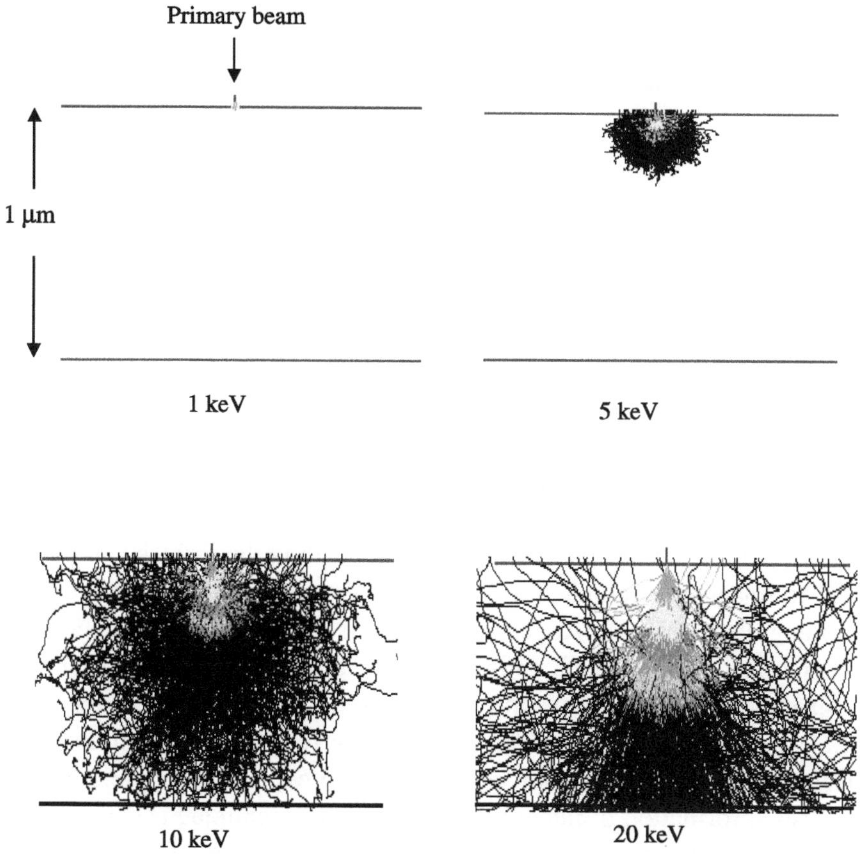

(b)

An approximate formula for estimating R has been given by Kanaya and Okayama (1972)

$$R = \frac{0.0276AV^{1.67}}{Z^{0.889}\rho} \quad (\mu\text{m}) \tag{1.1}$$

where V is the primary beam voltage, A is the atomic weight, ρ is the mass density, and Z is the atomic number. For silicon, R = 0.03 μm at 1 keV, and 4.72 μm at 20 keV.

There are a variety of other signals that can be imaged in a SEM (Reimer, 1998a). The primary beam/specimen interaction generates photons, and in the X-ray range, this allows for elemental identification on the specimen through a technique known as Energy Dispersive Spectroscopy (EDS). Emitted light, generally known as cathodoluminescence, can be used to image semi-conductor defects and is useful in the failure analysis of integrated circuits. Auger electrons also provide material contrast information. They need, however, an ultra high vacuum (UHV) environment to be useful, and are generally not detected inside most SEM specimen chambers. The electrical current created from absorbed electrons into the specimen can be monitored, and can yield failure analysis information of integrated circuits. This technique is called EBIC (Electron Beam Induced Current). For thin specimens, the transmitted electrons through the specimen can be imaged, and help to understand the defects and basic crystal properties of many materials. Voltage contrast, a technique that can monitor the specimen electric potential by analyzing the energies of the SE electrons, can also be used for integrated circuit specimens (Thong, 1993).

EDS is a standard attachment that can be fitted on most conventional SEMs. Although it is a very powerful technique, widely used in SEMs, it has some notable limitations, such as the primary beam energy having to be typically between 10 to 20 keV, and the lateral resolution being relatively low, usually lying between 1–5 μm (Goldstein *et al.*, 2003). These limitations come from the fact that X-rays are generated deep within a comparatively large interaction volume. The standard EDS attachment therefore cannot operate at low primary beam voltages (< 5 kV) and has considerably poorer spatial resolution than the SEM operating in its normal topographical mode (by over an order of magnitude). Another limitation of EDS is that it has considerably reduced sensitivity for lighter elements. For this reason, EDS is limited to probe materials that have an atomic number of 4 and above.

1.2 Basic principles of electron optics

The beam of electrons that emerges from the gun inevitably has a certain size in the transverse plane associated with it (perpendicular to their direction), known as the source size. Each electron lens demagnifies this source size by a factor of typically 10 to 40. The total demagnification, from the source to the specimen in a three-lens SEM system lies typically between 1,000 to 64,000. The objective lens, not only demagnifies the transverse electron beam size, but also imposes aberration limits on the spatial resolution of the SEM. To a good approximation, the condenser lenses only change the size of the projected source at the specimen.

A magnetic electron lens focuses an electron beam by introducing a gap in a magnetic circuit on the optic axis, creating a leakage field where the beam passes. The lens is axisymmetric, where the coil current flows in the azimuthal direction and the flux path is defined in the *r*-*z* plane. The flux lines and the axial fringing field distribution for a three-lens SEM layout are shown in Fig. 1-4. The magnetic flux of a coil is concentrated within a small volume by iron pole-pieces and the stray field across the gap forms the magnetic field of a bell-shaped distribution on the optical axis. As the result of the Lorentz force, off-axis electrons will travel in a screw trajectory (due to the radial component B_r) and curve towards the optic axis and cross it (due to the axial component B_z).

The condenser lens is usually air-cooled since it is a relatively weak lens and the heat generated by the modest current flowing through it is easily dissipated. Since the objective lens is the strongest lens in the SEM, with

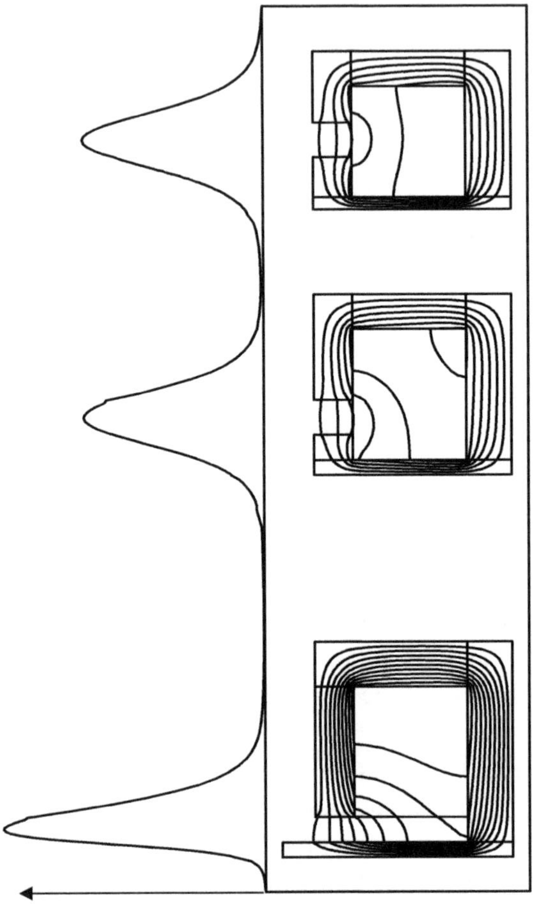

Fig. 1-4. Flux lines and magnetic axial field distribution of an SEM column.

the largest current, flowing through its windings, it is usually liquid-cooled. The design of this lens often allows for space to house the scanning coils, the stigmator, and the beam-limiting aperture.

The general relativistic equation of motion for an electron moving through electric and magnetic fields is given by

$$\frac{d}{dt}\left\{\frac{m\mathbf{v}}{\left(1-\mathbf{v}^2/c^2\right)^{1/2}}\right\} = -e(\boldsymbol{E} + \mathbf{v}\times\boldsymbol{B}) \tag{1.2}$$

where **v** is the velocity of the electron, *c* is the speed of light, ***E*** is the electric field intensity, and ***B*** is the magnetic flux density, *e* is the electron charge and *m* is the electron rest mass. This equation must be satisfied at every point on an electron's trajectory path, whether the electron is in the primary beam or is scattered back from the specimen surface. In practice, the above equation is rarely solved in this form. Electron lenses usually have rotational symmetry in their field distributions and primary beam electrons are confined to a small region around the lens optic axis. This means that the fields ***E*** and ***B*** that focus the primary beam can be expressed in terms of a polynomial in, *r*, the off-axis distance in the transverse plane. The electric potential $U(r,z)$, and magnetic field components $B_r(r,z)$ and $B_z(r,z)$ through which the primary beam electrons travel are usually expressed in the following form

$$\begin{aligned}
U(r,z) &= U(z) - \frac{r^2}{4}\frac{d^2U(z)}{dz^2} + \frac{r^4}{64}\frac{d^4U(z)}{dz^4} - \frac{r^6}{2304}\frac{d^6U(z)}{dz^6} + \ldots\ldots \\
B_z(r,z) &= B(z) - \frac{r^2}{4}\frac{d^2B(z)}{dz^2} + \frac{r^4}{64}\frac{d^4B(z)}{dz^4}\ldots\ldots\ldots\ldots \\
B_r(r,z) &= -\frac{r}{2}\frac{dB(z)}{dz} + \frac{r^3}{16}\frac{d^3B(z)}{dz^3}\ldots\ldots\ldots
\end{aligned} \tag{1.3}$$

where $U(z)$ is the axial electric potential, and $B(z)$ is the z-component of the axial magnetic field (see Appendix 1.0 for the full derivation). These expressions are derived from Maxwell's equations for static field distributions. Numerical methods are usually employed to solve for $U(z)$ and $B(z)$ (Khursheed, 1999).

If the transverse distances and velocities of the electron beam are assumed to be orders of magnitude smaller than their longitudinal counterparts, as is usually the case, then a simplified "paraxial equation" can be used to describe the first-order optical properties of the primary beam. The paraxial equation (see Appendix 1.1) is given by

$$\frac{d^2r}{dz^2}+\frac{1}{2U}\frac{dU}{dz}\frac{dr}{dz}+\left(\frac{1}{4U}\frac{d^2U}{dz^2}+\frac{e}{8mU}B^2\right)r=0. \qquad (1.4)$$

For a SEM that has only magnetic electron lenses, *U(z)* is constant along the path of the primary beam (except inside the electron gun), and its kinetic energy is given by *eU*.

The paraxial equation, being a second-order ordinary differential equation, has two independent solutions, *h(z)* and *g(z)*, so that the general solution *r(z)* is given by

$$r(z)=Cg(z)+Dh(z) \qquad (1.5)$$

where *C* and *D* are constants that can be solved by specifying the initial conditions for the ray, typically by using the initial value of *r* and its slope *dr/dz*. Two independent ray path solutions are shown in Fig. 1-5. Normally the slope of *g(z)* and the starting point of *h(z)* are specified to be zero. The point at which a paraxial ray focuses is called the Gaussian plane. Here the Gaussian plane is located at the focal point of ray *h(z)* and the linear magnification of the lens *M* can be found from the value of ray *g(z)* on the Gaussian plane. Typical paraxial rays for the SEM are shown in Fig. 1-6. These rays were calculated by solving Eq. (1.4) numerically. The axial field distribution is indicated by the fainter line. The path of the electron in *r* is usually measured in microns, whereas in *z*, distances are typically measured in centimeters.

The numerical solution of the paraxial equation is straightforward, and provides the ray paths, *g(z)*, *h(z)*, the magnification *M*, the Gaussian plane position z_i and the focal length *f*. It can be proved from the paraxial equation, that the primary beam after traversing a purely magnetic lens field or purely electrostatic lens field always converges (Dinnis, 1993).

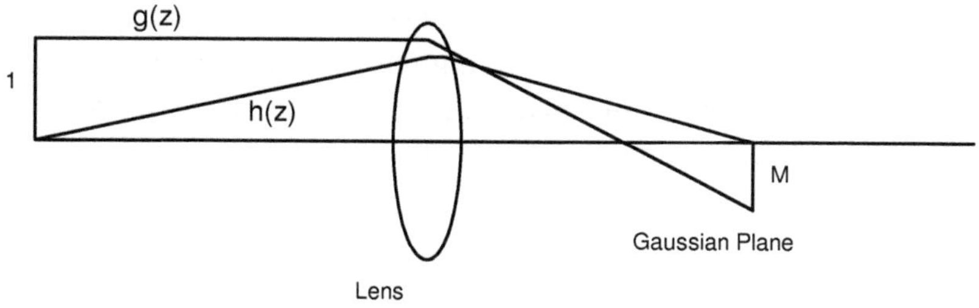

Fig. 1-5. Two independent paraxial ray solutions.

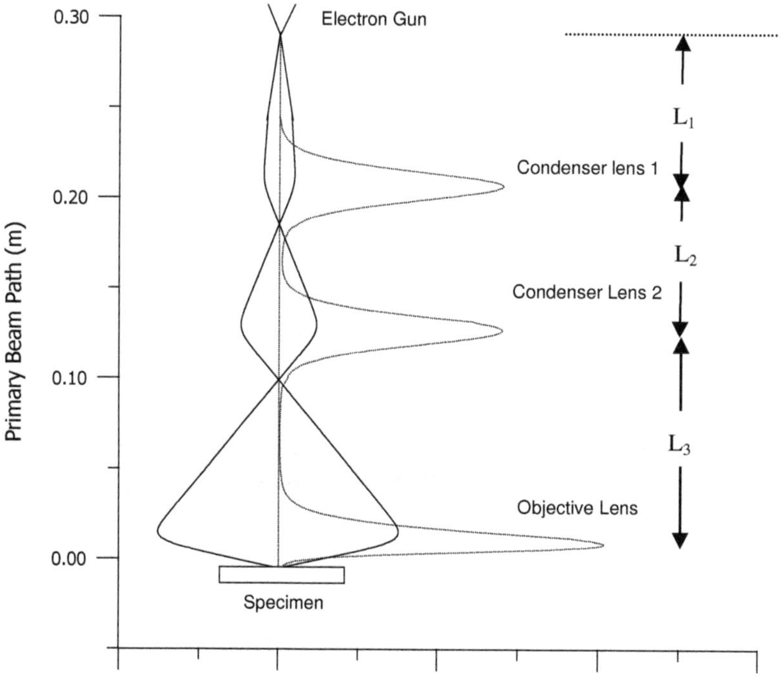

Fig. 1-6. Paraxial ray trajectory in a SEM column.

Useful estimates of the focal length of an objective lens can be obtained in certain circumstances, without solving the paraxial equation. It can be shown for instance for a "thin magnetic lens", that is, where the source and Gaussian planes, z_0 and z_i, lie outside the lens field distribution, the focal length is approximately given by

$$\frac{1}{f} = \frac{e}{8mU} \int_{z_0}^{z_i} B^2(z)dz. \tag{1.6}$$

Consider an idealized step function field distribution where $B(z) = B_0$ over the lens length L and zero every else. The focal length is then given by

$$f = \frac{8mU}{eB_0^2 L}. \tag{1.7}$$

Typical values of f for a conventional SEM range from 5 to 30 mm. For a 10 kV primary beam focused by a 50 mT field strength spanning 1 cm, the focal length is 18.2 mm.

The above expression shows that the objective lens focal length is proportional to the beam voltage, inversely proportional to the square of the axial field strength, and inversely proportional to the length of the lens field distribution. These relationships are of critical importance in electron optics. In conventional SEMs, the focal length of the objective lens is well approximated by the above weak lens formula, since the specimen is located below the final lens pole pieces. The above expression explains why the primary beam is focused on to the specimen by adjusting the lens coil current. For a non-saturated objective lens, B_0 is proportional to NI, the excitation strength, so the focal length depends inversely on the square of the lens coil current. The most effective way of shortening the focal length is to increase the axial field strength. The dependence of f on L demonstrates an important property about scaling down the size of the lens. If the lens is made smaller by a factor of k while keeping the beam energy and axial flux density constant, then the focal length will increase by a factor of k. To maintain the same focal length for a smaller lens, or to have it proportionally scale down, B_0 needs to be increased.

By Ampere's circuital theorem, it is easy to show that if the lens size is scaled down by k and the source excitation strength (NI) is also scaled down by the same amount, the axial field strength $B_0(z)$ remains constant and the focal length will be scaled up by k (Eq. (1.7)). On the other hand, if the same current is maintained through the lens windings as it is scaled down, $B_0(z)$ will in principle be scaled up by a factor of k and the focal length will be scaled down by k. However, maintaining the same coil current when the size of a magnetic lens is scaled down by a factor of k leads to the coil current density going up by a factor of k^2, and since even normal lenses require special cooling precautions, it is often impractical for a normal electromagnetic lens to be scaled down in size. Also, saturation effects limit the degree to which B_0 can be increased.

Another important consequence of the above formula is that as the primary beam voltage is increased, to keep the primary beam focused on the specimen, the lens axial field strength must be correspondingly increased. If for instance, the beam voltage is increased by a factor of 25, from say 1 kV to 25 kV, then the axial field strength needs to be increased by a factor of 5 to maintain the same focal length. These kinds of considerations are important to objective lens design. Most conventional SEMs are designed so that the beam voltage can typically vary between 1 kV to 30 kV.

For a thin electrostatic lens, it can be shown that the focal length is approximately given by

$$\frac{1}{f} = \frac{3}{16}\left(\frac{U(z_0)}{U(z_i)}\right)^{1/4} \int_{z_0}^{z_i} \left(\frac{U'}{U}\right)^2 dz \qquad (1.8)$$

A simple electrostatic lens can be created by two electrodes at two different voltages, $U_1(z_o)$ and $U_2(z_i)$. Assuming that the voltage on the axis varies linearly over a length L, as shown in Fig. 1-7, then it is straightforward to show that the focal length is given by

$$\frac{1}{f}=\frac{3}{16}\left(\frac{U_1}{U_2}\right)^{1/4}\frac{(U_2-U_1)}{L}\left(\frac{1}{U_1}-\frac{1}{U_2}\right). \tag{1.9}$$

If the primary beam is retarded from 10 kV to 1 kV over a distance of 1 cm, then the focal length is approximately given by 3.7 mm. Although conventional SEMs do not usually employ electrostatic lenses, the strong focusing effect produced by retarding the primary beam has been the source of some recent innovations in SEM objective lens design. Note that the focal length of an electrostatic lens scales down with its length.

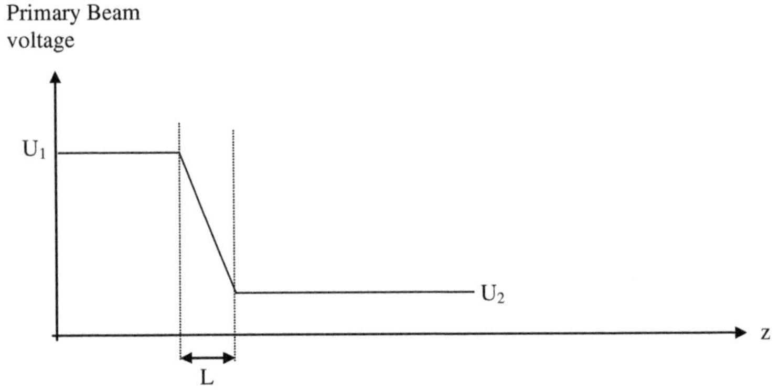

Fig. 1-7. Idealized two-electrode electrostatic lens.

From first-order optics considerations, the demagnification action of an electron lens can be related to its focal length and the distance from the source or beam crossover to the lens center. Let the distance from the source to the center of the first lens be L_1, the distances between subsequent lens centers L_2, and L_3, and the focal lengths for each lens f_1,

f_2, and f_3, as shown in Fig. 1-6. The total source demagnification, M_s, is approximately given by

$$M_s = \frac{f_1}{L_1}\frac{f_2}{L_2}\frac{f_3}{L_3}. \tag{1.10}$$

The beam probe diameter at the specimen d_0, in terms of the effective gun source diameter, d_G, is thus given by

$$d_0 = M_s d_G = \frac{f_1}{L_1}\frac{f_2}{L_2}\frac{f_3}{L_3} d_G. \tag{1.11}$$

The source demagnification parameter, M_s, is different from the image demagnification M_I. Most SEM users and manufacturers refer to "demagnification" or "magnification" to refer to the size of the specimen scanned area to image size, and the source demagnification is sometimes referred to as "spot size", usually measured on a normalized scale. The lens currents in all lenses will change their respective focal lengths, and hence vary the overall source demagnification. Since f_3, the focal length of the final lens, is fixed by the position of the specimen, it is the lens currents in the condenser lenses that are usually adjusted to vary the size of the projected source on the specimen plane.

The source demagnification action of an electron lens can also be understood with respect to the beam semi-angles, before and after traversing the lens. In the paraxial case, the final semi-angle at the specimen, α_0, is given by

$$\alpha_0 = \alpha_G / M_s = \frac{L_1}{f_1}\frac{L_2}{f_2}\frac{L_3}{f_3}\alpha_G. \tag{1.12}$$

Apertures are usually placed before each lens so that wider angle electrons are prevented from reaching the specimen. This means that at high source demagnifications, the electrons in the final probe are ones that have small semi-angles at the source. Unlike the case for light optics, aberrations in the final lens of a SEM require that the semi-angles at the specimen be relatively small, typically in the range of 1 to 10 mrad. For a source demagnification of 5,000, this means that the electrons which reach the specimen start off from the source with initial semi-angles of 0.2 to 2 μrad. The semi-angle at the specimen is kept small by the Final Aperture unit, usually located at the entrance of the objective lens bore. Final aperture sizes typically range from 30 to 100 μm in diameter. Seen in this way, source demagnification is a way of selecting the more parallel electrons in the primary beam, and improving SEM spatial resolution. Better spatial resolution inevitably reduces the available probe current.

Another important parameter for the SEM is its depth of focus. Compared to the light microscope, the SEM has a relatively large depth of focus, and this has been one the reasons why it has become so popular. The depth of focus is defined to be the distance *D*, within which height variations on the specimen are still in focus in the SEM image. Obviously, there is no definite point at which an image suddenly defocuses. The usual definition is given in terms of the probe resolution δ at the specimen and the semi-angle α_0

$$D = \frac{\delta}{\alpha_0}. \tag{1.13}$$

Given for instance a probe resolution of 10 nm, and a final semi-angle of 5 mrad, the depth of focus is 2 μm. Large depth variations in the surface topography can be observed, compared to the feature sizes that can be resolved. Smaller apertures will have the desirable effect of increasing the depth of focus. The main disadvantage of decreasing the aperture size is of course, reduction of the primary beam probe current. In practice, a compromise between depth of focus and signal to noise considerations must be found.

1.3 The electron gun

The performance of an SEM is critically dependent on the type of electron gun used. The column optics is adjusted so that the projected size of the source on the specimen is small. The energy spread of electrons emitted from the gun limits the resolution of the SEM at low voltages, and the final image quality depends upon how much current the gun can provide. The types of electron guns used in SEMs are the thermionic tungsten gun, LaB_6 gun and field emission (cold and Schottky types) gun.

Thermionic guns, such as the tungsten or LaB_6 cathode source, form a real crossover point between the cathode and anode electrodes, as shown in Fig. 1-8(a). They do this by employing a Wehnelt electrode placed in front of the cathode. The Wehnelt electrode is biased negatively with respect to the cathode voltage and has the effect of focusing the emitted electrons, effectively reducing the source size. A current is passed through the cathode and heats it to high temperatures so that electrons are emitted from the cathode surface. The electric field strength at the cathode surface is relatively low, and the emission current from the cathode is usually limited by a space charge effect, where the slower moving emitted electrons form a negatively charged cloud above the cathode tip. The emission current naturally comes to its own equilibrium.

Field emission guns use a negatively biased suppressor electrode, which is placed behind a finely pointed cathode, as shown in Fig. 1-8(b). The radius of curvature of the cathode tip in this type of gun can be as low as 0.01 μm. The suppressor electrode has the effect of pushing the emitted electrons forward, and a high electric field strength at the cathode surface is created (typically around 10^9 V/m or greater). Electrons leave the cathode tip via quantum tunneling and do not form a real crossover. They seem to emanate from a "virtual source", which is found by projecting back the emitted rays from the anode plane. The virtual source position typically lies several nanometers behind the cathode electrode tip and is around 15 nm in diameter.

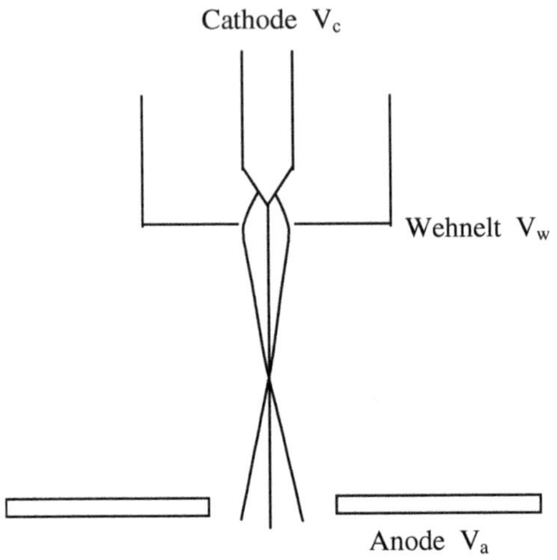

(a)

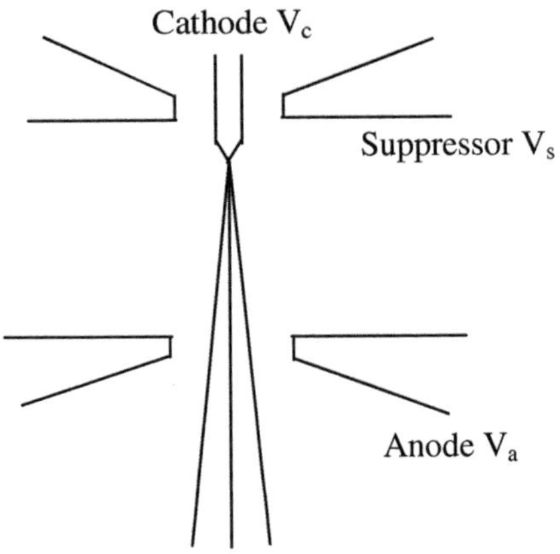

(b)

Fig. 1-8. Electron guns: (a) Thermionic gun, (b) Field emission gun.

The emission characteristics of both thermionic and field emission guns are schematically illustrated in Fig. 1-9. Curve A represents the situation where the electric field at the cathode tip is very low, such as in the thermionic gun case. The emitted electrons in this case are ones that are able to surmount the potential barrier between the Fermi level and vacuum level (work function). Curve B represents the Schottky situation with moderate extraction field strength. The potential barrier is effectively lowered. Curve C depicts the situation of a high extraction field (around 10^9 V/m or more). In this case, electrons are able to tunnel through the potential barrier. This last situation corresponds to the type of emission in a cold field emission gun, where electrons leave the cathode surface through a process of quantum mechanical tunneling.

A formula useful for estimating the current density at the cathode, J_c, where the extraction field strength is low, as in the case for thermionic guns, is given by Swanson and Schwind (1997)

$$J_s = \frac{4\pi m(kT)^2}{h^3} \exp\left(\frac{e^{3/2} F^{1/2} - \Phi}{kT} \right) \tag{1.14}$$

where h is Planck's constant, e and m is the respective charge and mass of an electron, k is Boltzmann's constant, T is the cathode temperature, Φ is the cathode work function, and F is the extraction field strength at the cathode tip surface.

When the electric field strength $F = 0$, this expression reduces to the well-known Richardson–Dushman equation describing emission for thermionic guns. As the extraction field strength increases, quantum mechanical tunneling starts to occur. A parameter measuring the fractional amount of tunneling current in the total current is given by q

$$q = 3.72 \times 10^{-3} \frac{F^{3/4}}{T}. \tag{1.15}$$

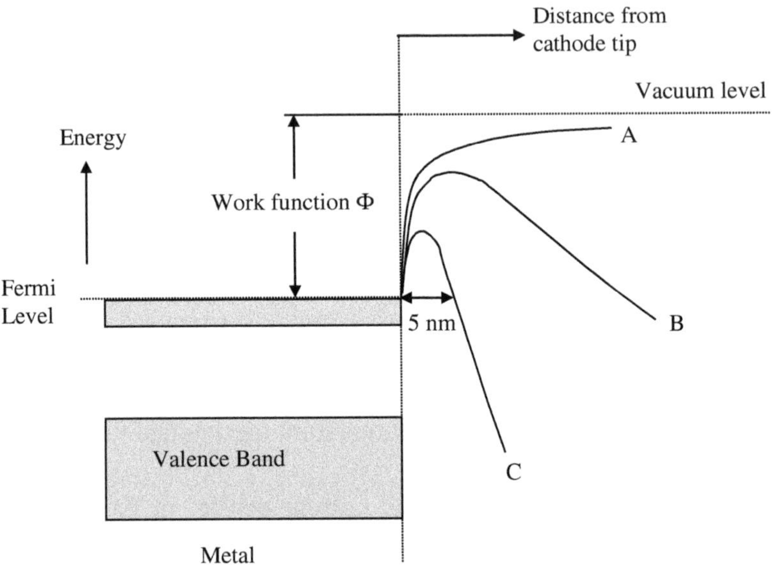

Fig. 1-9. Electron gun emission characteristics.

When for instance q = 0.5, tunneling current make up half the total current. As the extraction field strength increases, q also increases. Beyond $q > 0.15$, the gun is defined to be in an "extended Schottky emission" region, and the emission current density in this case, J_{EC}, is then better approximated by

$$J_{EC} = J_C \frac{\pi q}{\sin(\pi q)}. \tag{1.16}$$

This formula is valid for $q < 0.7$, and it describes the type of emission that occurs for most field emission guns. Clearly, from the above formulas, the current density for field emission guns is predicted to be much greater than for thermionic guns.

The brightness of an electron gun, β, is an important parameter that characterizes how much current it can provide at the specimen. Brightness β is defined by

$$\beta = \frac{J}{\pi \alpha^2} \tag{1.17}$$

where J is the current density and α is the beam semi-angle at the cathode. For a given SEM column, the brightness is a constant along the primary beam path. The current density at the cathode usually falls off sharply with increasing α. Since the brightness of a gun approximately rises linearly with the anode voltage (primary beam voltage U_p), it is sometimes quoted as the "reduced brightness", $\beta_R = \beta/U_p$.

At the specimen, for a projected source diameter of d_0 and semi-angle of α_0, the primary beam current, I_p, is given by

$$I_p = \frac{\pi^2}{4} \beta d_0^2 \alpha_0^2 . \tag{1.18}$$

The available probe current is directly proportional to the gun brightness. From this expression, it is clear that high probe currents cannot be obtained with small probe diameters (required for high resolution) and that a compromise between the two must be found. A minimum requirement on the probe current comes from inherent signal-to-noise limitations in SEM detection and display systems.

The tungsten filament is reliable, relatively inexpensive, and well understood. For many SEM applications where high brightness is not necessary, or where stable high currents are required, the thermionic tungsten filament may be used without loss of performance, and may, in fact, be the best choice. LaB_6 filaments are more costly, and need to be operated at a higher vacuum level, which is normally maintained by an additional pump. However, the increased current for a given probe size and significantly longer filament lifetime justifies the greater cost for many applications. The advantages of cold field emission guns include their higher brightness and lower energy spread. But they have the disadvantages of short-term fluctuations (noise) of the order of 2 to 5%, and a long-term drift of the emission current. The vacuum requirement for this type of gun is even higher. Although thermal field emission guns such as the Schottky electron emitter have slightly lower brightness than cold field emission guns, they are much more stable, and are widely used in practice. Fransen *et al.* (1999) have carried out a recent survey on the performance of Schottky electron emitters. Each type of gun varies in the amount of current it can produce, its effective projected size, its intrinsic energy spread, the stability of its emitted current, and its lifetime. Table 1-1 presents some typical values of electron gun parameters as reported by Delong (1993) and Postek (1997). It should be noted that the values in Table 1-1 are only approximate, and that some of them depend on the primary beam voltage.

Table 1-1. Electron gun parameters.

Source	Brightness (A/cm^2 sr)	Lifetime (hours)	Source size	Energy spread ΔE (eV)	Current stability	Vacuum pressure (Pa)
Tungsten hairpin	2×10^4–2×10^5	40–100	20–50μm	1–3	1%	$<10^{-3}$
LaB_6	10^6	200–1,000	5–20μm	0.5–2	1%	$<10^{-5}$
Field Emission						
Cold	10^8–10^9	>1,000	<5nm	0.1–0.4	5%	$<10^{-8}$
Schottky	10^7–10^8	>1,000	<15nm	0.4	2%	$<10^{-7}$

1.4 Lens aberrations and primary beam probe size

From first-order optics considerations alone, the size of the electron beam probe at the specimen should vary only as a function of the total projected source size but, in practice, the spatial resolution to which the primary beam can be focused at the specimen is limited by objective lens aberrations, schematically illustrated in Fig. 1-10. These aberrations are critically dependent on the final probe semi-angle α_0, given by $\alpha_0 = r_a/f$, where r_a is the final aperture radius and f is the objective lens focal length. For a conventional lens, the focal length is of the order of 10 to 30 mm and the lens aberrations usually limit α_0 to less than 5 mrad, which means that final apertures are typically smaller than 50 μm in radius.

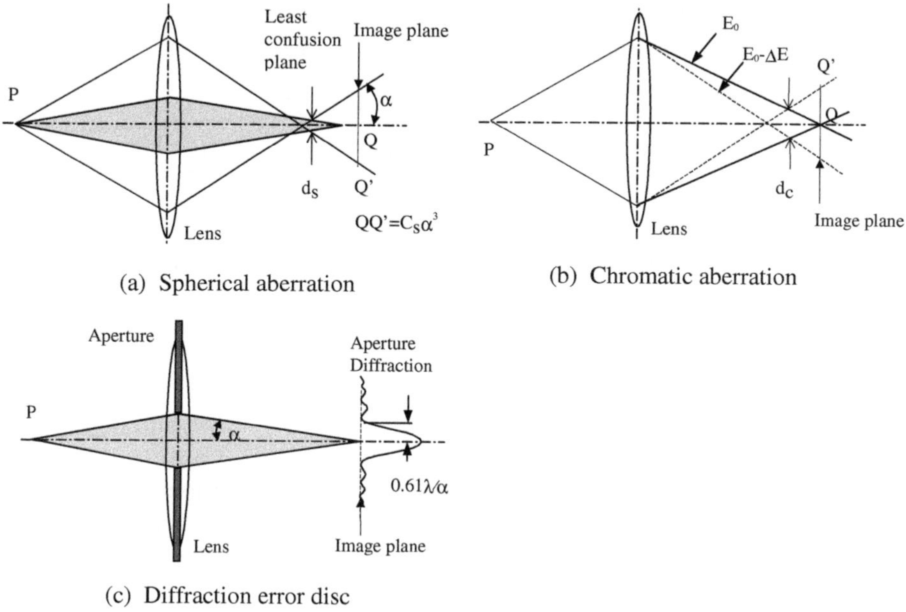

Fig. 1-10. Schematic representation of lens aberrations: (a) Spherical, (b) Chromatic, (c) Diffraction.

For a beam of electrons entering an electron lens, the focal point distance depends on its off-axis position through the lens. As the off-axis distance increases, the point of focus moves closer to the centre of the lens, as shown in Fig. 1-10(a). The resulting beam diameter in the Gaussian plane is $2C_s\alpha^3$, and in the plane of least confusion the diameter d_s is given by

$$d_s = 0.5C_s\alpha^3. \tag{1.19}$$

Here, C_s is the spherical aberration coefficient of the lens and α is the aperture-limited semi-angle, which can be changed by varying the final aperture size or the working distance. Note that the effect of spherical aberration can always be reduced by using small apertures.

Derivation of the expression for C_s is based upon taking into account higher-order terms in the field expansion expressions, and retaining up to third-order terms in r in the general equation of motion. For the non-relativistic case, the third-order expression for C_s is given by (see Appendix 1.2)

$$C_s = \frac{1}{16U_0^{1/2}} \int_{z_0}^{z_i} h^4 U^{1/2} \Big(\frac{5}{4}\left(\frac{U''}{U}\right)^2 + \frac{5}{24}\left(\frac{U'}{U}\right)^4 + \frac{14}{3}\left(\frac{U'}{U}\right)^3 \frac{h'}{h} - \frac{3}{2}\left(\frac{U'}{U}\right)^2 \frac{h'^2}{h^2}$$

$$+ \frac{eB'^2}{mU} + \frac{3e^2B^4}{8m^2U^2} - \frac{eB^2}{mU}\frac{h'^2}{h^2} + \frac{35eU'^2B^2}{16mU^3} - \frac{3eU'BB'}{mU}$$

$$- \frac{2eB^2U'}{mU^2}\frac{h'}{h} \Big) dz. \tag{1.20}$$

h is the paraxial ray solution with slope unity at the object plane. The above integral is usually calculated numerically. For the case of magnetic fields only ($U =$ constant and $B \neq 0$)

$$C_s = \frac{1}{16} \int_{z_0}^{z_i} \left(\frac{eB'^2}{mU} h^4 + \frac{3e^2B^4}{8m^2U^2} h^4 - \frac{eB^2}{mU} h'^2 h^2 \right) dz. \tag{1.21}$$

This expression has great importance in the subject of electron optics. It can be shown that this expression is always positive definite, that is, C_s for a magnetic round-lens electron optical system can never be zero and will always have a residual value (Hawkes and Kasper, 1989a). The same is true for the more general expression of C_s with combined rotational electric and magnetic field distributions. For round lens systems, C_s cannot be zero, and has at best, a minimum value. Since practically all SEMs use round lens systems, spherical aberration effects play an important role in limiting their resolution.

The focal position also depends on the primary beam energy. Due to an energy spread of ΔE at the electron gun, the probe size is broadened (see Fig. 1-10(b)). The disc of least confusion with diameter d_c is given by

$$d_c = C_c(\Delta E / eU_s)\, \alpha_0 \tag{1.22}$$

where C_c is the chromatic aberration coefficient and U_s is the primary beam voltage at the specimen. For weak lenses, C_c is approximately equal to the focal length. Note the linear dependence of the size of the chromatic aberration spot on ΔE, due mainly to the energy spread of the electrons at the electron gun. This brings out the importance of electron gun design for high resolution scanning electron microscopy. Apart from the linear dependence on C_c, it is obvious that the chromatic aberration effect will be worse for low primary beam voltages (low eU_s). As with the spherical aberration effect, chromatic aberration can be reduced by using smaller apertures. It should also be noted that the fluctuations of the acceleration voltage U and variations in the lens current I also cause chromatic aberration.

The expression for C_c can be derived by considering small perturbations in the beam energy. The equation of motion can then be reformulated, and the expression for C_c is given by (see Appendix 1.3)

$$C_c = \frac{3\Delta U}{8U_0^{1/2}h'(z_0)}\int_{z_0}^{z_i}\frac{U'^2h^2}{U^{5/2}}dz + \frac{\Delta Ue}{8mU_0^{1/2}h'(z_0)}\int_{z_0}^{z_i}\frac{B^2h^2}{U^{3/2}}dz$$
$$-\frac{e}{4mU_0^{1/2}h'(z_0)}\frac{\Delta B_{\max}}{B_{\max}}\int_{z_0}^{z_i}\frac{B^2h^2}{U^{1/2}}dz\,. \qquad (1.23)$$

This expression, like the previous one for spherical aberration, is very important in understanding the resolution limits of round lens systems. Since terms in the above integral for C_c are always positive, the chromatic aberration effect for round lens systems can be minimized, but not completely eliminated. This expression has critical importance for the development of SEMs today, since there has been a steady trend over the last decade towards low primary beam voltages, precisely the regime where the SEM's resolution is limited by chromatic aberration.

Several proposals and experiments have been made based upon using non-round lens fields to correct for the chromatic and spherical aberration effects (Zach, 2000). The search for SEM objective lenses that have a low chromatic aberration coefficient has been the subject of considerable research, and some proposals coming from this work will be described in a later chapter of this book.

Due to the quantum wave nature of an electron beam, Fraunhofer diffraction at the beam-limiting aperture in the final lens causes an Airy disc (see Fig.1-10(c)) with a half-width given by

$$d_{\mathrm{d}} = 0.6\lambda/\alpha \tag{1.24}$$

where the DeBroglie wavelength $\lambda = h/mv = 1.226/\sqrt{U}$, U is the primary beam voltage. Here λ is measured in nanometers and U is in volts. Note that the size of the diffraction blur is inversely proportional to α, varying in the opposite way to how it changes for the other lens aberrations. This means that from considerations of image resolution alone, that is, apart from signal to noise considerations, the size of the final aperture cannot be made too small. In practice, there is an optimum semi-angle that minimizes the image resolution and the final aperture size is usually selected so that the SEM operates at this point. The spatial resolution of an electron optical system, unlike in light optics, is not limited purely by diffraction.

The wavelengths for a beam of electrons at 1 and 25 keV are 38.7 pm and 7.74 pm, respectively, according to the above formula. This is typically 2 to 3 orders of magnitude smaller than the contributions from spherical and chromatic aberrations. But since the other lens aberrations limit the final semi-angle to only a few mrad, the diffraction contribution to the final spot size may in some cases reach a few nano-meters.

The asymmetrical pinhole lens, shown in Fig. 1-11, is the most common type of objective lens used in a SEM. The final pole-piece, operated relatively far away from the specimen, has a very small bore that keeps most of the magnetic field within the lens. This type of lens was used in the Cambridge Instruments S100 SEM, and has a bore diameter of 9 mm.

It provides space for various types of detectors below it. This space however, increases the aberrations on the objective lens, and therefore leads to a larger electron-probe size. The distance from the lens lower bore to the specimen, known as the working distance, lies typically in the range of 5 to 30 mm.

Figure 1-12 shows graphs of how C_s and C_c depend on working distance for the conventional pinhole lens shown in Fig. 1-11. They demonstrate that the on-axis aberration coefficients increase strongly with working distance. At a working distance of 5 mm and beam energy of 20 keV, f, C_s, and C_c are 15.69 mm, 30.6 mm, and 13.29 mm respectively. In general, for short working distances, both spherical and chromatic aberration coefficients are of the same order as the focal length. The probe diameters associated with spherical, chromatic, and diffraction aberration effects are shown in Figs. 1-13(a) and 1-13(b) for 20 keV and 1 keV beam energies respectively, as a function of the probe semi-angle for an energy spread of 1.5 eV. Obviously diffraction effects work in opposition to chromatic and spherical aberrations, that is: they are large for small semi-angles and small for large semi-angles. Clearly chromatic aberration is important at low primary beam voltages (1 keV), while spherical aberration dominates at high voltages (20 keV).

For most SEM users, the resolution of their final image is determined by signal-to-noise considerations. The excitation strengths of the condenser lenses and size of all apertures are first selected to ensure that an image can be obtained. The minimum primary beam current at the specimen needed to do this is typically of the order of a few pico-amps. A smaller signal than that will produce excessively noisy images. This minimum current requirement naturally sets limits on the spatial resolution of the SEM. For a given probe current I_p, the projected diameter of the source at the specimen, given by Eq. (1.18), is

$$d_0^2 = \frac{4I_p}{\beta\pi^2\alpha_0^2}. \tag{1.25}$$

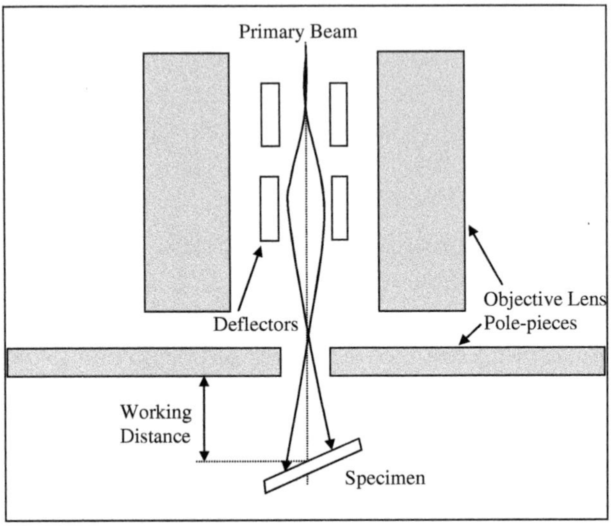

Fig. 1-11. A conventional SEM objective lens.

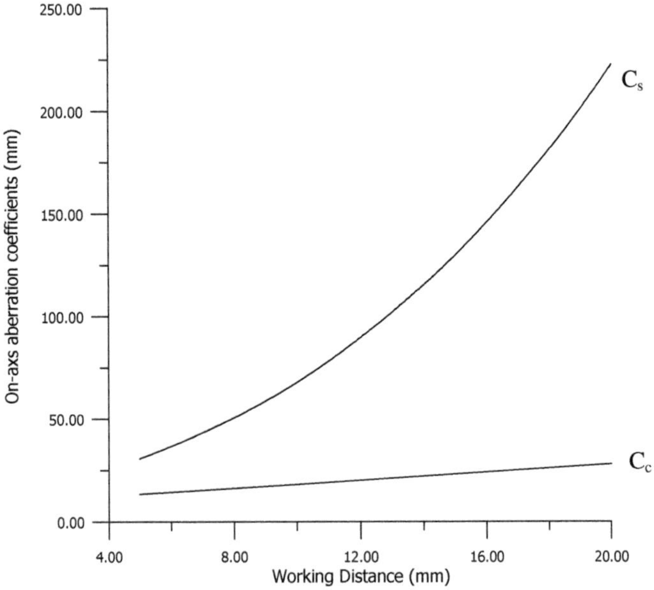

Fig. 1-12. On-axis aberration coefficients and their dependence on the working distance for a conventional objective lens.

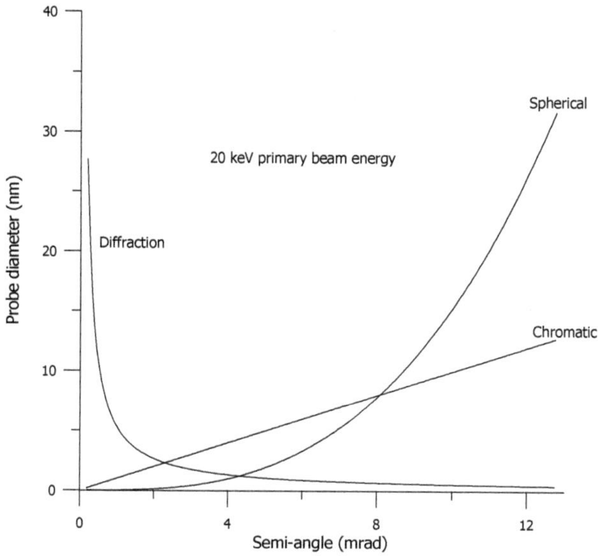

(a)

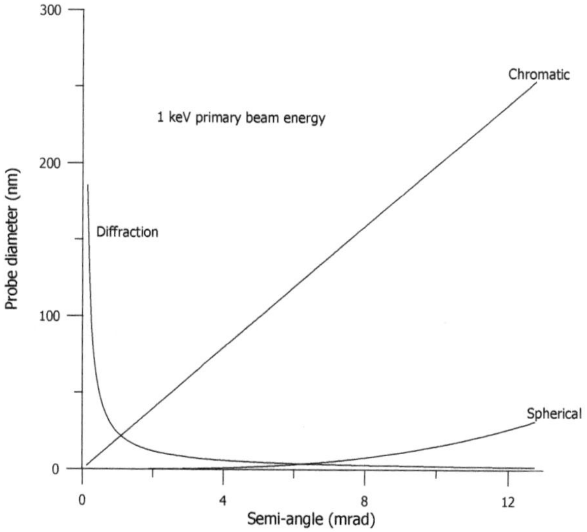

(b)

Fig. 1-13. Predicted probe diameter limited by aberrations and their variation with the final beam semi-angle for a conventional objective lens (S100 lens with an energy spread of 1.5 eV): (a) 20 keV primary beam energy, (b) 1 keV primary beam energy.

Note that as with the diffraction effect, the diameter of the probe increases as the final semi-angle decreases, which is in opposition to the way chromatic and spherical aberrations behave. A rough estimate for the total primary beam diameter d_p at the specimen can be obtained by adding all contributions that broaden the beam at the specimen in quadrature

$$d_p^2 = \left[C_0^2 + (0.6\lambda)^2\right]\alpha_p^{-2} + \frac{1}{4}C_s^2\alpha_p^6 + \left(C_c\frac{\Delta E}{E}\right)^2\alpha_p^2 \qquad (1.26)$$

where $$C_0^2 = \frac{4I_p}{\pi^2\beta}.$$

This formula has been widely used to estimate the primary beam probe size at the specimen (Cosslet, 1972; Reimer, 1998b), and is also taken to estimate the SEM image resolution (Zach, 1990, 2000; Frank and Mullerova, 1999). However, it overestimates the probe size by around a factor of two (Plies and Elstner, 1989). At high resolution, say a probe diameter of less than 5 nm, the quadrature formula becomes even more inaccurate.

Barth and Kruit (1996) have proposed a root-power-sum method, based upon using 50% of the beam current-enclosing diameter. This formula was tested against the exact wave optical approach and found to be more accurate than the quadrature sum method. The formula derived by Barth and Kruit is given by

$$d_p^2 = \left[\left(d_{sph}^4 + d_{df}^4\right)^{1.3/4} + d_G^{1.3}\right]^{2/1.3} + d_{chr}^2$$

$$d_{sph} = \left(\frac{1}{2}\right)^{5/2} C_s\alpha_p^3$$

$$d_G = \left(\frac{2}{\pi}\right)\left(\frac{I}{\beta}\right)^{1/2}\frac{1}{\alpha_p}$$

$$d_{df} = 0.54\frac{\lambda}{\alpha_p}$$
$$d_{chr} = 0.34C_c\frac{\Delta E}{E}\alpha_p. \qquad (1.27)$$

The Barth–Kruit root-power-sum method is used in aberration calculations throughout this book. There are other factors that also limit the spatial resolution. They include the effects of thermal and mechanical vibrations, stray electromagnetic fields, electron-electron interactions within the primary beam and the interaction volume at the specimen surface. All these effects become important at high resolution (typically less than 5 nm).

For a tungsten source, usually $C_0 >> \lambda$ and the probe resolution is limited by the projected source size (related to the total lens demagnification and the primary beam current through the gun brightness), chromatic, and spherical aberration effects. Consider, for instance, a probe current of 1 pA, which lies at the threshold of most SEM detection and display systems. The brightness of a tungsten source operated at a 20 keV primary beam energy is typically 2×10^5 Acm^{-2}sr^{-1}, and it falls to around 0.2×10^5 Acm^{-2}sr^{-1} at 1 keV. For these conditions, C_0 is 14.2 pm and 45 pm respectively, which is around a factor of 2 to 3 greater than 0.6λ (5.19 pm at 20 keV and 23.22 pm at 1 keV). However, for a field emission gun, whose brightness at 20 keV is around typically 10^8 Acm^{-2}sr^{-1}, and 10^7 Acm^{-2}sr^{-1} at 1 keV, C_0 is 0.636 pm and 2 pm respectively, which is an order of magnitude smaller than its corresponding 0.6λ values. So for field emission systems, the probe size is limited by diffraction, chromatic, and spherical aberration effects. The image resolution of field emission systems is usually **aberration limited**, while for tungsten gun SEMs, it is usually **beam current** limited.

Graphs demonstrating how the probe diameter varies with probe current for the S100 objective lens are shown in Figs. 1-14(a) and 1-14(b) for a tungsten and field emission source using the Barth and Kruit root-sum formula. Here the specimen is placed at a working distance of 5 mm, where f, C_s, and C_c are 15.69 mm, 30.6 mm, and 13.29 mm, respectively.

An energy spread of 1.5 eV is assumed for the tungsten gun, while a 0.4 eV spread is taken for the field emission source. Both the aberration coefficients and the gun energy spread are assumed to be independent of the primary beam voltage. The probe diameters given here are their minimum values taken over a wide range of beam semi-angles. Figures 1-14(a) and 1-14(b) clearly show that the probe diameter for the field emission SEM is around 3 times lower than its corresponding value for a tungsten source SEM at high voltages (20 keV), and around four times lower at low voltages (1 keV). They also demonstrate that the probe diameter for the tungsten gun SEM varies significantly with probe currents above 1 pA, whereas in the field emission case, the probe diameter remains approximately constant for probe currents less than 100 pA. These results are valid for high resolution applications where the probe diameter at 20 keV lies below 50 nm. For larger probe diameters however, typically above 100 nm, the field emission SEM gives less probe current than a thermionic source. This is important when using the SEM as an analytical tool, such as for cathodluminescence, X-ray and Auger microanalysis, EBIC, and acoustic and thermal-wave microscopy. For these kinds of applications, primary beam currents up to several hundreds of nano-amperes may be required, and a thermionic source SEM may be more effective than a field emission SEM.

If the source current and size are not included in the quadrature sum formula for the probe size, an aberration limited probe size estimate is obtained. Often, this is further simplified at either low or high primary beam voltages to be a combination of only two aberrations, chromatic and diffraction at low voltages, and spherical and diffraction at high voltages. An optimum semi-angle can then be found. Crewe (1997) has reported formulas to estimate the best resolution attainable under these conditions, borrowed from theory developed in light optics. The minimum resolution δ is derived for an optimum convergence semi-angle α_{op}, and is given by the following formula for high primary beam voltages

$$\delta = 0.61\frac{\lambda}{\alpha_{op}} = 0.15 C_s \alpha_{op}^3, \quad \alpha_{op} = \left(\frac{4\lambda}{C_s}\right)^{1/4}. \tag{1.28}$$

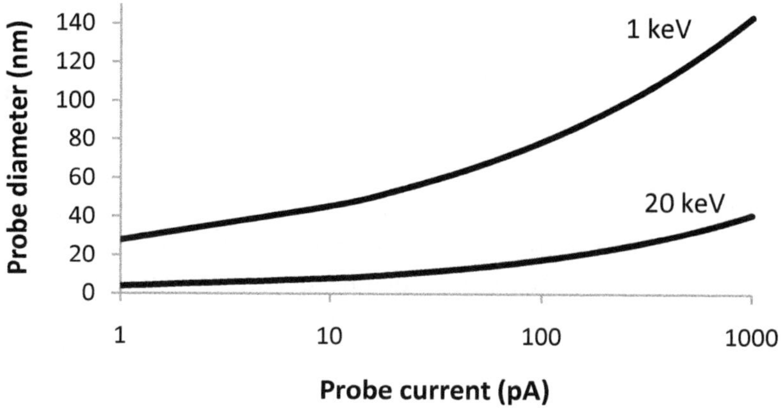

(a)

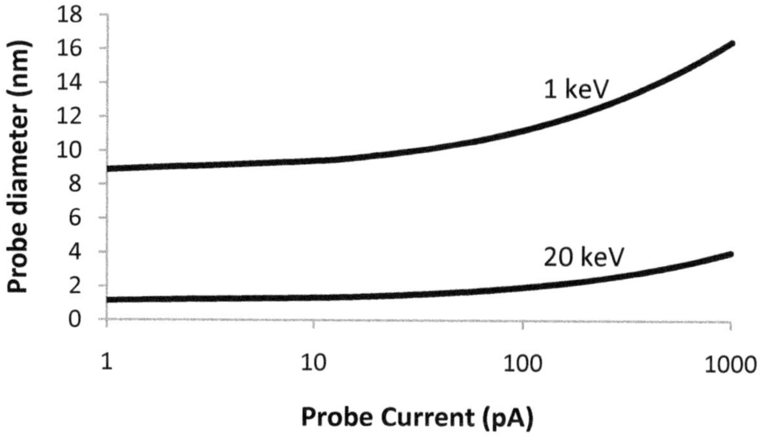

(b)

Fig. 1-14. Predicted probe diameter as a function of probe current for a conventional objective lens operated at a working distance of 5 mm (f =15.69 mm, C_s=30.6 mm, C_c=13.29 mm): (a) Tungsten gun (ΔE=1.5 eV, β=2×10^4 Acm^{-2}sr^{-1} at 1 keV, β=2×10^5 Acm^{-2}sr^{-1} at 20 keV), (b) Field emission gun (ΔE=0.4 eV, β=10^7 Acm^{-2}sr^{-1} at 1 keV, β=10^8 Acm^{-2}sr^{-1} at 20 keV).

For low primary voltages

$$\delta = 0.61\frac{\lambda}{\alpha_{op}} = C_c\alpha_{op}\frac{\Delta V}{V_p}, \quad \alpha_{op} = 2.74\times10^{-5}\left(\frac{V_p^{1/4}}{C_c^{1/2}\Delta V^{1/2}}\right) \tag{1.29}$$

where V_p is the primary beam voltage and ΔV is the voltage spread at the gun. These estimates are around a factor of 2 smaller than those calculated from the corresponding quadrature sum formula. The quadrature sum formula assumes that the aberrations are statistically independent and contribute equally to increase the probe size. However, this is not a valid assumption since the aberrations occur in quite different places and are different in character. Their combined effect is to slightly change the shape of the beam.

At low primary beam voltages, under high resolution conditions, the spherical aberration significantly broadens the probe size and cannot be simply omitted, as done in the above formula. Although formulas based upon the combination of only two aberrations have been widely used in the past to estimate the resolution of SEM images, they may provide misleading results in many instances, and are not recommended here. Also, certain amount of beam current will always be required, and it is therefore necessary to include the effect of the source.

1.5 Deflection systems

Deflection systems are needed for two purposes: one is deflection of the focused probe over the specimen in order to form an image or to position it at a particular point in order to make a measurement, and the other is for beam-blanking purposes. The former function is invariably carried out by magnetic deflection, whereas the latter is usually achieved through electrostatic deflection. The reasons for this come from the need to have precision deflection in scanning the probe with minimum spot aberration, while for beam blanking, the priority is speed.

A parallel-plate capacitor arrangement is often used for beam blanking. For scanning, the number of electrodes is usually a multiple of 4, the most common arrangement being either a quadruple or an octopole deflector, as shown in Fig. 1-15(a). Note that for the plates in this diagram, the applied voltages on the electrodes produce an even symmetry line on the *x*-axis and an odd symmetry line on the *y*-axis, therefore, the electric field will be mainly in the *x*-direction. The electrode arrangement can be regarded as an *x*-deflector. Similarly, *y*-deflectors are obtained by rotating the deflector by 90°. This rotation can, of course, be achieved electronically by suitably transposing the driving voltages.

A magnetic deflector usually has current-carrying conductors arranged on a surface of revolution about the *z*-axis in the form of a cylinder or a tapered horn. The windings may be in the form of either a saddle or toroidal type, and the deflector itself can be placed next to rotationally symmetric magnetic materials. A schematic diagram for toroidal and saddle deflection yokes are shown in Fig. 1-15(b). The toroidal coil is shown on the left, while the saddle is depicted on the right. The arrows indicate the direction of current flow. In this figure, the current direction in the coils along the main body of the deflectors has odd symmetry with respect to the *x*-axis. The resulting magnetic field will be mainly in the *x* direction, and they are often referred to as *x*-deflectors. *Y*-deflectors (not shown in the figure) are of identical design, and are merely rotated 90° relative to *x*-deflectors.

A significant difference between electrostatic and magnetic deflection is their relative deflection sensitivity. Electrostatic deflection has a low deflection sensitivity and thus needs high driving voltages, while magnetic field has a high deflection sensitivity and thus requires low driving currents. Scan generators for magnetic coils are simpler to design than electrostatic ones.

The major disadvantage of a magnetic coil comes from its eddy currents, which limit its scan frequency. Though the eddy currents in the magnetic

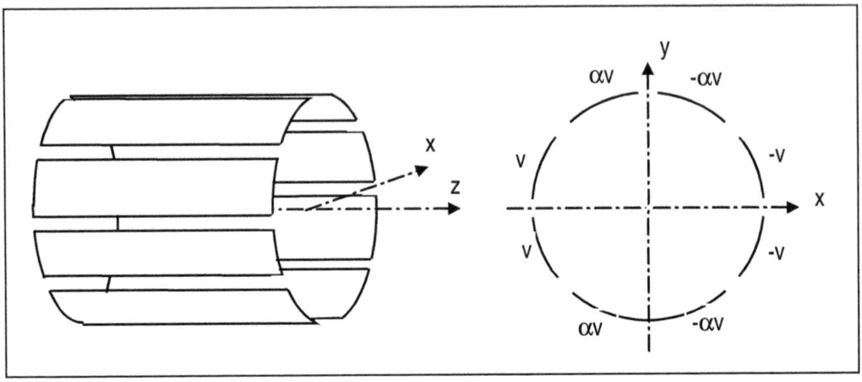

(a)

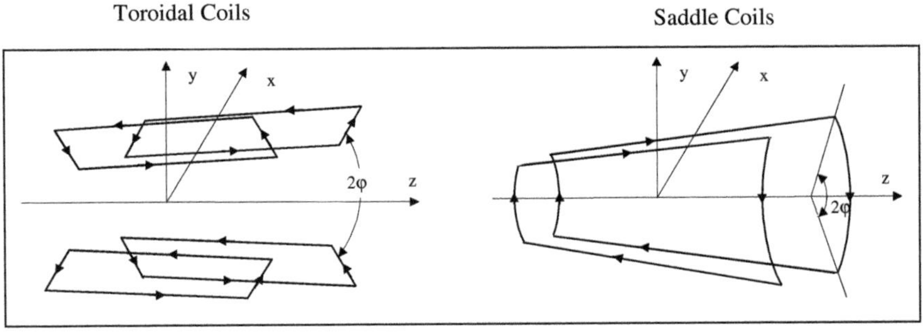

(b)

Fig. 1-15. Schematic drawing of primary beam scan deflectors: (a) Electrostatic, (b) Magnetic.

core can be overcome by using ferrite cores or air coils, in many cases where magnetic coils are placed in the vicinity of other iron components, say magnetic pole-pieces, the system will suffer from serious eddy current errors at high scanning speeds.

Taking into account deflector aberration effects on the primary beam spot is a complex task and, here, only the general form of the aberrations of a focusing/deflection system, purely electrostatic or purely magnetic is given. A detailed description of them can be found elsewhere (Ohiwa *et al.*, 1971; Munro, 1974; Zhu *et al.*, 1997).

Table 1-2 shows the general form of the primary beam aberrations for a magnetic focusing/deflection system. They have geometrical as well as chromatic contributions. In this table, α is the beam semi-angle, r is the off-axis deflection distance, and d is the primary beam diameter at the image plane z_I, as predicted by first-order optics. The aberrations are classified into four groups:

1. Axial aberrations, which depend only on the beam half-angle α,
2. Pure deflection aberrations, which depend on the deflection distance r
3. Shaped beam aberrations, which depend on shaped beam size d,
4. Hybrid aberrations, which depend on both r and d.

Table 1-2. General form of primary beam aberrations of a shaped-beam focusing/deflection system. Terms in bold type indicate aberrations that can be reduced or eliminated by dynamic correction. α is the beam half-angle, r is the off-axis deflection distance and d is the projected first-order shaped-beam size at the image plane.

	Axial	Pure deflection	Shaped beam	Hybrid
Spherical	$C_s\alpha^3$			
Axial Chromatic	$C_c\alpha\Delta E/E$			
Coma Radius		$K_R\alpha^2 r$	$K_{SR}\alpha^2 r$	
Coma Length		$K_L\alpha^2 r$	$K_{SL}\alpha^2 r$	
Field curvature		$\mathbf{K_F\alpha r^2}$	$K_{SF}\alpha d^2$	$K_{HF}\alpha rd$
Astigmatism		$\mathbf{K_A\alpha r^2}$	$K_{SA}\alpha d^2$	$K_{HA}\alpha rd$
Distortion		$\mathbf{K_D r^3}$	$K_{SD}d^3$	$K_{HD1}r^2d+$ $K_{HD2}rd^2$
Transverse Chromatic		$K_T r\Delta E/E$	$K_{ST}d\Delta E/E$	

As the deflection distance r and beam shape size d increase, these aberrations become more serious, and it is important to apply dynamic correction. Coma and transverse chromatic aberrations, if present, cannot be corrected dynamically, because the required correction potential does not obey Laplace's equation and, therefore, cannot be realized in practice. Fortunately, however, it is possible to design deflection systems which are free of coma and transverse chromatic aberration (Goto and Soma, 1977).

The three correctable aberrations are shown in bold type in Table 1-2. They are deflection field curvature (**K_F**), deflection astigmatism (**K_A**), and deflection distortion (**K_D**). Deflection field curvature can be corrected by a dynamic focus coil. The dynamic coil is energized in opposition to the main lens excitation, and its current should be proportional to the square of the deflection strength. Deflection astigmatism can be corrected by a quadruple field, produced from either an electrostatic or a magnetic octopole stigmator of suitable strength and orientation. The field strength should be proportional to the square of the deflection.

Deflection distortion is caused by the deflection distance at the image plane being slightly nonlinear relative to the deflection signal. It may include a radial error (isotropic distortion) and a circumferential error (anisotropic distortion). These distortions can be corrected by applying appropriate correction signals to the deflection currents.

Because "distortion" does not blur the spot and can be easily corrected, it is not included in the probe size estimate. The overall aberration before dynamic correction is evaluated as the quadrature sum of all the aberrations other than distortion, and the overall aberration after dynamic correction is defined similarly, except that two more dynamic correctable aberrations (astigmatism and field curvature) are not taken into account.

In-lens deflection is found effective in suppressing deflection aberrations. The concept behind it is the moving objective lens (Ohiwa *et al.*, 1971). With the aid of the MOL concept, Pfeiffer and Langner (1981) first achieved a practical normal incident scanning system or telecentric system. Other lens/deflection systems based on this approach are known as the VAL (variable axis lens) or VAIL (variable axis immersion lens).

The MOL concept is illustrated in Fig. 1-16. A point source of electrons, emitted from z_o, is imaged to z_i by a lens. The first deflector deflects the beam so that it enters the lens off-axis. The second deflector, placed inside the lens, shifts the electrical center of the lens off-axis, so that the deflected beam still passes through the optical center of the lens. The MOL reduces the effect of the off-axis lens aberrations. The spherical and chromatic aberrations can be kept small by having a short working distance L_2. At the same time, the deflection aberrations can be kept small by having a large distance L_1 from the first deflector to the lens (Heritage, 1975).

In order to establish the condition of a MOL system, the radial expansion of magnetic flux density is required. From Eq. (1.3), in which magnetic field component expansions in cylindrical coordinates up to the fourth order are given, it is clear that to a first-order approximation, the required deflection field is given by

$$D(r,z) = \frac{1}{2} B'(r) r(z). \tag{1.30}$$

This deflection field effectively shifts the lens axis by a distance r, and the off-axis lens aberrations can be greatly reduced. This is the MOL condition. Although MOL systems are more popularly associated with lithography, they have been used for specialist SEM applications, such as quantitative voltage contrast (Richardson and Muray, 1988).

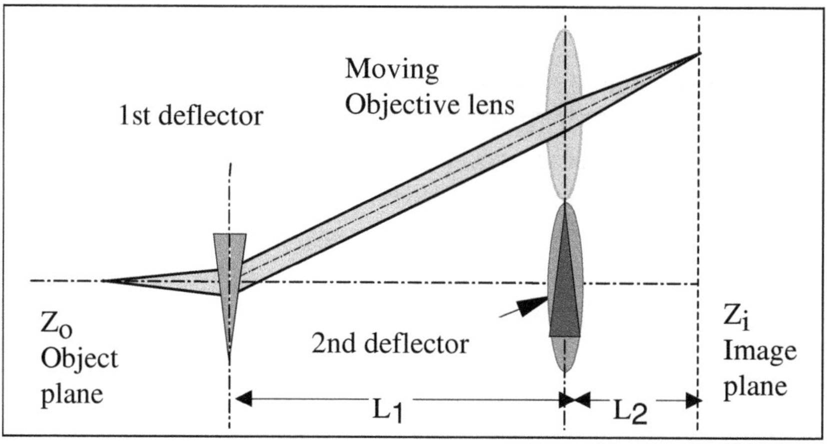

Fig. 1-16. The moving objective lens (MOL) concept.

1.6 Quadrupole stigmators

In general, the shape of the primary beam at the specimen is not perfectly round. Due to many factors, such as beam misalignment and mechanical asymmetry of lens pole-pieces, it has an elliptical shape. In order to correct for this, additional lenses placed above the objective lens, known as "stigmators" are used. In its simplest form, stigmatic correction consists of quadrupole field distributions, one to shape the beam in the x-direction, and the other to shape the beam in the y-direction (orthogonal to the x-direction). Quadrupole fields can be electric or magnetic, and have wide application in many branches of applied physics for focusing charged particle beams. In the SEM however, they are mostly used as weak focusing lenses, reshaping the primary beam, rather than bringing it to a focus on the axis. It is therefore sufficient, in the present context, to discuss only their first-order optical properties. The following summary is based upon the work of Banford (1966a).

Figure 1-17 depicts the cross-section of an idealized magnetic quadrupole field distribution, where solid lines represent magnetic scalar equiptentials, and dotted lines denote lines of magnetic field. The magnetic scalar potential is fixed (at pole pieces) along hyperbolae, whose nearest distance to the origin O is $\sqrt{2}a$ (distance a from both the x and y axis). It is assumed here that the primary beam of electrons travel perpendicularly into the page and that the quadrupole field strength and distribution are constant along the stigmator length (along the z-direction), and that there are no components of the magnetic field in the z-direction (perpendicular to the page), in practice, small fringe fields will inevitably be present at the edges of a stigmator unit.

The idealized magnetic scalar potential, $\psi(x,y)$, is given by the simple analytical function

$$\psi(x, y) = \frac{\psi_0}{a^2} xy \, . \tag{1.31}$$

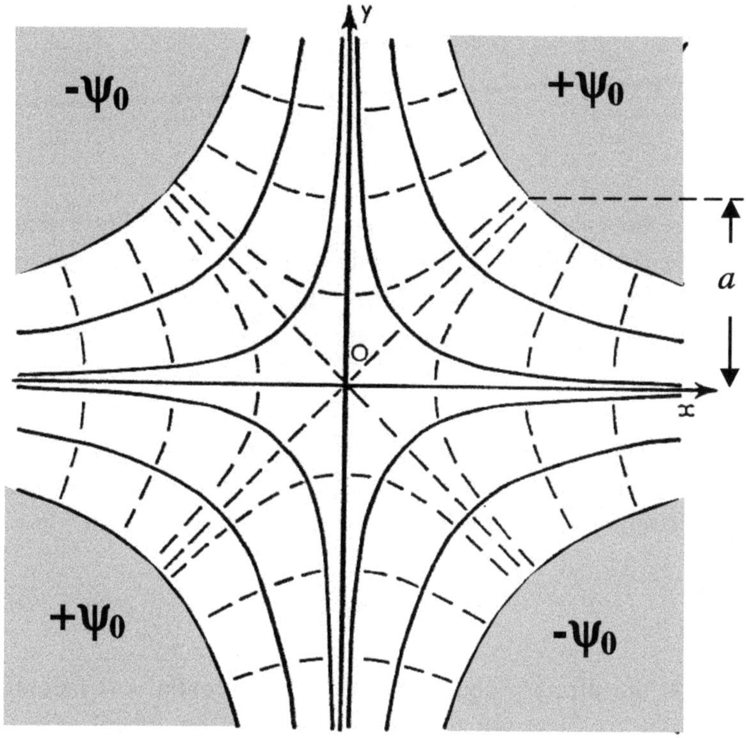

Fig. 1-17. Idealized magnetic quadrupole field distribution. Solid lines represent magnetic scalar equipotentials, while dotted lines denote lines of magnetic field.

This distribution obviously satisfies Laplaces's equation, and the components of the magnetic field are given by

$$B_x = -\mu_0 \frac{\psi_0}{a^2} y$$
$$B_y = -\mu_0 \frac{\psi_0}{a^2} x \tag{1.32}$$

where μ_0 is the magnetic permeability of free space. Substituting these expressions into the equations of motion for an electron travelling into the plane, gives the following equations

$$\ddot{x} = \mu_0 \eta \dot{z} \frac{\psi_0}{a^2} x$$
$$\ddot{y} = -\mu_0 \eta \dot{z} \frac{\psi_0}{a^2} y \tag{1.33}$$

where η is the charge to mass ratio for an electron. These equations clearly indicate that there is a focusing action in the y-direction (giving simple harmonic motion solution), and a diverging force in the x-direction. The action of the quadrupole lens is therefore to focus the beam in one direction, while defocusing it in the orthogonal direction, precisely the kind of action required to transform an elliptical beam cross-section into a circular one. If one assumes that the velocity of the beam in the z-direction is many orders of magnitude greater than the transverse components, the following approximation can be made

$$\dot{z}^2 \approx 2\eta\phi_p \tag{1.34}$$

where ϕ_p is the primary beam voltage. This formula will naturally be more accurate for electrons that travel closer to the optical axis, which have smaller transverse velocity components. The paraxial (near axis) equations for the magnetic quadrupole can then be written as

$$x'' = \mu_0 \sqrt{\frac{\eta}{2\phi_p}} \frac{\psi_0}{a^2} x$$
$$y'' = -\mu_0 \sqrt{\frac{\eta}{2\phi_p}} \frac{\psi_0}{a^2} y\,. \tag{1.35}$$

Solutions to these equations describe the first-order optical properties of the magnetic quadrupole. It is usual to decouple the quadrupole field variation in the z-direction from its transverse variations, and represent the action of the quadrupole over an effective length L. Solutions of these equations are of course, straightforward, and usually expressed in matrix form, linking initial values of x, y and their gradients at the quadrupole entrance, at say point 1, to their predicted values at the quadrupole exit, point 2, by

$$\begin{pmatrix} x_2 \\ x_2' \end{pmatrix} = \begin{pmatrix} \cosh\beta L & \dfrac{1}{\beta}\sinh\beta L \\ \beta\sinh\beta L & \cosh\beta L \end{pmatrix} \begin{pmatrix} x_1 \\ x_1' \end{pmatrix}$$
$$\begin{pmatrix} y_2 \\ y_2' \end{pmatrix} = \begin{pmatrix} \cos\beta L & \dfrac{1}{\beta}\sin\beta L \\ -\beta\sin\beta L & \cos\beta L \end{pmatrix} \begin{pmatrix} y_1 \\ y_1' \end{pmatrix} \tag{1.36}$$

where $\beta^2 = \mu_0(\psi_0/a^2)(\eta/2\phi_p)^{1/2}$.

Electrostatic quadrupoles have similar first-order optical properties to magnetic ones. For the same focusing action as the magnetic quadrupole just presented, a focusing action in the y-direction, and defocusing in the x-direction, the idealized electric potential distribution $\phi(x,y)$ is given by

$$\phi(x,y) = -\frac{\phi_0}{a^2}\left(y^2 - x^2\right). \tag{1.37}$$

The equipotential lines and field lines of such a distribution are shown in Fig. 1-18. In this case, the distance a denotes the distance between the origin O and electrodes along each axis. The transverse electric field components in this case are given by

$$E_x = -2\frac{\phi_0}{a^2}x$$
$$E_y = 2\frac{\phi_0}{a^2}y. \tag{1.38}$$

The transverse equations of motion are therefore given by

$$\ddot{x} = 2\eta\frac{\phi_0}{a^2}x$$
$$\ddot{y} = -2\eta\frac{\phi_0}{a^2}y. \tag{1.39}$$

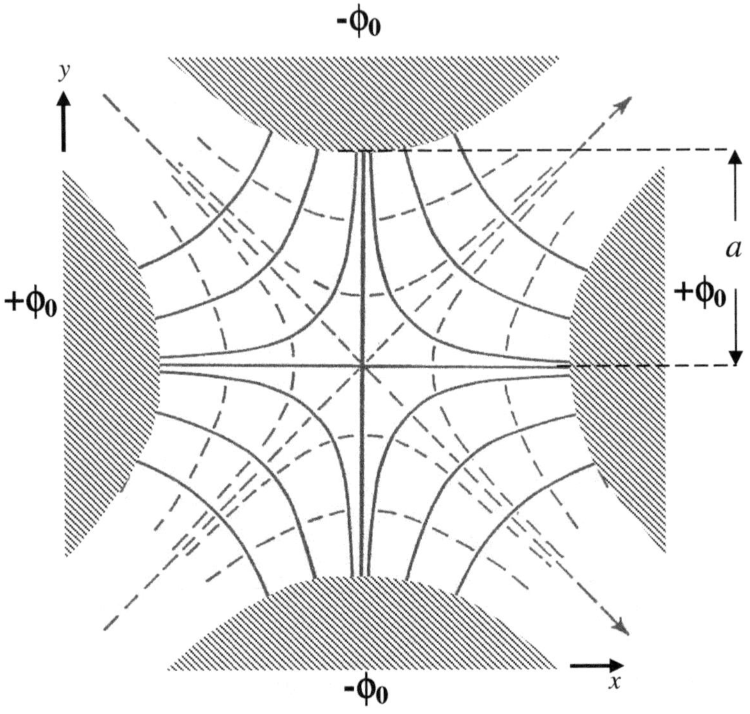

Fig. 1-18. Idealized electric quadrupole distribution. Solid lines represent field lines, while dotted lines denote equipotential lines.

Similar to the magnetic case, the paraxial equations of motion are derived making the assumption that the velocity in the z direction is many orders of magnitude greater than the transverse velocity components

$$
\begin{aligned}
x'' &= \frac{\phi_0}{a^2\phi_p} x \\
y'' &= -\frac{\phi_0}{a^2\phi_p} y.
\end{aligned}
\tag{1.40}
$$

Clearly focusing will occur in the y-direction, and defocusing in the x-direction. Solutions to these equations are of the same form as in the magnetic case, with $\beta^2=(\phi_0/\phi_p a^2)$. Note that these paraxial equations are independent of η, charge-to-mass ratio, and therefore, just as with purely round lens electric fields, apply equally to electrons and ions.

A quadrupole field distribution is often approximated by using straight or circular plates, which are much easier to manufacture than hyperbola shapes. Since in a SEM, these kinds of field distributions are only used for small corrections in the beam shape, an exact match to hyperbola electrode/pole piece shapes is not required. Figures 1-19(a)–1-19(d) depicts some common types of stigmator field distributions used in the SEM.

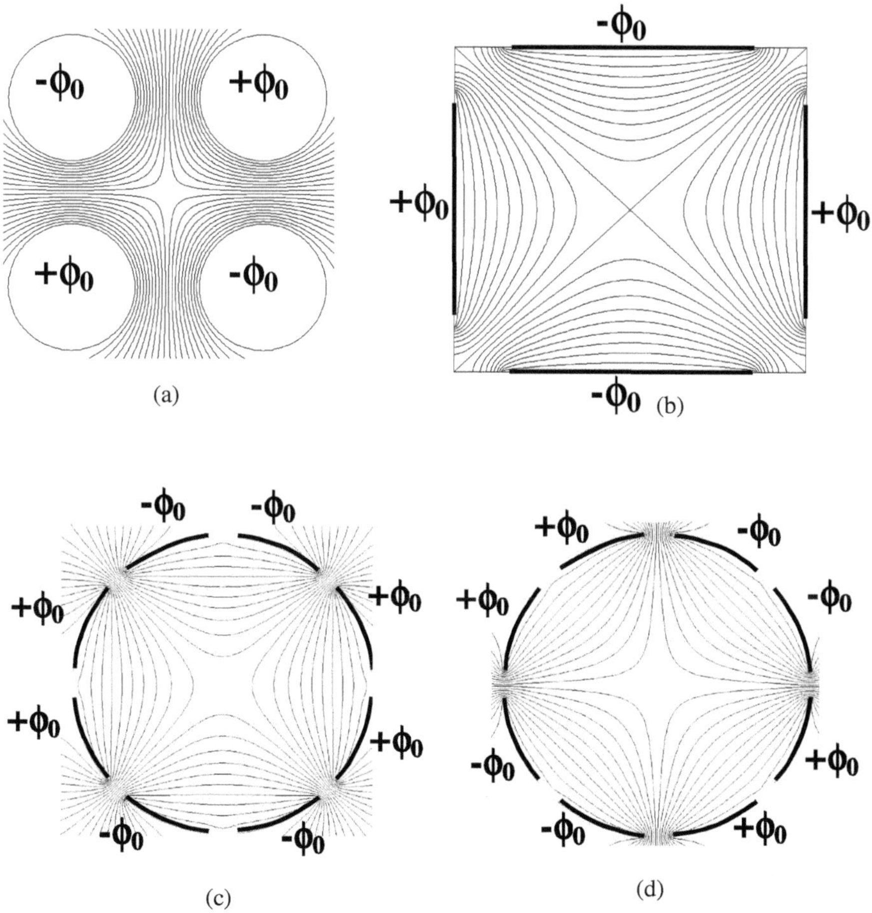

Fig. 1-19. Stigmator field distributions: (a) Circular rods, (b) Straight plates, (c) Octopole cylindrical plates, (d) Octopole cylindrical plates with 45° field rotation.

Figures 1-19(c) and 1-19(d) illustrate how a single layout of 8 cylindrical plate electrodes (octopole) can be used to rotate the field distribution by 45°, in this way, the direction of stigmatic correction can be varied by changing the voltage excitation on the electrodes.

For magnetic stigmators, a coil is wound on to rounded pole-pieces that is attached to an outer casing, as shown in Fig. 1-20. Symmetry planes are created along lines OA and OB.

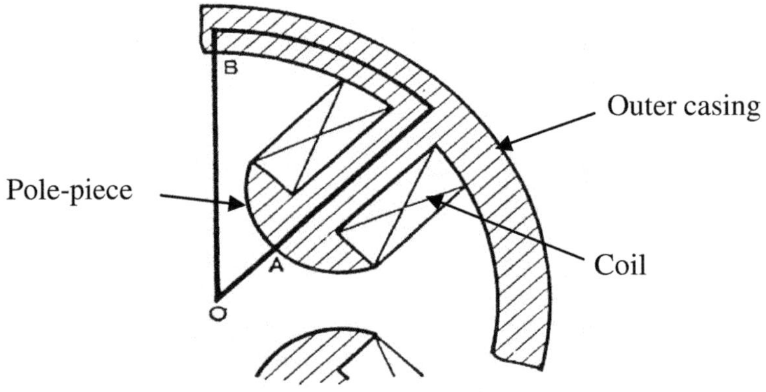

Fig. 1-20. Pole-piece inside a magnetic stigmator unit.

1.7 SEM output signals

Figure 1.21 is a schematic diagram illustrating the relative positions of the secondary electron and backscattered electron detectors as used in most SEMs. The backscattered detector, placed just below the final lens lower pole-piece, may be a silicon p-n junction type or a micro-channel plate which has a small central hole, typically around 5–10 mm in diameter, so that the primary beam can pass through it. The secondary electron detector in Fig. 1-21 was first proposed by Everhart and Thornley (1960) and is by far the most widely used secondary electron detector in SEMs. It is often referred to simply as the Everhart–Thornley (E-T) electron detector. The E-T detector consists of a scintillator, a light pipe, and a photomultiplier tube (PMT). The E-T is electrically isolated from the rest of the microscope and has a wire mesh screen in front of it at a potential typically +200 to +300 V, in order to draw some of the low energy SEs on to the scintillator. A small fraction of the BSEs from the specimen also directly enter the E-T detector. All electrons entering the E-T detector are accelerated by a voltage on the scintillator, typically between 10 to 12 kV. Once inside the scintillator cage, the SEs strike the scintillator material at high energies (typically 10–12 keV) and produce light that travels through a light pipe to the PMT, which converts the light into amplified electrical signals.

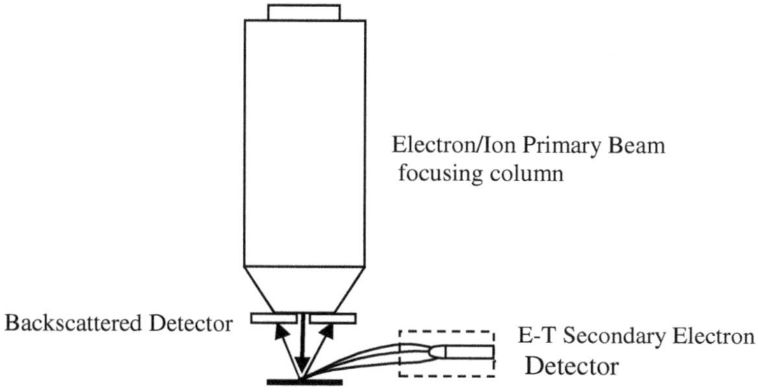

Fig. 1-21. Conventional detector layout.

The secondary electrons that are collected by the E-T detector are generated in different ways, as shown in Fig. 1-22. SE_1 are secondary electrons that are generated by the primary beam within a t_{SE} = 0.5–1.5 nm thick layer below the specimen surface, while SE_2 type secondaries emanate from BSEs traversing this region. SE_3 type secondaries come from backscattered electrons that scatter on external surfaces (away from the specimen), such as the lower lens pole-piece and specimen chamber walls.

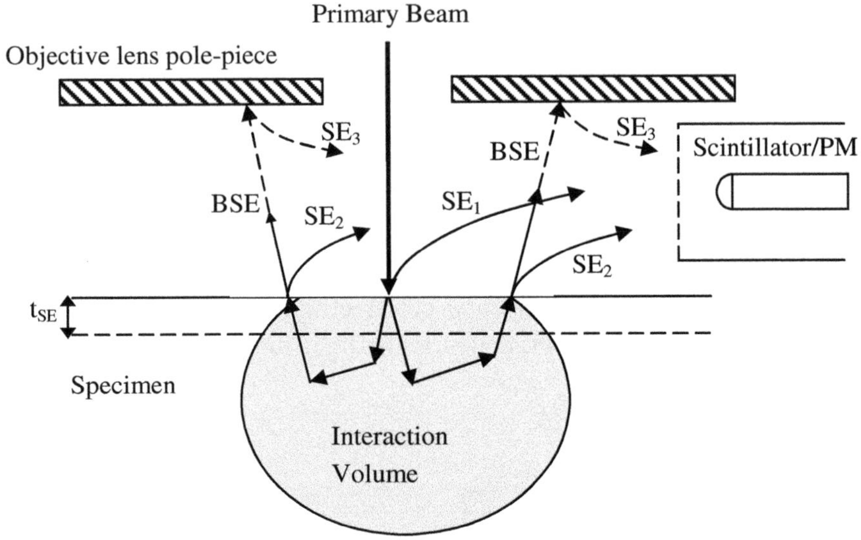

Fig. 1-22. Secondary electron signal generation.

Only some 20–50% of the total SE signal can be attributed to the primary beam, SE_1, (Reimer 1998c), and because they emanate from a shallow escape depth, they are capable of providing high resolution information in the final SE image. The other parts of the SE signal (formed from SE_2 and SE_3) carry information relating to the greater interaction volume that penetrates much deeper into the specimen (since they are indirectly generated by BSEs), and in general, they degrade the final image resolution. These different parts to the SE signal are in general, not separable, and are mixed together to form one bulk SE signal.

The underlying cause for both secondary and backscattered topographic contrast is related to the change in path length that scattered electrons below the specimen surface experience before emerging from the specimen at different polar angles θ (with respect to the surface normal). These electrons must travel a path length proportional to $1/\cos\theta$, and assuming the degree of energy loss is proportional to path length within the specimen, the intensity of these electrons is diminished by $\cos\theta$. To a good approximation, the polar angular distribution of scattered electrons emerging from the specimen surface is therefore proportional to $\cos\theta$, obeying Lambert's Law (Reimer, 1998d). On the other hand, if there are local changes in the specimen slope with respect to the horizontal, let us say by θ, the intensity of SE or BSE emission rises by a factor of $(1/\cos\theta)$, coming from the fact that the path length for scattered electrons generated beneath the specimen surface is shortened. It is these types of variations in SE and BSE intensity that gives rise to SE and BSE topographic contrast, making surface imaging possible inside the SEM.

1.8 The emission hemisphere and BSE collection

There is an important distinction to be made between the angular distribution of both SEs and BSEs at the point of emission, on the specimen surface, and their subsequent angular distribution from the point of view of detection. On leaving the specimen surface, scattered electrons start to occupy a volume of space that is hemispherical in shape, as shown in Fig. 1-23. The relative intensity at any point on the hemisphere's surface is obviously independent of the azimuthal angle, ϕ, and proportional to $\cos\theta$, due to the path length changes beneath the specimen.

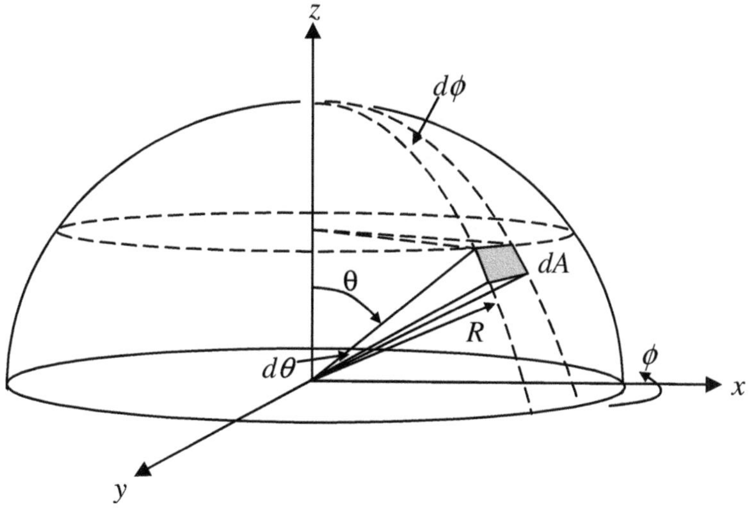

Fig. 1-23. The emission hemisphere of scattered electrons.

The signal current at the detector will not only depend on the relative size and shape of the detector, but also on the trajectory paths electrons take, which in turn, depends on the electric/magnetic field distributions present. If the detector is hemispherical in shape, centered at the point of emission, and there are no electric/magnetic fields present, all scattered electrons will be collected. Let us say they generate a total current of I_0. The area dA on a small part of the hemispherical surface, represented by the shaded area in Fig. 1-23, is obviously given by

$$dA = (Rd\theta)(R\sin\theta d\phi) = R^2 \sin\theta \, d\theta \, d\phi. \tag{1.41}$$

If the initial cosine distribution is now taken into account, the fraction of current on the emission hermisphere is therefore given by

$$\frac{dI}{I_0} = \frac{R^2 \cos\theta \sin\theta \, d\theta \, d\phi}{\pi R^2} = \frac{1}{2\pi} \sin 2\theta \, d\theta \, d\phi. \tag{1.42}$$

Note that the total current I_0 is proportional to πR^2, found from integrating dA for θ varying between 0 to $\pi/2$ and ϕ from 0 to 2π. The emission current, according to this formula, is therefore a maximum for a polar angle of 45°. In the case of BSEs, their trajectory paths can be well approximated by straight lines in a conventional SEM, due to them having several orders of magnitude higher energies than secondary electrons. For the rotationally symmetric overhead BSE detector, the relative detected current is therefore given by

$$\frac{I}{I_0} = \frac{1}{2}\left[\cos 2\theta_1 - \cos 2\theta_2\right] \tag{1.43}$$

where the BSE current is detected for θ between θ_1 and θ_2. Consider the case for instance, of a BSE detector having an inside diameter of 5 mm and an outside diameter of 20 mm, located 10 mm above the specimen, so that θ_1 = 14° and θ_2 = 45°, then the relative amount of BSE current collected is 44.1%.

For the case of a BSE detector being placed to one side of the specimen, where detection is defined by azimuthal and polar angles, ϕ_1 to ϕ_2 and θ_1 and θ_2, then the relative detection current is given by

$$\frac{I}{I_0} = \frac{(\phi_2 - \phi_1)}{4\pi}\left[\cos 2\theta_1 - \cos 2\theta_2\right]. \tag{1.44}$$

This formula can be applied to estimate the number of BSEs that are inadvertently detected by the SE detector. In this case, detected BSEs typically range from 75 to 85° in the polar direction, and over a 10° interval in the azimuthal direction, resulting an estimated 0.164% of the BSE emission being collected by the SE detector. For relatively small detector opening angles, $\Delta\phi$ and $\Delta\theta$ around θ_0, the above expression is simplified to

$$\frac{I}{I_0} = \frac{\Delta\phi\,\Delta\theta}{2\pi}\sin 2\theta_0. \tag{1.45}$$

1.9 The scattered electron energy distribution

Another important factor related to the estimation of scattered electron collection efficiency is the emission energy distribution of BSEs and SEs. For the case of BSEs, the energy distribution can be predicted from Monte-Carlo simulation programs that model the way primary beam electrons scatter inside the specimen, involving multiple inelastic collisions, already mentioned earlier in this chapter. The energy distribution is obtained by monitoring the energies of those electrons that scatter back up from the specimen surface. Figure 1-24 shows the energy distribution from such a Monte-Carlo simulation for specimens of different materials. Specimens are 1 μm thick and the primary beam voltage is 3 kV. The beam interaction with the specimen for these conditions is a bulk one, and variations in the thickness of the specimen do not affect the BSE spectrum. These results were generated from a program called McforVB written by Joy (1995). Figure 1-24 indicates

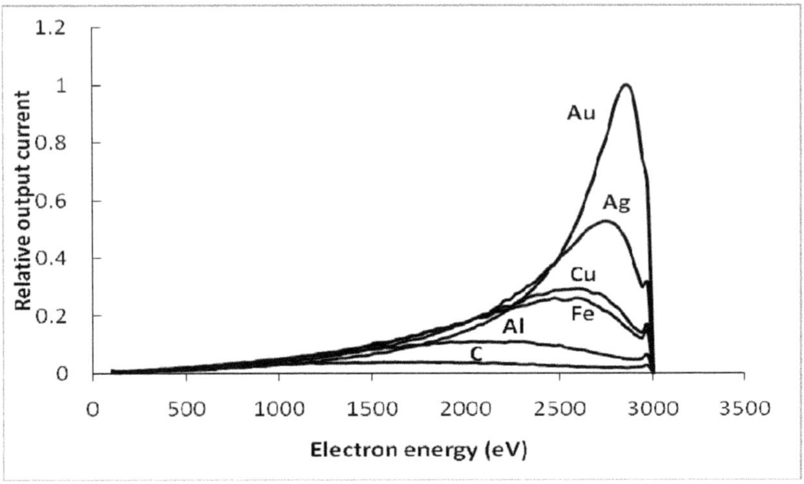

Fig. 1-24. Monte-Carlo simulated BSE energy spectrum for different materials.

that the shape of the BSE spectrum changes considerably with atomic number. For higher atomic numbers, the BSE spectrum contains a pronounced peak near the primary beam energy, while for lower atomic numbers, the BSE spectrum height drops and its width broadens. In general, the kind of simulation results depicted in Fig. 1-24, correlate well with what has been measured experimentally (Reimer 1998e).

The energy distribution for secondary electrons in most cases can be fairly well approximated by the Chung–Everhart distribution, given by

$$\frac{dN}{dE} = \frac{6W^2 N_{SO} E}{(E+W)^4} \tag{1.46}$$

where E is the kinetic energy of the SE leaving the specimen, W is the specimen work function, and N_{SO} is the total number of secondaries that are emitted. This distribution is plot in Fig. 1-25 for $W = 4$ eV (Al) and energies up to 20 eV. The distribution rises sharply, but falls off slowly, peaking at an energy of $W/3$.

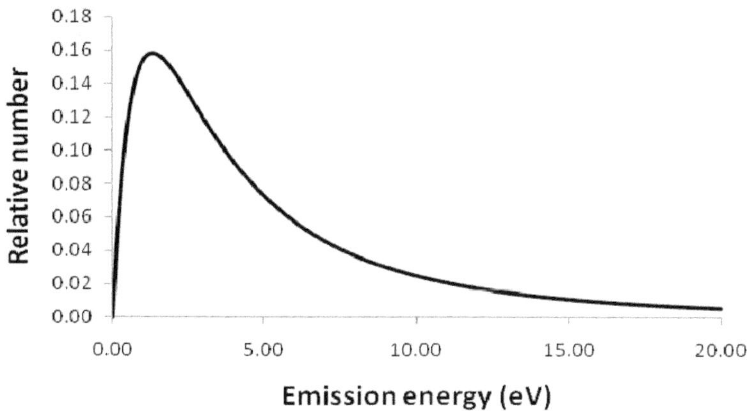

Fig. 1-25. The SE Chung–Everhart energy distribution.

Figure 1-26(a) plots this distribution for the work functions of 4, 4.5 and 5 eV, while Fig. 1-26(b) plots it for the specimen potentials of -1 to -5 volts in steps of 1 volt. The latter graph was plot by using the fact that the initial kinetic energy of the secondary electrons, E_0, is directly lowered or raised by the specimen voltage V_S, so that its kinetic energy is subsequently, $E = E_0 - eV_S$. These graphs indicate how work function (linked to material type) and surface specimen potential information is embedded in the SE energy spectrum.

1.10 The SE collection efficiency

The specimen/scintillator secondary electron collection efficiency in most conventional SEMs is typically less than 15%. This is because the scintillator detector captures scattered electrons over limited angles of emission. Figures 1-27(a)–1-27(d) show equipotentials and trajectory paths for scattered electrons leaving the specimen in 20° polar anglular intervals and initial energies of 3, 10, and 1 keV. The simulation results depicted in these figures were obtained by using the KEOS finite-element field solving and Runge–Kutta ray tracing programs, Khursheed (1999). In this simple two-dimensional model of the

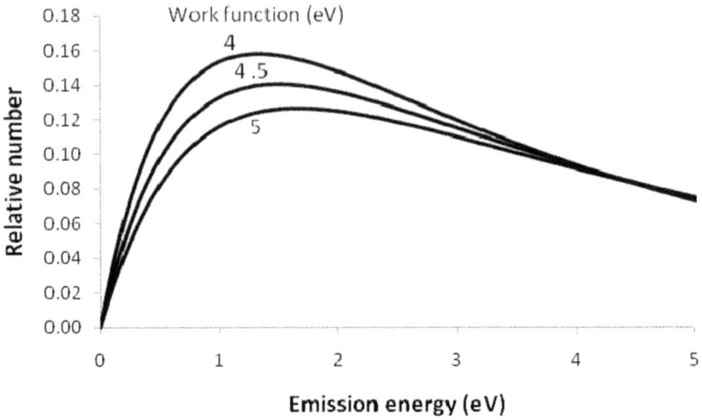

(a)

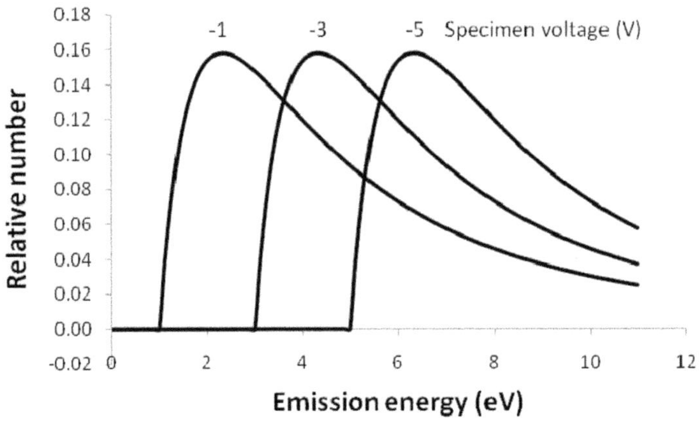

(b)

Fig. 1-26. SE energy distribution: (a) Work function variations, (b) Specimen voltage variations.

scintillator/specimen layout, the scintillator cage has a voltage of 250 V, and is assumed to lie in the plane of emission. The working distance here is 13 mm. These ray tracing results show that less than half the secondary electrons are collected. For other emission planes, that is, ones that do not intersect with the scintillator cage, the collection efficiency will be much lower. The scintillator collection angle in the azimuthal angle is typically less than 90°. This gives a total angular collection of less than 12.5%. As seen from Figs. 1-27(b)–1-27(d), the collection efficiency decreases as the emission energy increases. In some SEMs, the scintillator detector is mounted at an angle, alongside the conical pole-piece of the objective lens, as shown in Figs. 1-28(a) and 1-12(b). This arrangement gives better collection of the secondary electrons when the specimen is tilted.

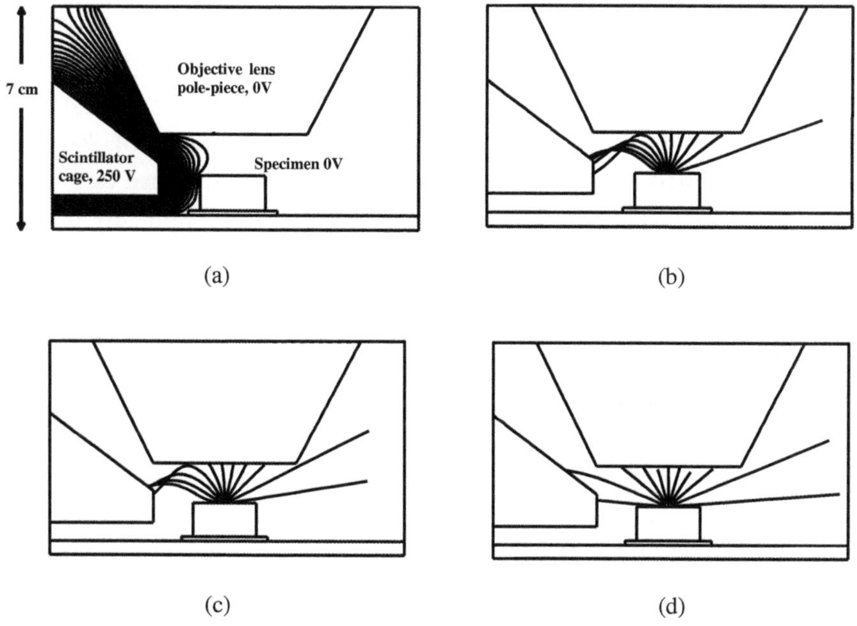

Fig. 1-27. Simulation of secondary electron collection: (a) Equipotential lines, (b) Trajectory paths for 3 eV energy secondary electrons over 20° polar angular intervals, (c) Trajectory paths for 10 eV energy secondary electrons, (d) Trajectory paths for 1 keV energy backscattered electrons.

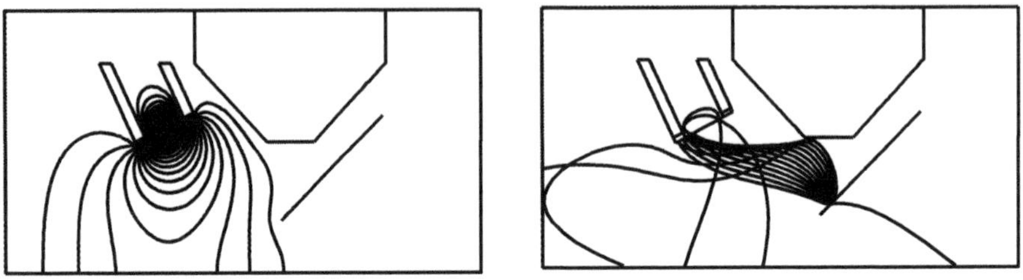

(a) (b)

Fig. 1-28. Simulation of secondary collection with a tilted specimen and scintillator cage: (a) Equipotential lines, (b) Trajectory paths of 3 eV secondary electrons over 20° polar angular intervals.

More comprehensive ray tracing of the SEs in three dimensions confirms the conclusions made from the two-dimensional simulations. Figure 1-29 depicts direct ray tracing of 1, 5 and 10 eV secondary electrons emitted at 10 polar angles ranging from -90 to 90° in uniform steps through a fully 3D solved field distribution that models the E-T detector, using the Lorentz 3D program, Integrated Engineering Software (2009). Figure 1-29(a) shows SE ray paths for a working distance of 20 mm, while Fig. 1-29(b) depicts SE ray paths for a working distance of 5 mm. Collection efficiencies for 1, 5 and 10 eV SEs are approximately 68, 40 and 30% respectively for a working distance of 20 mm. For a working distance of 5 mm, the collection efficiency is calculated to be 24, 22, and 18% (for 1, 5 and 10 eV respectively). Similar simulations performed by Balasubramanyam *et al.* (1995), Konvalina and Mullerova (2006) and Mullerova and Konvalina (2009) also report similar predictions for SE transport efficiency. It should be noted however, that collection efficiency of the scattered electrons is only part of understanding the resolution in the final image. Obviously, a higher collection efficiency is desirable, more collected electrons will reduce background image noise. On the other hand, the sensitivity of the contrast mechanism that gives rise to topographic contrast is also important, that is, there is a certain sensitivity with which variations in specimen topography are translated into changes in SE intensity, and this sensitivity will vary according to

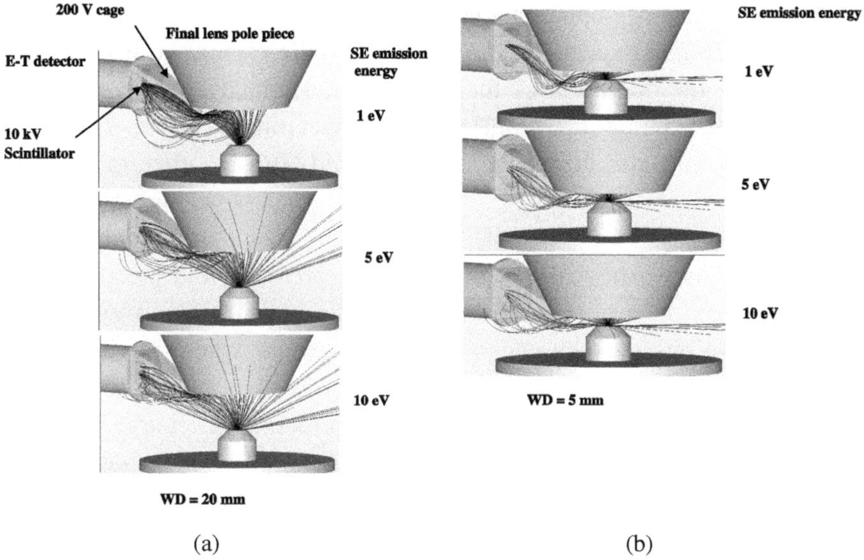

Fig. 1-29. Direct ray tracing of secondary electron trajectory paths from specimen to scintillator at emission energies of 1, 5 and 10 eV for 20° intervals in the polar and azimuthal angular directions . The scintillator potential is set to 10 kV and its surrounding cage is biased to 200 V. SE ray paths for (a) WD = 20 mm, (b) WD = 5 mm.

SE emission energy and angle. Lower energy SEs are expected to be more sensitive to topographic specimen changes, therefore, it may even be advantageous to deliberately filter out some higher energy SEs from detection, retaining the more topographically sensitive ones. The E-T detector, in its normal mode of operation, has reduced collection efficiencies for higher energy secondaries, and this may be one of the reasons why it has remained the most popular of all SE detectors for over 50 years.

1.11 Specimen charging

Severe charging up of insulator specimen surfaces can occur when irradiated by the primary beam of a SEM, and this often creates the appearance of fictitious artifacts in the SEM image. It may typically give rise to localized dark/bright areas in the image, and in some circumstances, may even reflect back the primary beam so that an image of the lens pole-piece/upper chamber is obtained, instead of the specimen surface. Since most specimens have non-metallic regions, it is therefore important to the operation of the SEM to understand how an insulator specimen can charge up. For biological specimens, the specimen is usually dried out and coated with a thin gold film in order to make it metallic.

Figure 1-30 shows how the total scattered electron yield (secondary + backscattered) σ, varies as a function of the primary beam energy E. There are two energies, known as cross-over points, where $\sigma = 1$ (at $E = E_1$ and $E = E_2$) that is, where for every primary electron going into the specimen, a scattered electron leaves it, creating a charge balance condition at the specimen surface. Since E_1 generally lies between 100–200 eV, it does not have much significance for SEM imaging. However, E_2 typically ranges from 1 to 3 keV (depending on the specimen), and knowledge of its precise value can often help reduce erroneous charging effects.

Consider the situation where the primary beam energy lies between the two cross-over points, $E_1 < E < E_2$, indicated by point A in Fig. 1-29, in this case, $\sigma > 1$, so the specimen will charge up positively (more electrons leaving it than entering it), this is a self balancing stable situation, because as the specimen charges up, it attracts some of the low energy scattered electrons back to its surface, reaching a small equilibrium positive voltage, typically of a few volts. On the other hand, if the primary beam energy is greater than E_2, indicated by point B in Fig. 1-30, in this case $\sigma < 1$, so the specimen charges up negatively. The negative build up of charge, unlike the positive build up of charge, does not have a self-stabilizing mechanism, so in time, there is nothing to stop the surface of the specimen charging up to negative voltages that are comparable in magnitude to the primary beam voltage. In fact, when the specimen charges up to negative voltages exceeding the magnitude of the primary beam voltage, it will act like an electric mirror, reflecting the primary beam back up the column/chamber.

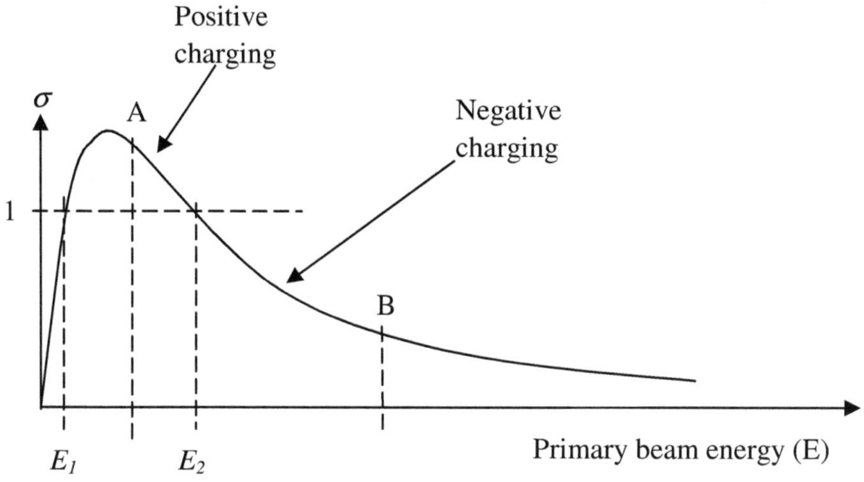

Fig. 1-30. Scattered electron yield (σ), as a function of primary beam energy (E) for insulators specimens.

For non-metallic specimens, the primary beam voltage is often adjusted to be around the second-cross-over point (E_2), a point of approximate charge balance. It must be noted however, that charging of the specimen varies from point to point on the specimen, since local specimen tilt also affects the number of scattered electrons that leave the specimen surface, so severe transverse fields can also arise. How fast an insulating specimen charges up also depends upon whether there are conductive paths down to ground, often, in order to minimize charging effects, the specimen is mounted on to a conductive plate. One way of monitoring the degree to which a specimen charges up negatively, is to monitor the surface potential via the use of an energy spectrometer.

1.12 Elastic BSE imaging

The shape of the BSE energy spectrum varies considerably with the polar angle range of emission. Figure 1-31 shows Monte-Carlo simulations for the BSE spectrum for a range of different 1° polar angle intervals on a gold specimen. This graph shows that the wide-angle BSEs, that is, BSEs emitted within the polar angular range 89–90°, have an energy spectrum consisting largely of elastic BSEs, where the energy spectrum is confined to a sharp peak around the primary beam energy. As the polar angle interval decreases, the shape of the BSE spectrum broadens and covers a larger range of energies. For the conventional BSE detector, usually placed under the objective lens, the polar angular range typically varies between 5 to 45°, the energy spectrum is therefore a relatively broad one, usually taken to span 50 eV to the primary beam energy.

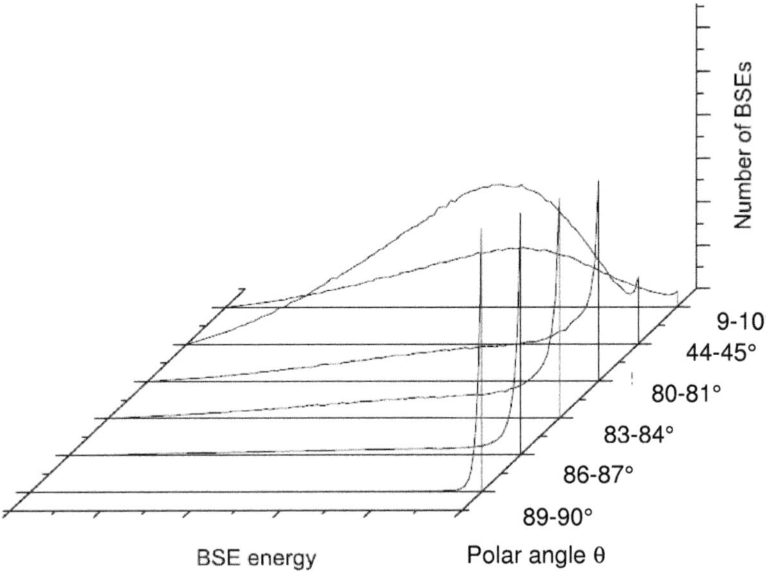

Fig. 1-31. Simulated Monte-Carlo BSE spectrum as a function of polar detection angle for gold.

Since BSE trajectory paths in a conventional SEM column are approximately straight (the SE detector does not significantly affect them), additional BSE detectors can be placed at the sides of the specimen in order to capture wide-angle BSEs. Wells (1971) pointed out that detection of elastic BSEs provides more information than is possible through the conventional BSE detector. This is because elastic BSEs are emitted closer to the specimen surface, and therefore have a smaller interaction volume, providing greater resolution in the BSE image. Figure 1-32 depict Monte-Carlo simulations of the scattering interaction inside a gold specimen, confirming this point. Wide-angle BSEs, emitted in the 89° to 90° polar angle interval, come from a much smaller interaction volume located close to the specimen surface (within several nano-meters), while BSEs over the whole polar angle range, 0° to 90°, come from a much larger interaction volume (reaching more than 20 nm below the specimen surface).

Elastic BSEs can either be obtained by energy filtering BSEs over narrow angles of emission (conventional detection), as proposed by

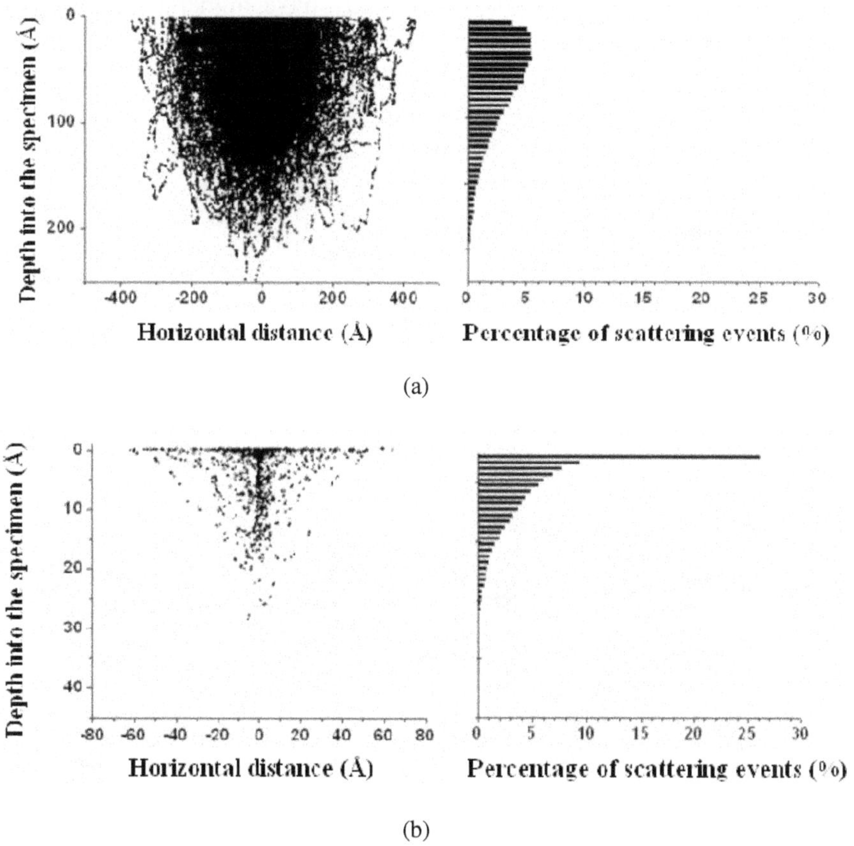

Fig. 1-32. Simulated BSE scattering events depth distribution of gold substrate for a primary beam of 5 keV for (a) polar emission angles between 0° to 90°, (b) emission polar angles between 89° to 90°.

Wells (1971), or by the detection of wide-angle BSEs, as proposed by Luo and Khursheed (2007). Rau and Robinson (1996) describe a BSE energy spectrometer for the SEM which is able to provide useful depth information about a multi-layer specimen by action of energy filtering.

1.13 Selected SEM image examples

Example of tungsten gun SEM images are shown in Fig. 1-33 at image magnifications ranging from 100 to 12,000. Compared to optical microscope images, these images not only have higher resolution (are much shaper), they have more depth information, made possible by the relatively small semi-angle at the specimen in the SEM (3–10 mrad).

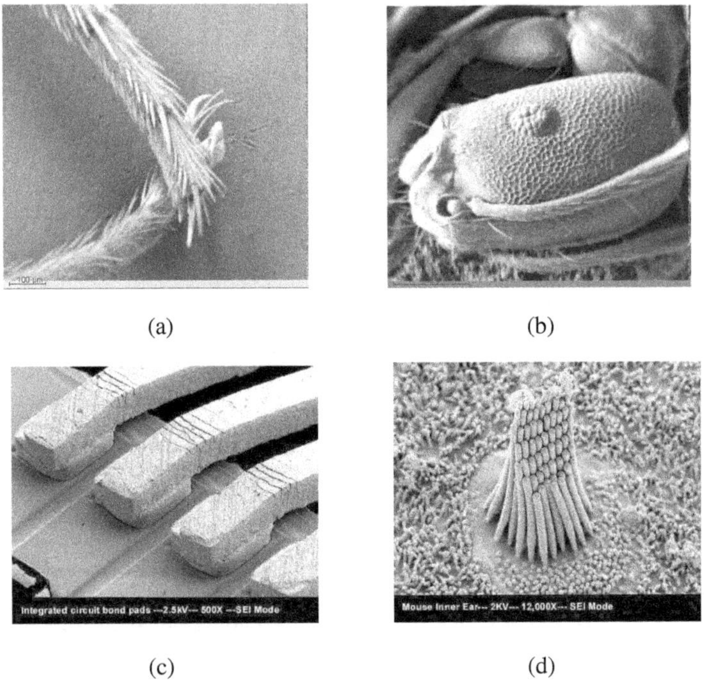

Fig. 1-33. Tungsten gun SEM images: (a) Fly leg, ×100, 15 kV (Courtesy of MEBI Pte Ltd, Singapore), (b) Ant head, ×200, 15 kV (Courtesy of MEBI Pte Ltd, Singapore), (c) Integrated circuit board, ×500, 2 kV (Courtesy of JEOL Inc., USA), (d) Mouse inner ear, ×12,000, 2 kV (Courtesy of JEOL Inc., USA).

Figure 1-34 shows some field emission gun SEM image examples for magnifications ranging from 27 to 200,000. In general, for image magnifications beyond 100,000, field emission gun SEMs are preferred. The better gun brightness and smaller energy spread of a field-emitter compared to the tungsten source, means that it is able to provide higher quality images at high resolution (typically below 5 nm), particularly at low primary beam voltages (≤ 3 kV).

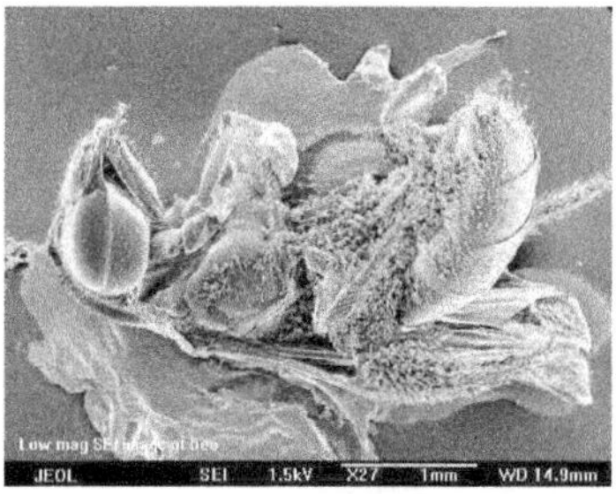

(a)

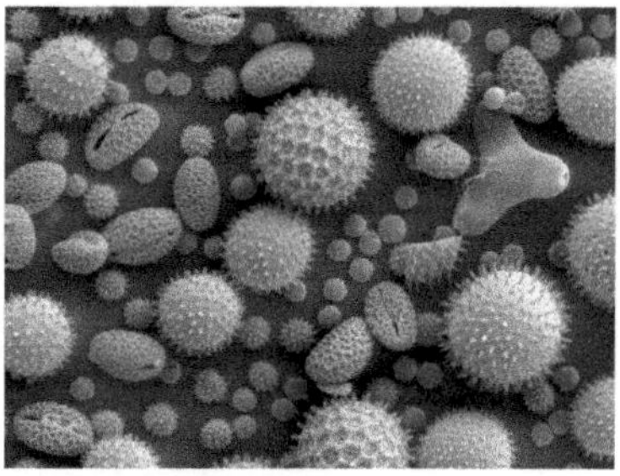

(b)

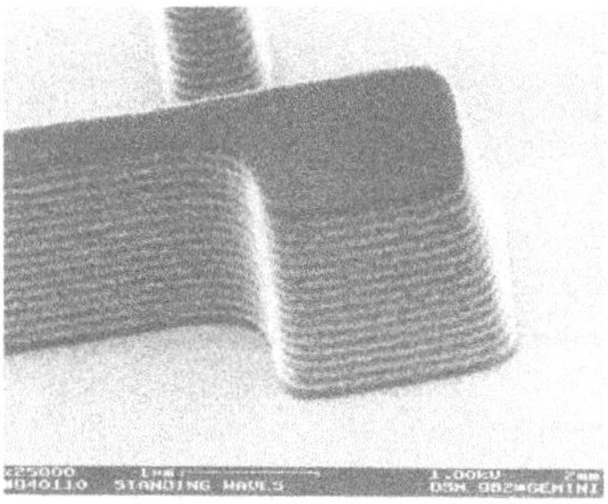

(c)

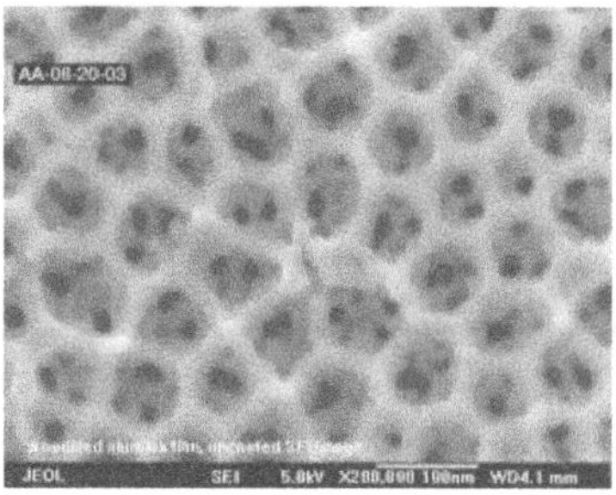

(d)

Fig. 1-34. Field emission gun SEM images: (a) House bee, ×27, 1.5 kV (Courtesy of JEOL Inc., USA), (b) Pollen from a variety of common plants, ×500, (c) Photoresist, ×100,000, 1 kV, (Gemini column, Courtesy of Carl Zeiss), (d) Anodized alumina film, ×200,000, 5 kV (Courtesy of JEOL, USA).

Figure 1-35 shows an example of how a SEM BSE image is able to provide material contrast information. The bright particles in the image correspond to a palladium catalyst, while the dark area comes from the SiO_2 carrier. Although images like these indicate a difference in material on the specimen, it is not possible to identify them. In order to subsequently perform quantitative material analysis in the SEM, an EDS is usually used (Reimer, 1998f). However, as discussed earlier, the BSE energy spectrum varies significantly with specimen material type, and in principle, with the use of an appropriate energy spectrometer, BSEs could also be used to provide some degree of quantitative material contrast information.

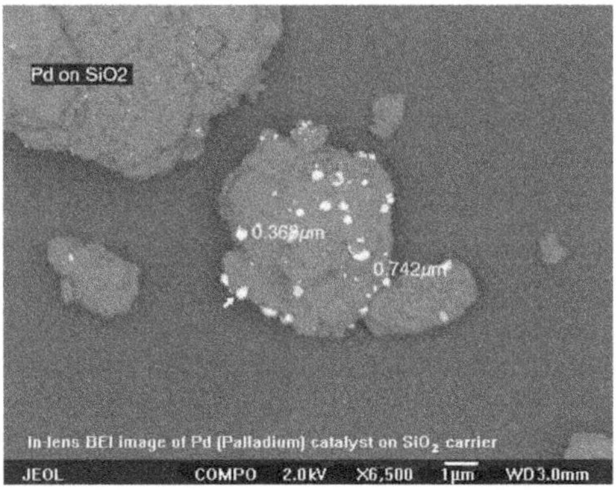

Fig. 1-35. BSE material contrast image of palladium catalyst on SiO_2 carrier, ×6,500, 2 kV (Courtesy of JEOL, USA).

Figure 1-36 show a series of BSE images taken at different polar detection angles (Luo and Khursheed, 2007) over a 1° collection interval, illustrating how BSEs can be used to provide specimen depth information. At one extreme, the wide-angle BSEs, shown in Fig. 1-36(a), reveals surface information, which in this case, shows the presence of various particles on the surface of an aluminium substrate specimen. At the other extreme, the narrow angle BSEs, shown in

(a) (b) (c) (d)

Fig. 1-36. BSE images of a buried gold line 100 nm below the surface of an aluminum substrate at different polar detection angles: (a) 89–90°, (b) 82–83°, (c) 44–45°, (d) 9–10°.

Fig. 1-36(d), reveals the presence of a gold buried layer track 100 nm below the specimen surface. This type of rudimentary tomographic information can also be obtained by the use of a BSE energy spectrometer (Rau and Robinson, 1996; Rau *et al.*, 2002). The elastic BSE image, shown in 1-36(a), has a higher image resolution than the conventional BSE image, which typically integrates BSEs emitted over polar angles ranging from 15 to 45°.

Figure 1-37 shows a variety of different SEM images for a gold on carbon test specimen, often used to estimate SEM image resolution. Figure 1-37(a) illustrates how the performance of a tungsten gun SEM degrades as the primary beam voltage is decreased. The image both darkens and becomes blurred as the primary beam voltage is lowered, this is due to the gun brightness decreasing, and probe aberrations becoming larger. For this reason, the tungsten gun SEM at a primary beam voltage of 1 kV is rarely operated at high image magnifications. The field emission gun SEM for the same conditions, gives much better results, as shown in Fig. 1-37(b). Low voltage SEM operation is preferred, since it has better SE yield, minimizes charging, and causes less specimen damage. Figure 1-37(c) shows how increasing the working distance degrades SE image resolution. The image taken at the working distance of 4 mm clearly shows many more fine features than the one taken with a working distance of 20 mm.

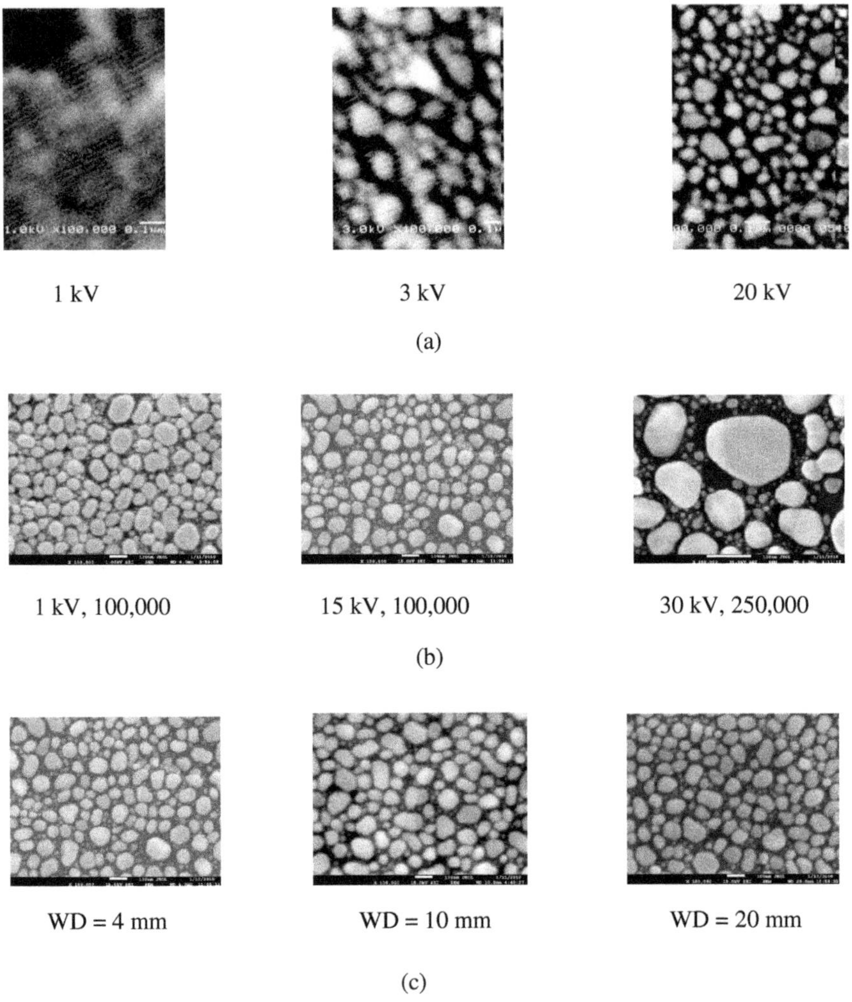

Fig. 1-37. Image resolution comparison for a gold on carbon test specimen: (a) JEOL 5600 Tungsten gun SEM at different primary beam voltages for a demagnification of 100,000, (b) JEOL 7100F Field emission gun SEM at different primary beam voltages and demagnifications, (c) JEOL 7100F Field emission gun SEM at different working distances for a primary beam voltage of 15 kV.

Figure 1-38 compares SE and BSE images for the test gold on carbon specimen. The BSE detector is usually divided into two different parts (halves), A and B. When the signals on A and B are added, the conventional BSE is obtained, one that is sensitive to material contrast. On the other hand, if the signals on the segments are subtracted, the image enhances topographic information and suppresses material information. The signal formed for A + B is termed BSE composition, while the signal A - B is referred to as BSE topographic. Figures 1-38(a) and 1.38(b) show that the SE image is of higher resolution than the BSE composition image, this comes from that fact that the interaction volume for BSEs is much larger than the one from which SEs are generated. Although the resolution of the BSE topographic image is also poorer than the SE image, it provides complementary information to the BSE composition image.

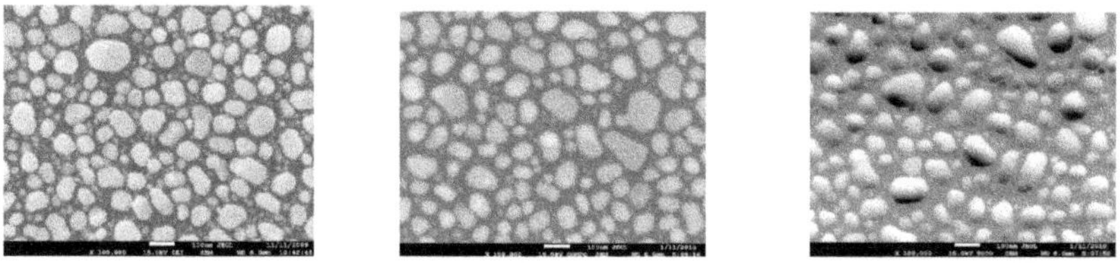

(a) (b) (c)

Fig. 1-38. Comparison of SE and BSE field emission gun images for a gold on carbon test specimen. A JEOL 7100F SEM is used at a demagnification of 100,000, a primary beam voltage of 15 kV and working distance of 6 mm: (a) SE image, (b) BSE composition image, (c) BSE topographic image.

Chapter 2

Spectrometer Design Principles

The motivation for acquiring secondary electron and backscattered electron energy spectra in the SEM comes from that fact that they contain valuable information about the specimen. Shifts in specimen potential, create corresponding shifts in the secondary electron energy spectrum, and change in specimen atomic number produces significant changes in the backscattered electron energy spectrum shape. The technique of quantitative voltage contrast, based upon shifts in the SE spectra, has been well established, and indeed, has historically been the main driving force behind the development of energy spectrometers for the SEM (Thong, 1993). This will be discussed much more fully in Chapter 5. On the other hand, the possibility of performing quantitative material analysis through the capture of the backscattered energy spectrum is only recently coming to light. Material identification has been proposed on the basis of either correlating experimental BSE spectra with their corresponding Monte-Carlo simulated ones (Luo and Khursheed, 2008), or by calibrating how the BSE spectrum's full-width at half-maximum height (FWHM) changes with atomic number (Hoang and Khursheed, 2009). This method might well be a useful companion method to the widely used Energy Dispersive X-Ray Spectrometry technique (EDS). The BSE method of elemental identification can in principle be used with LVSEMs (low voltage SEMs), overcoming important restrictions of standard EDS, namely, of being limited to operate with primary beam energies between 10 to 20 keV and having micron spatial resolution (Goldstein, 2003). Even for promising alternative low voltage EDS techniques (LVEDS), the LVSEM mode of operation (0.5 to 1 kV, with a few pA of probe current) is still far from possible to achieve (Boyes, 2002).

Backscattered spectrometers have also been used to provide depth information about multi-layer specimens (Rau and Robinson, 1996). Energy filtering for improving the resolution of backscattered electron images has long been recognized (Wells, 1971). Recent interest in capturing the secondary electron energy spectrum has come from attempts at quantifying specimen charging (Belhaj *et al.*, 2008), and for quantitative dopant concentration mapping on semi-conductor specimens (El-Gomati *et al.*, 2005, Kazemian *et al.*, 2006).

Before proceeding to discuss specific spectrometer designs in relation to the SEM, it is important to gain an overview of how they have been used elsewhere. The following summary of spectrometer design principles is largely drawn from their development in surface science and accelerator physics. A word of explanation about terminology is perhaps in order before proceeding. An "electron energy spectrometer", a term widely used in accelerator physics and other branches of applied physics, is usually referred to as an "electron energy analyzer" in surface science, and these two terms are used interchangeably throughout this book. Many of the principles summarized in this chapter can also be found in a recent review made by Yavor (2010).

2.1 Figures of merit

Figure 2-1 compares the different ways in which information is contained within SE, BSE and Auger electron spectral signals. The SE energy spectrum shifts as the specimen surface voltage (V_S) changes. The BSE spectrum changes shape for different atomic number (N). In the Auger electron energy spectrum, materials with different atomic number are represented by small peaks located at different energies. Auger spectrometers aim to reproduce the spectrum precisely, requiring high energy resolution in order to distinguish peaks that are located close to one another. On the other hand, SE and BSE spectrometers only need to detect changes in their respective spectra. Compared to Auger spectrometers, they detect relatively broad variations, and the most important criterion for them is to have a high signal to noise ratio, that is, changes to their spectra as the parameter of interest changes must be significantly higher than the background noise level. In the context of quantitative voltage contrast, a criterion of minimum resolvable specimen voltage change has been defined in terms of primary beam shot noise (Gopinath, 1977). This will be further discussed in Chapter 5.

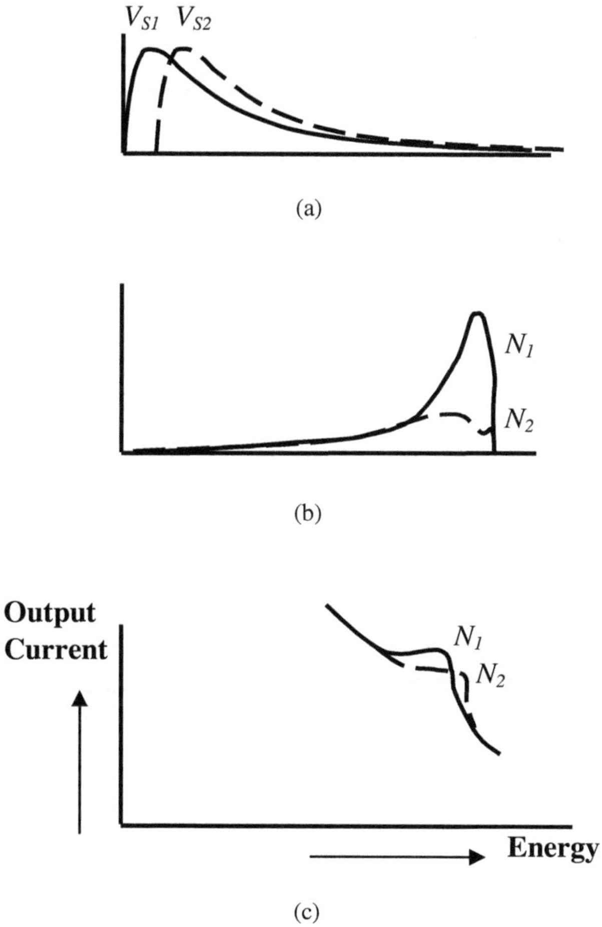

Fig. 2-1. Schematic comparison of scattered electron energy spectral information: (a) Secondary electron contrast of surface potential changes (V_{S1} to V_{S2}), (b) Backscattered electron of material contrast (N_1 to N_2), (c) Auger electron of material contrast (N_1 to N_2).

In this chapter, the performance of spectrometers will be summarized in relation to conventional figures of merit, of which the two most important are energy resolution and transmittance. There are many factors that limit the resolution of an energy analyzer. Some may come from imperfections of the analyzer's optics, that is, from aberrations. Others may be statistical in nature, say from the shot noise inherent in the primary beam and its interaction with the specimen. Other errors come from inevitable instabilities on the analyzer power supplies.

Let us assume that there is a monochromatic beam entering the analyzer, and that the effect of all these errors creates a truncated Gaussian shaped analyzer output signal, which would be, in the absence of analyzer errors, a perfect delta function. The finite size of an analyzer's output signal is usually referred to as its trace-width. For most analyzers, this signal width corresponds to the finite area of over which they arrive on the analyzer detection plane. Let the signal be centered at an energy of E_0, and have a width of ΔE_W at its base, as shown in Fig. 2-2. The energy resolution of the analyzer, defined by $\Delta E/E_0$, is usually taken to be approximately half the base resolution ($\Delta E_W/E_0$). This definition of energy resolution can be thought of as providing a reasonable estimate of how well two such signals can be separated, as indicated in Fig. 2-2. The validity of such an estimate is of course, dependent on the trace-width current distribution, which can vary greatly. In the following pages, spectrometers will be compared primarily in terms of their base energy resolution. The energy resolution will be taken as a percentage ($\Delta E/E_0 \times 100$), unless otherwise specified. Sometimes the reciprocal of energy resolution, called spectrometer resolving power will be used, defined by $E_0/\Delta E$.

Another important characteristic of an energy analyzer is its collection efficiency, or in spectrometer terminology, its "transmittance". Transmittance here, is defined to be the fraction of electrons at a given energy that arrive at the analyzer detector, compared to those that leave the specimen. The concept of an "emission hemisphere" was already outlined in the context of calculating the transport efficiency of scattered

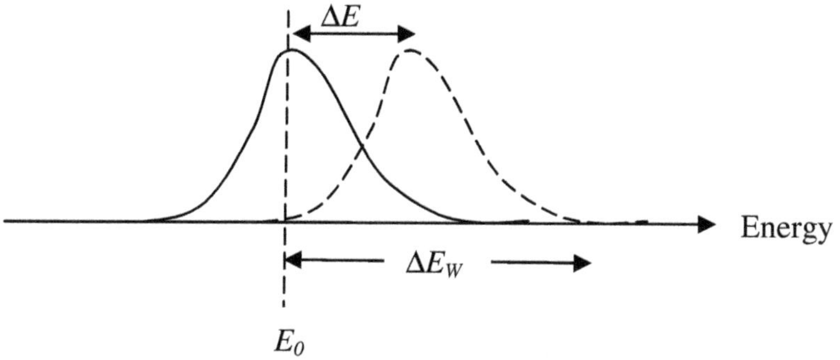

Fig. 2-2. Definition of analyzer resolution.

electrons in the previous chapter. The same concept will be used here to characterize the transmittance of an analyzer. In spectrometer literature, sometimes "transmittance" is taken to refer to the transport efficiency of electrons through the spectrometer alone, rather than from the specimen to the detector. However, for simplicity, transmittance will be taken to be the same as collection efficiency, the transport efficiency of electrons from the specimen to the plane or point of detection. The emission hemisphere for scattered electrons covers a solid angle of 2π steradians. If an analyzer, for a given energy, were able to detect all scattered electrons, and transfer them all to its detector (for a given energy), then it would have 100% transmittance. Most analyzers only detect a small portion of the full 2π steradian solid angle of emission, and their transmittance is usually limited by entrance apertures that define cut-off angles in the polar and azimuthal directions. In such cases, as with the estimate of BSE collection efficiency, the transmittance of the analyzer, T, can be calculated by

$$T = \frac{(\phi_2 - \phi_1)}{4\pi}[\cos 2\theta_1 - \cos 2\theta_2] \times 100 \tag{2.1}$$

where the analyzer azimuthal and polar entrance angles range from ϕ_1 to ϕ_2 and θ_1 to θ_2 respectively. Even in cases where analyzers are designed to capture the full 2π steradian solid angle of emission, electrons are usually scattered and absorbed within the analyzer. The transmittance of most analyzers lies well below 50%.

2.2 The SAM and the SEM

Auger Electron Microscopes (AESs), particularly Scanning Auger Electron Microscopes (SAMs), share much in common with SEMs. SAMs direct a focused primary electron beam on to a specimen, and capture electrons that are subsequently scattered back from it. Their electron beam columns have all the same kind of features as a standard SEM, a tungsten/field emission electron gun, condenser lenses, an objective lens, scanning/stigmator coils, a final aperture, and electron detectors, but in addition to all this, they have something that most SEMs do not usually have, namely, an electron energy analyzer.

Auger electrons emanate from the inner shell of atoms on and just below the specimen surface, and their kinetic energies are related to atomic shell energy differences, something which is of course, characteristic of the atoms being probed. The kinetic energies of Auger electrons can therefore be used to provide valuable elemental information about the specimen. The kinetic energy of Auger electrons typically range from 50 eV to 3 keV, depending on the type of sample material being probed, and their relative intensities are much lower than those for secondary or backscattered electrons: only one Auger electron is typically ejected from the sample surface for every 10,000 primary beam electrons, a yield of 10^{-4}. Compare this to SE yields of typically 0.35–1.6 for metals, and 1–10 for insulators, and 0.1–0.5 for BSE yields (Reimer, 1998c). However, Auger peaks in the energy spectrum are relatively narrow, having widths in the order of a few to tens of eV, and this makes it possible to locate them, despite the fact that they are submerged in a large background SE/BSE signal, as schematically illustrated in Fig. 2-3(a).

In order to distinguish Auger signals from their large SE/BSE background, experimental Auger measurements often operate in differentiated form, as depicted by the experimental Auger signal shown in Fig. 2-3(b). This brings out a very important difference between the energy spectrum to be acquired in the SEM, compared to spectra usually obtained in surface science, namely, what is considered to be an unwanted background signal in surface science, usually discarded, is the target spectrum to be detected in the SEM.

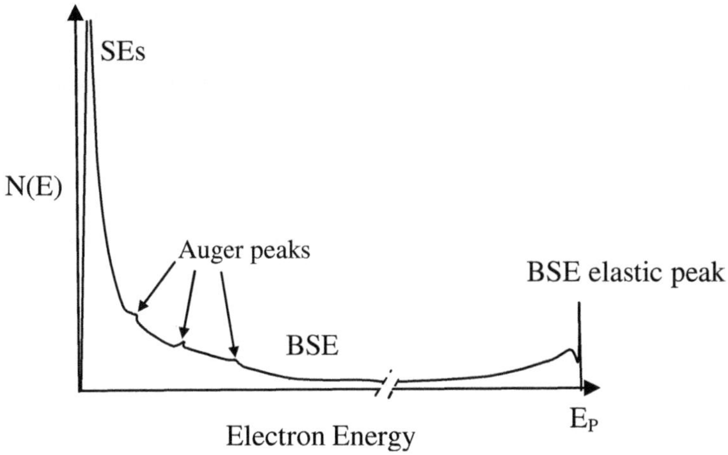

(a)

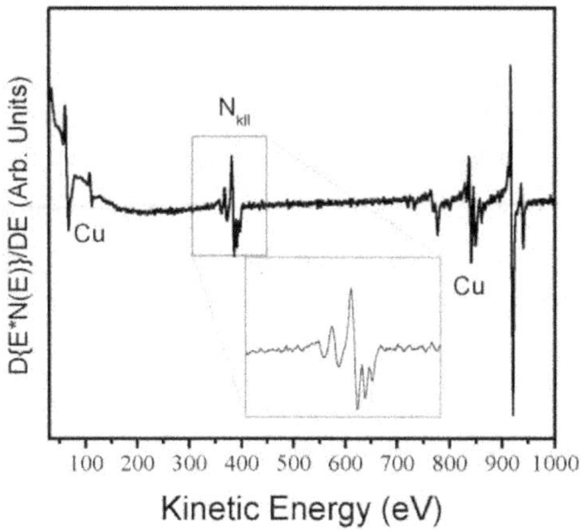

(b)

Fig. 2-3. Auger Electron spectra: (a) Schematic illustrating background SE/BSE signal, (b) Example experimental data of Copper Nitride.

Another important difference between Auger spectral analysis and the kind of energy spectrum that can be acquired in the SEM comes from the fact that the detection of Auger electrons requires an ultra high vacuum (UHV) environment, typically in the 10^{-10} Torr range. This is because the exit depth of Auger electrons is very small, only 1–10 mono-layers, and given that many mono-layers of contamination normally exist within the SEM chamber, whose vacuum level typically varies between 10^{-5} to 10^{-4} Torr, detection of Auger electrons in the SEM specimen chamber is not possible. The strongest Auger signal in the SEM chamber provides information about surface contamination, normally showing up as a carbon peak, and very little of anything else. For this reason, although scanning Auger electron microscopes (SAMs) have much in common with SEMs, their UHV requirement makes them very different instruments to SEMs in practice. As part of their operation, SAMs need an ion flood gun to clean the specimen in situ, and in general, the UHV chamber has a host of different devices/detectors, other than the electron beam column and SE detector. The specimen needs first to be prepared for examination in a Cleaning Chamber, and then transferred to an Analysis Chamber. Although a Scanning Auger Electron microscope is essentially a UHV Spectroscopic SEM, it is a very different instrument to the SEM in practice.

The working distance (distance of final lens to specimen) in the SAM is usually larger than it is for a SEM. The final primary beam probe size on the specimen in the SAM is generally a factor of 2–4 times greater than corresponding probe sizes inside the SEM. Since the intensity of the Auger signal is so small, it turns out that this is in many ways desirable from a signal-to-noise point of view, since a larger probe size at the specimen gives much larger detected currents (current of a non-aberration limited primary beam at the specimen is proportional to the square of its diameter).

2.3 The retarding field analyzer

One of the oldest methods to determine the kinetic energy of a charged particle is to use a retarding electric field, as shown in Fig. 2-4. The

Retarding Field Analyzer (RFA), as it subsequently became known in surface science, is essentially a high pass energy filter, where only particles with kinetic energies greater than a certain potential barrier, defined by eV_R, can reach the collector. Whether the grids are planar, as shown in Fig. 2-4(a), or circular, as shown in Fig. 2-4(b), depends on the angular distribution of the source and size of the analyzer entrance aperture. Obviously circular grids are preferred for scattered particles emanating from a point source, on the other hand, if the source is relatively far from the analyzer, and the entrance angles into it are relatively small, planar grids can be used. An entrance angular range of say, $\Delta\alpha$ with respect to the horizontal direction in the planar grid analyzer shown in Fig. 2-4(a) will give rise to an energy error of $KE_0 \sin^2\Delta\alpha$ in the final energy measurement, where KE_0 denotes the incident kinetic energy.

The reason that double grids at potential $-V_R$ are used is that they reduce the effect of non-uniformities of equi-potentials. Non-uniformities of this kind not only cause variations in the electric potential across the grid, but it also creates a weak focusing effect on electrons that pass through the grid. The potential in the space between grid-lines is inevitably lower than its value on the grid-line, and the magnitude of the drop is both proportional to the strength of electric fields on either side of the retarding grid, as well as the size of the space (hole) between grid-lines. By using double-grids (at the same potential), the electric field at least on one side of the retarding grid is zero. An estimated improvement of around one order of magnitude in the energy resolution of a RFA can be obtained by using double grids for Auger electron spectroscopy applications (Taylor, 1969).

Figure 2-5 shows a schematic diagram of how a RFA has been used both for Auger electron spectroscopy as well as Low-electron energy diffraction (LEED) (Talyor, 1969). Here the electron beam focusing column (not shown) is located above the spectrometer, and passes through a metal tube that shields it from the analyzer electrodes. The collector in this case is a fluorescent screen, which can image diffraction

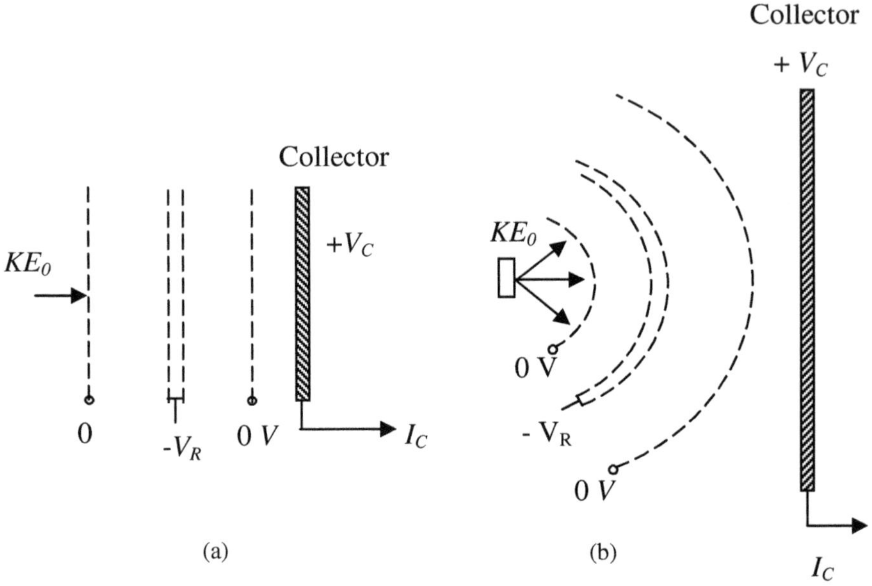

Fig. 2-4. Principle of a Retarding Field Analyzer: (a) Planar grid, (b) Circular grid shape.

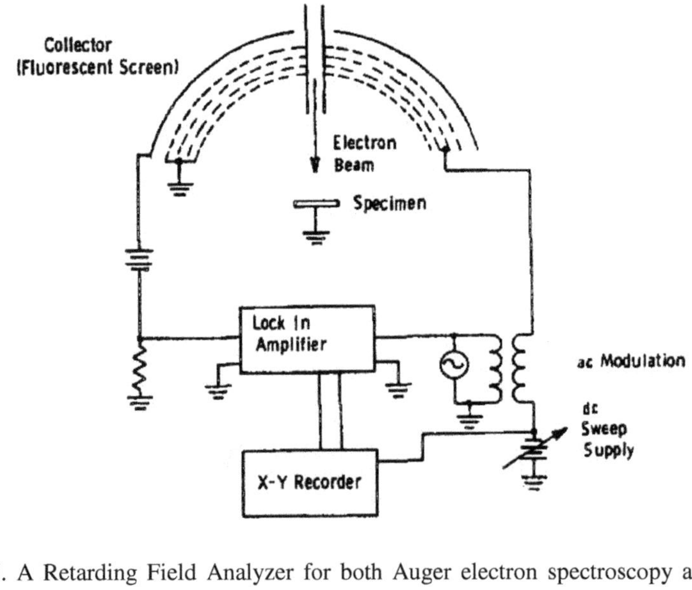

Fig. 2-5. A Retarding Field Analyzer for both Auger electron spectroscopy as well as Low-electron energy diffraction (LEED).

patterns, in the Auger electron spectroscopic mode, it will monitor a single bulk current. The primary beam energies for LEED, typically range from 20 to 200 eV, and do not require much focusing. The retarding potential, $-V_R$ is applied to the middle two grids, and has both AC and DC components. The technique of modulating an analyzer electrode voltage with a small AC signal, by say $k\sin\omega t$, and use of a lock-in amplifier at the spectrometer output, is quite a common technique in surface science in order to obtain the spectrum and its derivative. This technique is applicable to many different types of analyzers, not just to the retarding field analyzer. The DC bias on the retarding grid sets the potential barrier close to the energy range of interest, that is, close to a specific Auger peak energy. The amplitudes of the first and second harmonic of the output current $I(V_R + k\sin\omega t)$ is then monitored by the lock-in amplifier. By using a Taylor series expansion, it is straightforward to show that the amplitude of the first harmonic is predominantly proportional to the energy distribution $N(E)$, while the amplitude of the second harmonic is largely determined by its derivative, $dN(E)/dE$, as long as the amplitude of the AC signal is small compared to the DC component. This elegant technique of simultaneously obtaining the energy distribution and its derivative is only applicable when reproducing small variations in the energy spectrum. For signals that have relatively small widths, such as Auger electron peaks, it cannot provide the large scale variations of the SE or BSE spectrum.

In principle, if corrective steps are taken to reduce the field distortions imposed by grid non-uniformities, the RFA is capable of providing high energy resolution, of the order of 0.1% for electron energies that range between 500–1000 eV (Staib and Dinklage, 1977). The transmittance of the RFA is high in theory, since it can have an entrance solid angle of close to 2π steradians, but in practice, scattering at grids limits the transmittance to below 50%.

The most serious problem with the RFA is its relatively low output signal-to-noise ratio, which comes from the fact that only a small number of the electrons that reach the detector are sensitive to variations of the filter grid potential. Since the analyzer is essentially a high pass filter, all electrons having kinetic energies higher than the potential barrier are detected, therefore the small signal changes created by the filter potential changing take place over a large background signal, most of which does not carry any signal information. Of course, the AC modulation lock-in amplifier technique should in principle circumvent this, but in practice, a significant amount of noise is generated through scattering at the grids, which is greatly increased by allowing the high energy electrons to reach the detector. Several strategies to reduce the background noise at the output of retarding field analyzers have been proposed, most of which are based upon post-filtering the high energy electrons that pass through the potential barrier, effectively creating a band-pass filter. Figure 2-6 illustrates a method proposed by Hutchital and Rigden (1972), where a mirror grid is used to focus lower energy electrons (E_1) that surmount the potential barrier on to a small detector, while higher energy electrons (E_2) are not deflected sufficiently to be detected. The "Stop" electrode blocks the direct line-of-sight path to the detector.

Several other methods of post-filtering electrons that surmount the energy barrier in a retarding field analyzer have been proposed (Staib, 1972). Although RFAs are not widely used in surface science today,

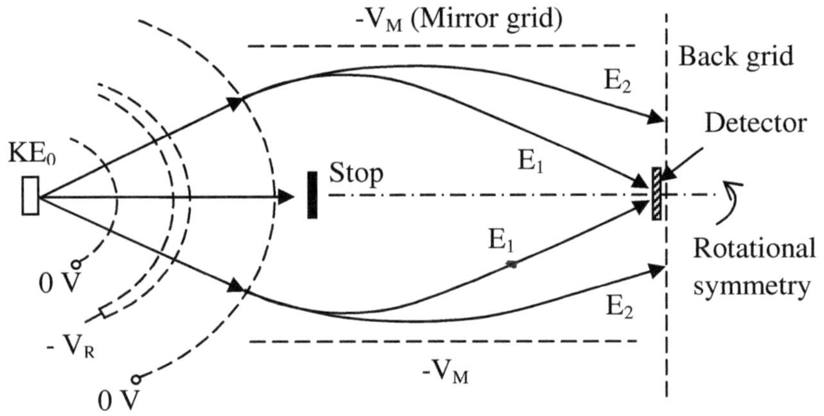

Fig. 2-6. Technique proposed to reduce background noise on a Retarding Field Analyzer.

experience derived from their use, and special techniques to improve their performance help to establish guiding principles for how such spectrometers might be incorporated into the design of the SEM. It turns out that RFAs are the most common type of spectrometers used to acquire the SE spectrum.

2.4 Deflection field analyzers

Most energy analyzers function by deflecting or bending electrons through angles that typically range from 45–180°. Normally, this action not only causes energy dispersion, but also focuses electrons on to a slit or detector plane, making it possible to separate or select electrons according to their initial energies. The simplest mode of operation for deflection analyzers is to use an output slit, so that electrons over a narrow energy range are detected. The deflection field strength within the analyzer is then ramped in time, so that the resulting time varying output signal represents the energy spectrum. This band-pass mode of operation is different from the retarding field analyzer, only a narrow range of energies reach the detector at any point in time, and it is inherently sequential, since it takes time to vary the analyzer deflection field strength. Deflection field analyzers therefore have better signal-to-noise characteristics than retarding field analyzers; however, the action of deflection imposes its own restrictions on the energy resolution that can be achieved.

2.4.1 *The parallel plate analyzer*

The focusing action inherent in magnetic or electric deflection fields is readily appreciated by understanding how the simplest of all such analyzers operates, the Parallel Plate Analyzer (PPA), which employs a uniform electric field to simultaneously disperse and focus an electron beam (Green and Proca, 1970). Figure 2-7 shows the layout of such an analyzer. Here electrons inside the analyzer start with an initial angle α relative to the horizontal (x) direction at $y = 0$. An electric field in the vertical (y) direction, V_m/d, slows down the incident electrons and

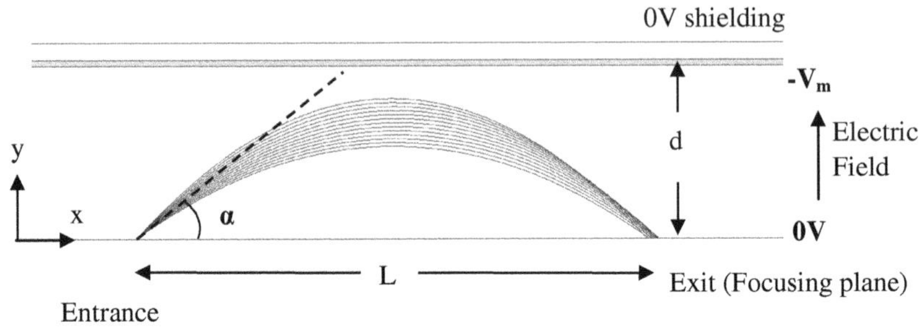

Fig. 2-7. Parallel Plate Analyzer layout.

mirrors them back to the $y = 0$ plane. Figure 2-7 also indicates trajectory paths derived by numerical simulation, through the use of the fourth-order Runge–Kutta method. The initial angles range from ±6° around a central ray of 45°, 13 trajectories are plot in uniform steps over this interval. The initial energy is 1 eV for all electrons, the mirror voltage (V_m) is -0.75 volts, and the plate separation is 15 cm.

The equations of motion for electrons moving inside the analyzer are given by

$$\begin{aligned} m\ddot{x} &= 0 \\ m\ddot{y} &= -e\frac{V_m}{d}. \end{aligned} \tag{2.2}$$

The solution to which is a well-known parabola governed by the following equation

$$y = -\frac{\eta}{2}\frac{V_m}{d}\left(\frac{x}{v_{x0}}\right)^2 + \frac{v_{y0}}{v_{x0}}x \tag{2.3}$$

where η is the electron charge to mass ratio, and v_{x0}, v_{y0} are initial velocities in the x and y directions respectively. If the initial kinetic energy of the particle is given in terms of energy E_0 (measured in electron volts), then the above expression can be expressed as

$$y=-\frac{\eta}{4}\frac{V_m}{d}\frac{\left(1+\tan^2\alpha\right)}{E_0}(x)^2+x\tan\alpha. \tag{2.4}$$

By solving the above equation for $y = 0$, the distance L (between entrance and exit focal points) is found to be

$$L=2d\frac{E_0}{V_m}\sin 2\alpha. \tag{2.5}$$

By fixing the initial angle α and length L, the mirror voltage V_m required to allow electrons of a certain pass energy E_0 is applied. It is scanned over a certain range, corresponding to the part of the energy spectrum that is of interest. Now consider small angular variations, $\pm\Delta\alpha$ around an initial angle α_0. Let the length L_0 be defined by no such angular variations, that is

$$L_0=2d\frac{E_0}{V_m}\sin(2\alpha_0). \tag{2.6}$$

It is straightforward to show that changes in the length due to initial angle variations give rise to a spherical aberration dispersion term, ΔL_A given by

$$\Delta L_A=L_0\left(\cos(2\Delta\alpha)+\frac{\cos(2\alpha_0)}{\sin(2\alpha_0)}\sin(2\Delta\alpha)\right)-L_0. \tag{2.7}$$

If $\Delta\alpha$ is small compared to α_0, then the sine and cosine functions can be approximated to give

$$\Delta L_A\approx L_0\left(1+\frac{\cos(2\alpha_0)}{\sin(2\alpha_0)}2\Delta\alpha-2\Delta\alpha^2\right)-L_0. \tag{2.8}$$

Now **first-order focusing** occurs where small variations with respect to $\Delta\alpha$ are zero, that is, to the first-order, trajectories that leave with slightly different initial angles focus at the same exit point. This condition is obviously achieved when $\alpha_0 = 45°$, so that

$$\Delta L_A \approx -2L_0\Delta\alpha^2 = -4d\frac{E_0}{V_m}\Delta\alpha^2. \tag{2.9}$$

Note that the change in distance here is negative, that is, small variations in angle from the central ray, whether negative or positive, result in electrons focusing before the original point of focus, this is similar to round lens spherical aberration.

Now consider the dispersion ΔL_D due to a change in initial energy, ΔE, assuming it to be much smaller than the initial energy E_0, it is simple to show that for the central ray, it is approximately given by

$$\Delta L_D \approx L_0\left(\frac{\Delta E}{E_0}\right). \tag{2.10}$$

Like round lenses, the point of focus moves linearly forward for larger energies, and linearly backwards for lower energies. The linear energy and quadratic angle dependence at the focal plane according to the above formulas are plot in Fig. 2-8. Also indicated are values derived by direct ray tracing, that is, numerical simulation using a fourth-order Runge–Kutta method. As expected, the analytical solution and numerical simulation agree very well (since the electric field in this simple idealized case is constant).

The width of the focal point at the output of an analyzer due to spherical aberration or energy dispersion is usually referred to as trace-width, W. For most spectrometers, the trace-width dependence with energy variations is linear and a single parameter known as dispersion D is usually sufficient to characterize it, normally linked to the gradient of the trace-width as it varies with energy

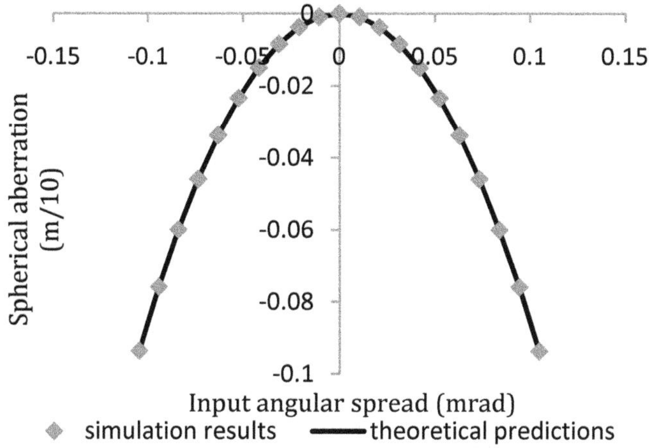

(a)

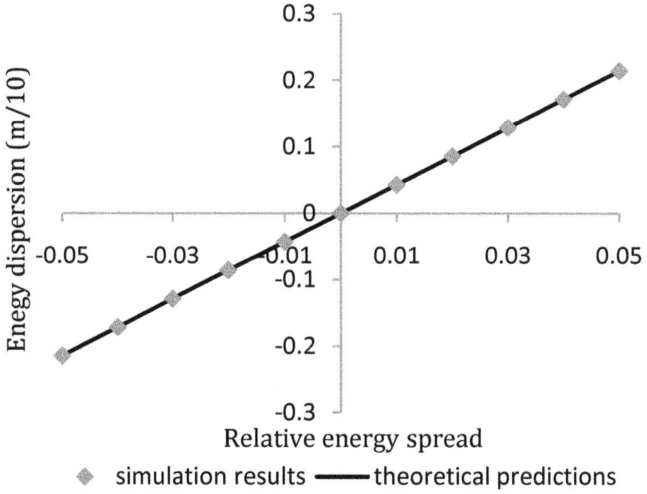

(b)

Fig. 2-8. Simulation results for a first-order focusing PPA: (a) Spherical aberrations at an input angular spread of ±6°, (b) Energy dispersion of the analyzer at a relative energy spread of ±5%.

$$D = E\left(\frac{dW}{dE}\right). \tag{2.11}$$

Once the dispersion of an analyzer is known, the trace-width at the analyzer output can be estimated for a given relative energy change

$$\Delta W = D\frac{\Delta E}{E}. \tag{2.12}$$

Or alternatively, a given trace-width can be mapped to an equivalent energy variation. This helps define the energy resolution of an analyzer. If ΔW_A is caused by spherical aberration, and is the parameter limiting the analyzer energy resolution, then the relative energy resolution is given by

$$\frac{\Delta E}{E} = \frac{\Delta W_A}{D}. \tag{2.13}$$

For the PPA analyzer at an entrance angle of 45°, the relative base energy resolution can therefore be estimated from

$$\begin{aligned} \Delta L_D &\approx L_0\left(\frac{\Delta E}{E_0}\right) = \Delta L_A \approx 2L_0\Delta\alpha^2 \\ &\Rightarrow \left(\frac{\Delta E}{E_0}\right) \approx 2\Delta\alpha^2. \end{aligned} \tag{2.14}$$

Here the dispersion $D = L_0$. In general, the relative energy resolution $\Delta E/E_0$ of an energy analyzer can be expressed in the following form

$$\frac{\Delta E}{E} = \frac{W_{SL}}{D} + \frac{W_A}{D} \tag{2.15}$$

where W_{SL} is the width of an output slit. For the purposes of estimating the ultimate energy resolution, the output slit term is usually omitted, and the aberration term is expanded in terms of a polynomial having spherical aberration coefficients (the first three of which are denoted by C_{S2}, C_{S3}, and C_{S4}

$$W_A = C_{S2}\Delta\alpha^2 + C_{S3}\Delta\alpha^3 + C_{S4}\Delta\alpha^4 + ... \tag{2.16}$$

It is assumed here that there is first-order focusing. The relative energy resolution of the analyzer can therefore be expressed in general terms by

$$\frac{\Delta E}{E_0} = \frac{C_{S2}}{D}\Delta\alpha^2 + \frac{C_{S3}}{D}\Delta\alpha^3 + \frac{C_{S4}}{D}\Delta\alpha^4 + ... \tag{2.17}$$

Both the energy dispersion D and spherical aberration characteristics of the analyzer (spherical aberration coefficients) need to be known in order to estimate the analyzer relative energy resolution.

Considerable effort in analyzer design usually goes into achieving the **second-order focusing**, that is, finding conditions where $C_{S2} = 0$, so that the focal point position does not change with respect to second-order variations in the input angle. For this situation, the focal point does not change with respect to $\Delta\alpha^2$, and the trace-width dependence with angular spread becomes a third-order one, resulting in significantly better energy resolution, or, making it possible to have greater entrance apertures (and therefore higher transmittance) for the same energy resolution. Either way, second-order focusing analyzer designs normally perform better than first-order focusing ones. Experience has shown that in order to achieve second-order focusing, greater flexibility in the analyzer design is required, where more free parameters can be varied.

It is instructive to see how this is achieved in the case of the PPA, one of the rare occasions where trajectory paths can be derived analytically. Figure 2-9 shows the second-order focusing layout. The point source and exit focal plane are now located outside the electric field a distance h

below the 0 volt lower plate), entrance/exit grids allow electrons to enter and exit the analyzer. Second-order focusing occurs for an entrance angle of 30° and $h = (1/4)\ (E_0/V_m)d = 5$ cm. Trajectory paths are numerically plot for the same angular spread (and analyzer plate voltage/separation distance as before) and indicate that a much sharper point of focus has been achieved, in comparison to the first-order trajectory paths shown in Fig. 2-7. All electrons are assumed to have an energy of 1 eV.

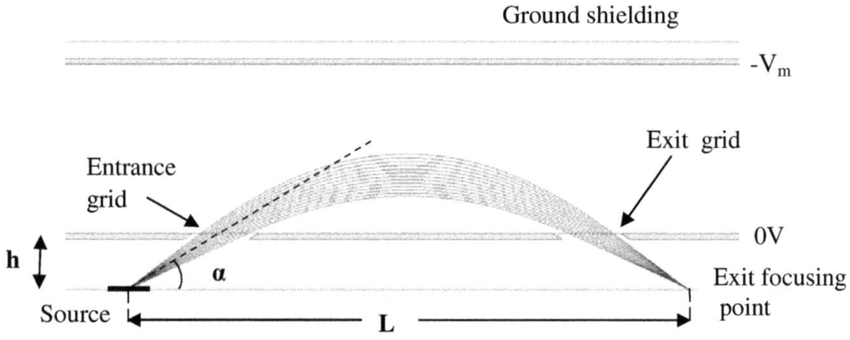

Fig. 2-9. Second-order focusing PPA layout (α = 30°).

With the addition of field-free regions at the analyzer entrance and exit, it is straightforward to show that the distance L between source and the exit focal point now becomes

$$L = \frac{2h}{\tan\alpha} + 2d\frac{E_0}{V_m}\sin 2\alpha. \tag{2.18}$$

Note that the entrance and exit angles are equal in magnitude (as it is for a mirror in light optics). Second-order focusing requires that the distance L not change with respect to first and second-order variations of the initial angle, that is

$$\frac{\partial L}{\partial \alpha} = \frac{\partial^2 L}{\partial \alpha^2} = 0. \tag{2.19}$$

If these constraints are enforced, the following conditions for second-order focusing can be derived (Schmitz and Mehlhorn, 1972)

$$\begin{aligned}
&\alpha = 30^\circ \\
&h = \frac{1}{4}\frac{E_0}{V_m}d \\
&L = 6\sqrt{3}h \\
&\Delta L_A = 32h\frac{\sin(2\Delta\alpha)\sin^2\Delta\alpha}{1-4\sin^2\Delta\alpha}.
\end{aligned} \tag{2.20}$$

The relative base resolution for second-order focusing is given approximately by

$$\frac{\Delta E}{E} = \frac{16}{\sqrt{3}}\Delta\alpha^3. \tag{2.21}$$

Figure 2-10 shows the trace-width dependence with energy and angular spread for these conditions, and as expected, the analytical prediction shows good agreement with numerical simulation (direct tracing using the fourth-order Range–Kutta method). The energy dispersion variation is still linear, but the spherical aberration dependence is now clearly cubic.

The relative base energy resolution of the second-order focusing PPA is only a factor of around 2 times better than the first-order focusing one, 1.054% compared to 2.18% respectively, but becomes better for smaller angular spreads, for $\Delta\alpha = 3^\circ$, it improves to around 4 (0.1317% compared to 0.545%).

Detailed investigation of trajectory paths around the exit focal plane for the second-order focusing PPA reveals that the best energy resolution does not occur at the Gaussian plane (at the distance h below the analyzer plate), but slightly above it, as shown in Fig. 2-11. Here,

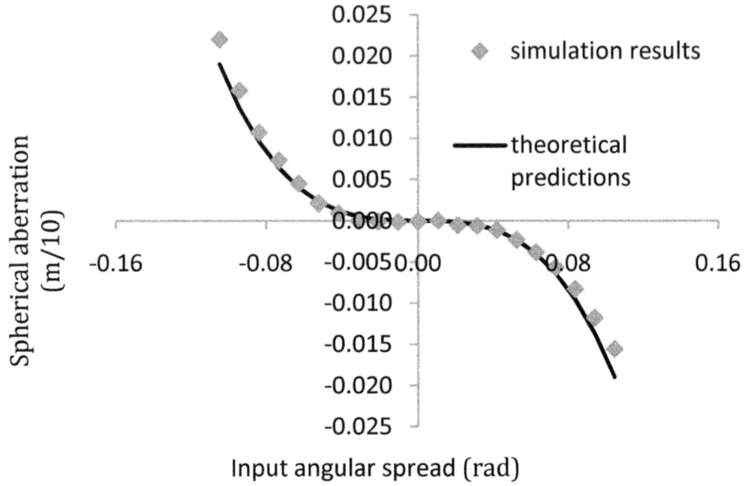

(a)

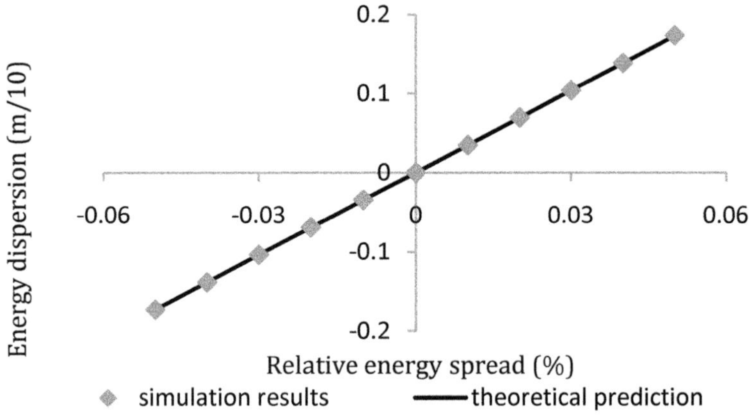

(b)

Fig. 2-10: Simulation results for a second-order focusing PPA: (a) Spherical aberrations at an input angular spread of ±6°, (b) Energy dispersion of the analyzer at a relative energy spread of ±5%.

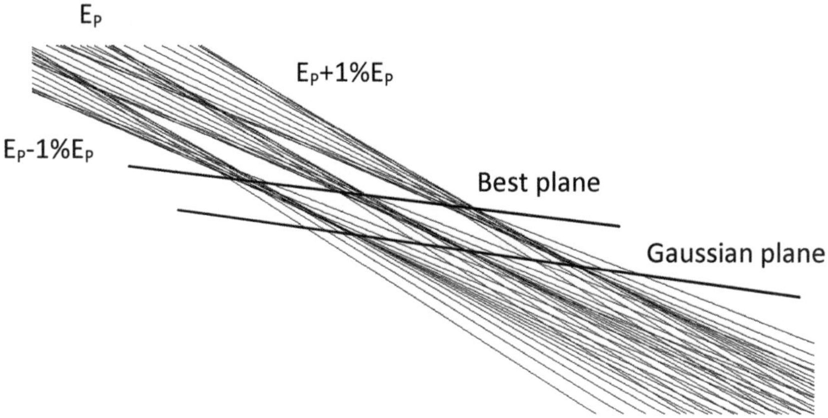

Fig. 2-11. Trajectory paths around the exit focal plane for the second-order Parallel Plate Analyzer for 3 different energies. The angular spread ranges from -6 to 6° in uniform steps, and the relative energy changes by 1%. E_P is the analyzer pass energy.

trajectory paths are plot for a relative energy difference of 1% around the pass energy (E_p) and an angular spread that varies from -6 to 6° around $\alpha = 30°$. The base energy resolution can be estimated visually. Figure 2-11 confirms that the base resolution (full-width) along the Gaussian plane is approximately 1%, since it is approximately equal to the same distance that separates the central rays of different energies (1%). Along the best focal plane, the full-width decreases by a factor of 3.3, reducing the base energy resolution to be around 0.3%, this type of improvement is well-known for second-order focusing systems, and will be referred to here as the best base energy resolution. For an angular spread of ±6°, the best base energy resolution of the second-order focusing PPA is around 7 times better than the first-order focusing one.

The transmittance of the PPA is relatively low, and this comes from the fact that in the out-of-plane direction, that is in the direction perpendicular to the plane of deflection (into the paper), the energy resolution is also limited. This is of course expected, since electrons that have a velocity component in the out-of-plane direction, will effectively have lower pass energies in the plane of deflection, for the same initial kinetic energy. If we denote β as the out-of-plane angle, then it can be shown that on the Gaussian plane, the relative base resolution for the first-order focusing PPA is given approximately by

$$\frac{\Delta E}{E_0} = 2\alpha^2 + \beta^2 \tag{2.22}$$

and for second-order focusing

$$\frac{\Delta E}{E_0} = \frac{16}{\sqrt{3}}\alpha^3 + \beta^2. \tag{2.23}$$

For second-order focusing on the best focal plane

$$\frac{\Delta E}{E_0} = \frac{4.8}{\sqrt{3}}\alpha^3 + \beta^2. \tag{2.24}$$

If the entrance aperture in the out-of-plane direction is adjusted so that the overall energy resolution drops by a factor of two, compared to its in-plane value, then the out of plane angular spread must be restricted to

$\beta = \sqrt{2}\alpha$ (first-order focusing) (2.25)

$\beta = 3.049\alpha^{3/2}$ (second-order focusing on the Gaussian plane) (2.26)

$\beta = 1.66\alpha^{3/2}$ (second-order focusing on the best focal plane). (2.27)

The out-of-plane angular spread must therefore be restricted to the same kind of values used for the in-plane angular spread. The transmittance for second-order focusing on the Gaussian plane for an angular spread of -6° to 6° around 30° in both directions is given by

$$\frac{I}{I_0} = \frac{\Delta\alpha\,\Delta\beta}{2\pi}\sin 2\alpha_0 = \frac{0.2094^2}{2\pi}\sin 60^0 \approx 0.6\%. \tag{2.28}$$

In order to keep the best base energy resolution below 0.6%, the out-of-plane angular spread must be restricted to be less than ±3.23°, bringing down the transmission to less than 0.323%. This transmission figure is further reduced by scattering on the entrance/exit grids. If it is assumed each grid has 90% transparency, then the transmittance falls to 0.261%. Compare this for the transmittance of hemispherical retarding field analyzers, which can have transmittances of up to 50%. The most serious drawback of the PPA is this relatively low transmittance value, generally only a few fractions of a percent.

One way of greatly enhancing the PPA transmittance is to make it rotationally symmetric, necessitating the use of a circular annular slit/detection plane, as shown in Fig. 2-12. This analyzer, known as the Fountain Analyzer, has energy resolution/focusing characteristics identical to the PPA, since like the PPA, it uses upper and lower parallel plates to create a uniform electric field. However, its transmission is greatly enhanced, since all electrons in the out-of-plane (azimuthal) direction reach the exit focal plane. Due to its inherent rotational symmetry, all electrons that pass through it experience in-plane deflection, and there are no out-of-plane electrons which compromise the energy resolution.

As shown in Chapter 1 (Eq. 1.43), the transmittance for rotationally symmetric detection with polar angles of entry between θ_1 to θ_2 can be estimated by

$$\frac{I}{I_0} = \frac{1}{2}\left[\cos 2\theta_1 - \cos 2\theta_2\right]. \tag{2.29}$$

If we substitute angular spreads of ±6° around 30° into this expression, the transmittance of a second-order focusing Fountain Analyzer having a best base energy resolution of 0.3% is 18%, without taking into account scattering on grids. If we again assume 90% transparency on each grid, this transmittance is lowered to around 14.6%, a transmittance of over 50 times better than the one for the PPA.

The Fountain Analyzer raises some practical difficulties. Firstly, it is only useful if electrons from a source are scattered over the full 2π sr solid angle. Secondly, it must be possible to place the source just below the analyzer on its rotational axis of symmetry. Thirdly, an annular slit or ring detector is required. These practical difficulties have limited the use of the Fountain Analyzer in practice. In the SEM, electrons do scatter over the full 2π sr solid angle, and the specimen can be placed just below the analyzer lower plate. However, the greatest difficulty for incorporating the analyzer in the SEM is that the primary beam needs to travel through the analyzer in order to reach the specimen, therefore the field inside the analyzer will disturb the primary beam, and may defocus it. Primary beam passing holes in the analyzer lower/upper plates in the presence of an electric field will certainly have a focusing effect on the primary beam. A later chapter will show that it is possible to redesign the Fountain Analyzer to screen this effect, and that the Fountain Analyzer design can be used inside the SEM.

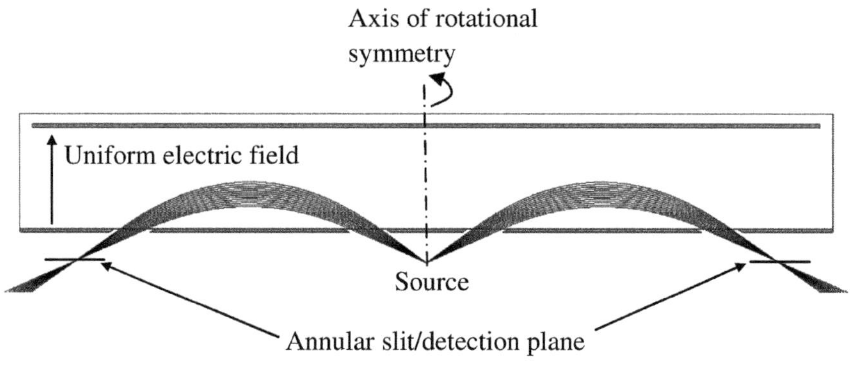

Fig. 2-12. Schematic for the Fountain Analyzer layout.

2.4.2 *The cylindrical mirror analyzer*

Another analyzer that uses the same kind of principle as the PPA, that of mirroring electron trajectory paths by the action of an electric field distribution, is the Cylindrical Mirror Analyzer (CMA). Figure 2-13 shows a schematic diagram of the CMA layout. In this case, the electric field distribution is created between concentric cylinders which are biased at different voltages, the inner one is usually grounded, located at radius R_1 from the rotational axis of symmetry, and the outer one, located at radius R_2 is biased to a mirror voltage ($-V_m$). An electron optical column is usually placed inside the analyzer, in a field-free central region, and like the PPA, there are entrance/exit grids. The source (specimen) and focal point lie on the rotational axis of symmetry. This analyzer has been widely used for Auger spectroscopy.

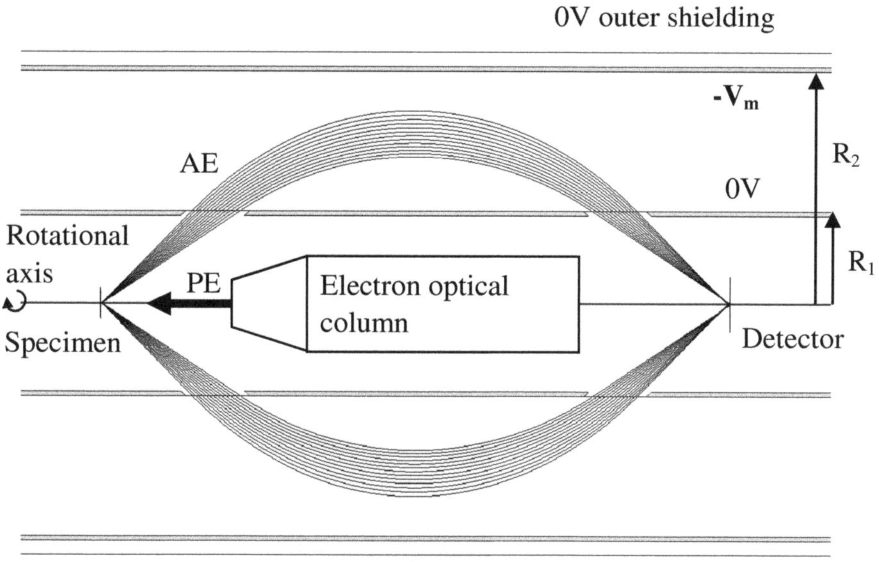

Fig. 2-13. Layout of the Cylindrical Mirror Analyzer.

Unlike the PPA, the electric field strength is not constant inside the analyzer, it is proportional to the inverse of the radius, as with all other coaxial cylinder geometries. In general, the pass energy, E_0 can be found from the following relation

$$E_0 = \frac{k}{R_1} \frac{V_m}{\ln(R_2 / R_1)} \tag{2.30}$$

where k is a constant that depends on the distance L separating the specimen and detector plane along the axis of rotational symmetry, which is fixed for a given CMA design, related to the initial take-off angle α_0. The range of scan for the mirror voltage V_m to reproduce a certain part of the energy spectrum can be calculated from this formula.

Figure 2-13 depicts direct ray tracing for a popular second-order focusing CMA layout, where the take-off angle α_0 is 42.3° (relative to the rotational axis), this is the second-order focusing condition as described by Sar-el (1967) and Hafner *et al.* (1968), for other take-off angles, the analyzer is characterized by first-order focusing. In Fig. 2-13, trajectory paths are plot for an angular spread $\Delta\alpha$ ranging from -6° to 6° in uniform steps, with R_1 = 1 cm, R_2 = 2.5 cm, E_0 = 1 eV, and V_m = 0.669 volts. Changes in the distance separating the specimen and detector, ΔL, are approximately given by

$$\Delta L = 5.6 R_1 \frac{\Delta E}{E_0} - 15.4 R_1 \Delta\alpha^3 + 10.3 R_1 \frac{\Delta E}{E_0} \Delta\alpha. \tag{2.31}$$

The first term describes energy dispersion, the second terms comes from spherical aberration, and the third terms comes from a combination of both energy and angle dispersion. If it is assumed that the limiting factor on energy resolution comes from angular dispersion, that a certain shift ΔL caused by $\Delta\alpha$ is equivalent to a ΔE as given by the above formula, then relative changes in energy can be found from equating ΔL to zero and solving for $\Delta E/E_0$

$$\frac{\Delta E}{E_0} = \frac{15.4 \Delta\alpha^3}{5.6 + 10.3 \Delta\alpha}. \tag{2.32}$$

For $\Delta\alpha$ ranging from -6° to 6°, the corresponding relative change in energy goes from -0.391% to 0.265%, giving a total base relative energy resolution (full-width) of 0.656%, around 1.5 times lower than that for the second-order focusing PPA/Fountain Analyzer.

Figure 2-14 shows detailed simulated trajectory paths around the exit focal plane for the CMA. Here 3 sets of 13 trajectories were plot with an angular spread ranging from -6° to 6° range in uniform steps, separated by relative energy differences of 1%. The relative base (full-width) energy resolution at the Gaussian plane from these simulations is 0.662%, close to the 0.656% predicted from the analytical approximation (Eq. 2.32). It is clear that like the PPA/Fountain Analyzer, the best resolution is located slightly above the Gaussian plane, by a radius r_c, where the full trace-width is reduced by a factor of around 3 as compared to its value on the Gaussian plane (on the rotational axis of symmetry). The best base resolution for an angular spread of ±6° in the second-order focusing CMA derived from these simulations is around 0.2%, slightly lower than the corresponding (in-plane) value of 0.3% for the PPA/Fountain Analyzer.

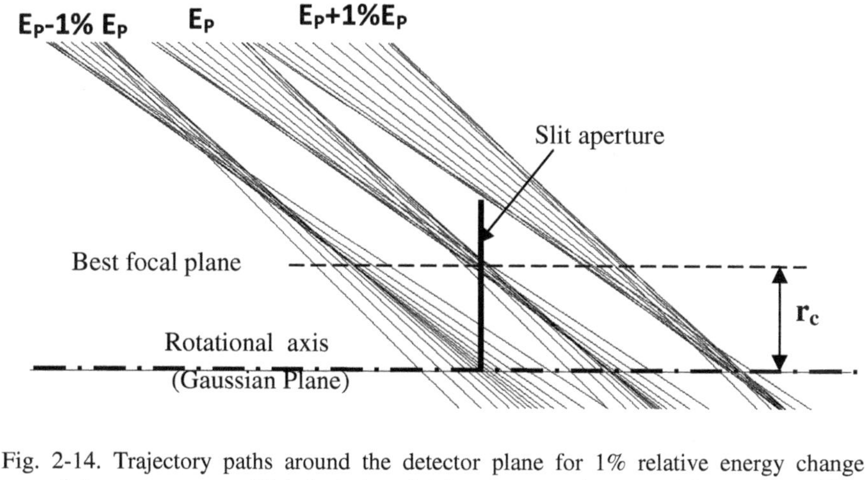

Fig. 2-14. Trajectory paths around the detector plane for 1% relative energy change around the pass energy. 13 trajectories plot between angular spread from -6° to 6° in uniform steps. E_P is the analyzer pass energy.

From a transmittance point of view, the CMA is much better than the PPA, like the Fountain Analyzer, it has rotational symmetry and all electrons that pass through it are deflected by the same in-plane field distribution, there are no out-of-plane trajectory paths (neglecting the finite size of the source). The transmittance of the second-order focusing CMA is estimated to be 16.84%, assuming that its entrance/exit grids have 90% transparency. This is slightly higher than that of the Fountain Analyzer, coming from the fact that its take-off angle of 42.3°, is closer to 45°, the angle of maximum transmittance for rotationally symmetric detection, compared to 30° for the second-order focusing Fountain Analyzer. The CMA is superior in energy resolution and transmittance to the Fountain Analyzer, and this maybe one of the reasons why it has been widely used in surface science. Another reason for its popularity comes from the fact that the electron column can be placed inside the analyzer, so that the electron focusing primary beam has a field free path to the specimen. This is not easy to achieve for the Fountain Analyzer. One disadvantage of the CMA is that it is sensitive to misplacement of the specimen. Since the aperture plane is usually fixed, misplacement of the specimen can result in a shift ΔL, which causes sizable errors in the final energy spectrum. From Eq. 2.31, if for instance R_1 = 15 mm, and the angular spread is ±6°, a misplacement error of 1mm in specimen position can generate a 12 eV shift in the final spectrum at a pass energy of 1,000 eV. This depth of focus limitation obviously becomes much more critical for smaller angular spreads, limiting the use of the CMA for high energy resolution spectroscopy applications (where a relative energy resolution of say 0.2% is required). In the present context, where it is broad SE and BSE spectral variations that are of interest, the CMA's depth of focus limitation is not a serious disadvantage.

Instead of placing the entire primary beam electron column inside in the CMA analyzer, it is possible to put in only the objective lens, and deflect the scattered electrons as they exit the CMA on to an off-axis multi-channel detector (Narum, 1993). Some commercial Scanning Auger Microscopes use this method. Figure 2-15 shows a schematic drawing of the PHI 700 Scanning Auger microscope made by Physical Electronics, which uses such a technique. The column can work in a SEM imaging mode, as well as a SAM analysis mode. The working distance in such columns must be fixed, and is inevitably longer than it would ordinarily be if the CMA were not present. This instrument is one working example of how electron energy analyzers can be combined with SEMs.

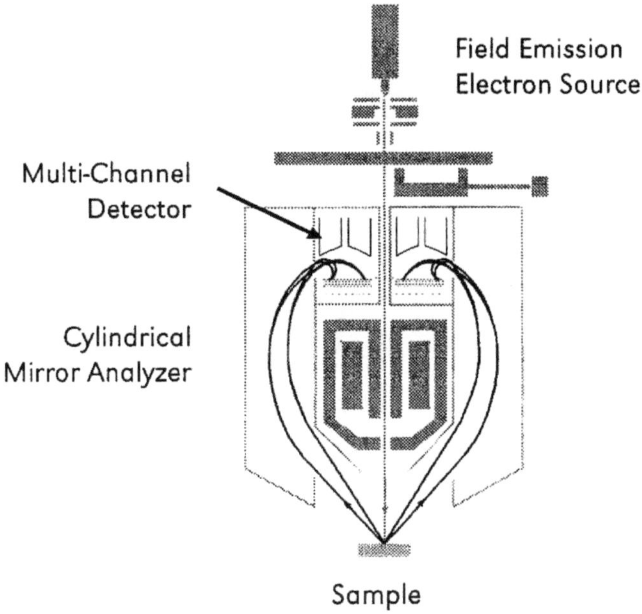

Fig. 2-15. Schematic diagram of a magnetic objective lens inside a CMA. Courtesy of Physical Electronics.

2.4.3 *Electric sector analyzers*

Electric sector analyzers function by using a radial electric field to bend a beam of electrons through a deflection angle Φ, as shown in Fig. 2-16(a). Inner and outer curved plates biased to voltages V_1 and V_2, having radii R_1 and R_2 about the point O, create the in-plane radial electric field (in the x-y plane), while in the out-of plane direction (in the z-y plane) the shape of the electrodes can take different forms. In general, toroidal deflection analyzers (TDAs) are characterized by curved electrodes in the out-of-plane direction, as shown in Fig. 2-16(b). For the special case where the in-plane and out-of-plane electrode curvature is the same, the analyzer is known as the Spherical Deflection Analyzer (SDA). For an out-of-plane parallel plate structure, the analyzer is known as the Cylindrical Deflection Analyzer (CDA), as shown in Fig. 2-16(c). The following summary is based upon the work of Banford (1966b).

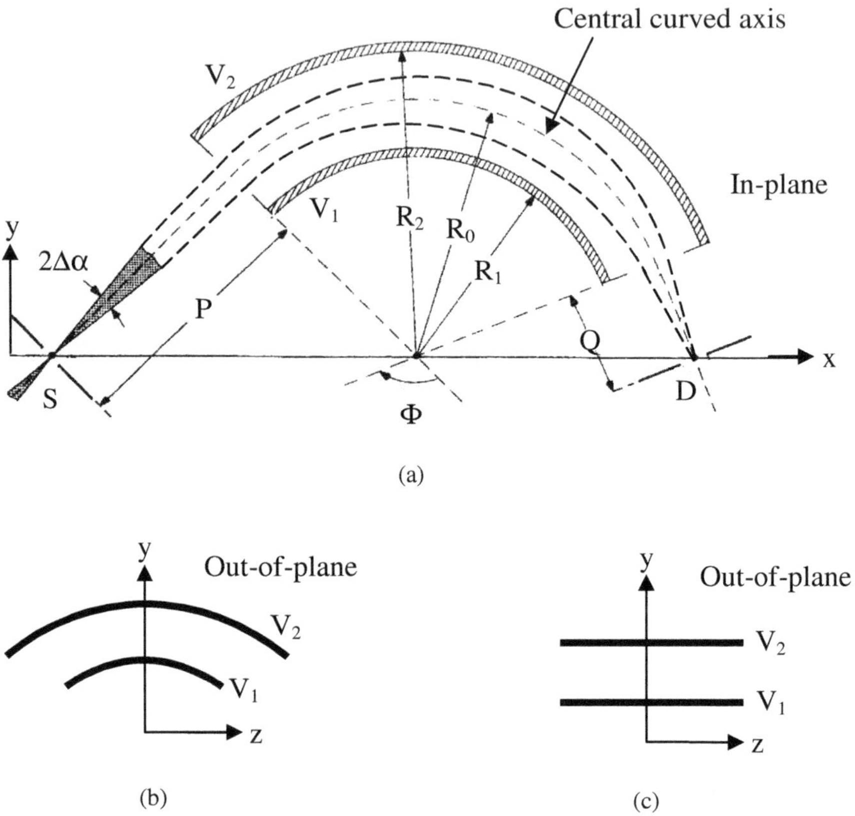

Fig. 2-16. Schematic diagram for the electric sector analyzer layout: (a) In-plane radial electric field deflection, (b) Toroidal/spherical deflection analyzer, (c) Cylindrical deflection analyzer.

Figure 2-17 shows a schematic representation of an electron travelling a small distance r from the central curved axis located at R_0. As an electron travels along the central curved axis, its path length can be expressed in terms of the angle θ measured with respect to the point O, and is θR_0. Let the radial component of the electric field strength, E_r, at point $R_0 + r$, have the following form

$$E_r = \frac{c}{(R_0 + r)^N} \tag{2.33}$$

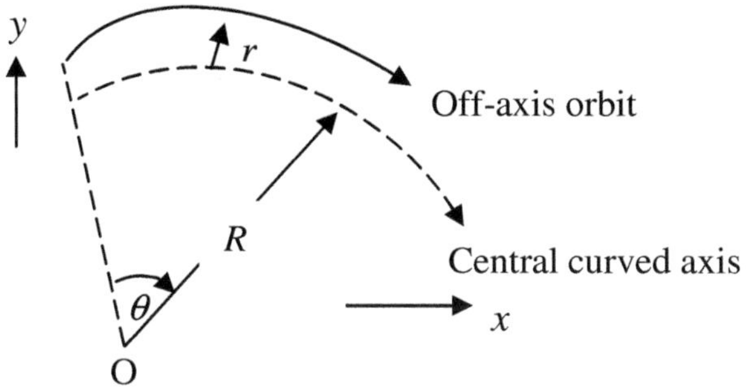

Fig. 2-17. Schematic diagram of an off-axis curved orbit inside the electric sector analyzer.

where c is a constant, and N = 1 for straight parallel plates in the out-of-plane direction (Cylindrical Deflection Analyzer, as depicted in Fig. 2-16(c), and N = 2 for the Spherical Deflection Analyzer, as shown in Fig. 2-16(b). Assuming that $r << R_0$, E_r can be expressed as

$$E_r \approx E_{R0}\left(1-\frac{Nr}{R_0}\right) \quad (2.34)$$

where E_{RO} is the electric field value on the central curved axis, given by $E_{RO} = c/R_0^N$.

The force and energy equations in the radial direction on the central curved axis are

$$\frac{mv_0^2}{R_0} = eE_{R0} \quad (2.35)$$

$$\frac{1}{2}mv_0^2 = eV_0 \quad (2.36)$$

where v_0 is the electron velocity along the central curved axis, on which the radial component of the velocity, v_r is zero. The small potential drop for an off-axis point at r can be equated to the kinetic energy in the radial direction

$$eV_0 - erE_{R0} \approx \frac{1}{2}mv_r^2. \tag{2.37}$$

Substituting for E_{R0} and V_0 in terms of the velocity v_0 in this equation, gives an expression for v_r

$$v_r^2 = v_0^2\left(1 - \frac{2r}{R_0}\right). \tag{2.38}$$

The radial force acting on an electron off-axis at point r is

$$m\ddot{r} = \frac{mv_r^2}{(r+R_0)} - eE_r. \tag{2.39}$$

Substituting in for E_r and v_r into this equation, approximating $(r + R_0)^{-1}$ with $(1-r/R_0)/R_0$ and neglecting terms in r^2, the above equation can be simplified to

$$\ddot{r} \approx \frac{v_0^2}{R_0} - 3\frac{v_0^2}{R_0^2}r - \eta E_{R0}\left(1 - \frac{Nr}{R_0}\right). \tag{2.40}$$

When substituting E_{R0} in the above equation in terms of v_0 (along the central curved axis, Eq. 2.35), this further simplifies to

$$\frac{\ddot{r}}{v_0^2} + (3-N)\frac{r}{R_0^2} = 0. \tag{2.41}$$

If it is assumed that the velocity along the path of each trajectory is much greater than its radial component, so that the overall path length $s \approx v_0 t$, where the time variation in r can be replaced by r', its variation with respect to s, the double derivative with respect to s is given by

$$r'' = \frac{\ddot{r}}{v_0^2} \tag{2.42}$$

The equation of motion then becomes

$$r'' + (3 - N)\frac{r}{R_0^2} = 0. \tag{2.43}$$

This equation explains why, for trajectories close to the central curved axis, there is a focusing effect. The above equation is a simple second-order harmonic differential equation which has sine and cosine function solutions of the following form

$$r = A\cos(\frac{Q_{re}s}{R_0}) + B\sin(\frac{Q_{re}s}{R_0}) \tag{2.44}$$

where A and B are constants and $Q_{re} = (3\text{-}N)^{1/2}$. The term, s/R_0 is equal to the angle θ, measured with respect to the point O. If the analyzer entrance values of r and r' are denoted by (r_1, r_1') and their subsequent values at angle θ are (r_2, r_2'), then the solution to trajectory paths through the analyzer can be expressed as

$$\begin{bmatrix} r_2 \\ r_2' \end{bmatrix} = \begin{bmatrix} \cos Q_{re}\theta & \dfrac{R_0}{Q_{re}}\sin Q_{re}\theta \\ -\dfrac{Q_{re}}{R_0}\sin Q_{re}\theta & \cos Q_{re}\theta \end{bmatrix} \begin{bmatrix} r_1 \\ r_1' \end{bmatrix}. \tag{2.45}$$

Some important conclusions about electric sector analyzer design can be made from the above trajectory path solution. If the source is located at the analyzer entrance, $r_1 = 0$, then

$$r_2 = \frac{R_0}{Q_{re}}\sin Q_{re}\theta. \tag{2.46}$$

Therefore, first-order focusing within the analyzer $(r_2 = 0)$ will occur at a deflection angle, $\theta = \pi/Q_{re}$, which for the CDA ($N = 1$) is $\pi/\sqrt{2}$ (127.279°), and for the SDA, it is π. Analyzers that use such conditions at their exit planes are illustrated in Figs. 2-18(a) and 2-18(b), and have been widely used in applied physics for a variety of different applications. In these figures, rays have been simulated using analytical field expressions. Figures 2-18(a) and 2-18(b) show that first-order focusing properties at the analyzer exit are typically maintained for an energy spread of approximately ±5% around the pass energy, which means that a limited degree of energy parallel detection is possible with analyzers of this type.

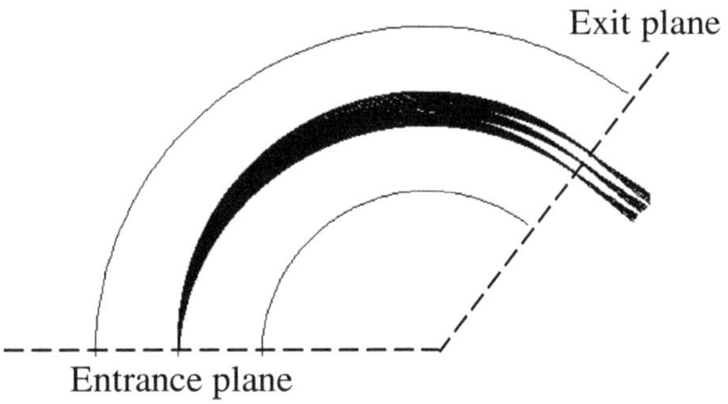

(a)

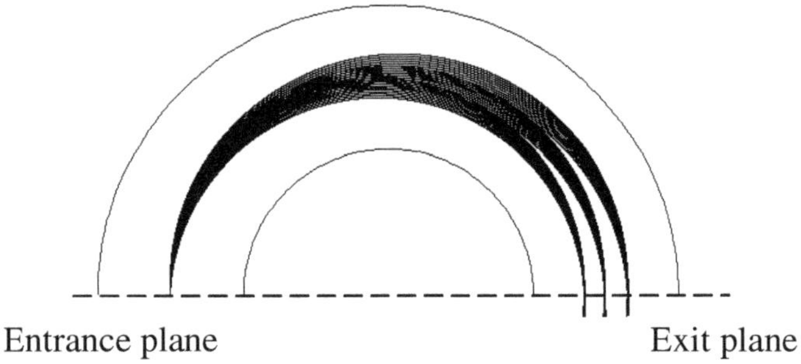

(b)

Fig. 2-18. Simulated ray tracing of commonly used CDA and SDA layouts. Rays have an input angular spread of -50 to 50 mrad in 10 mrad steps and initial energies of $0.95E_p$, E_p and 1.05 E_p, where E_p is the pass energy: (a) CDA, 127.3°, (b) SDA, 180°.

In order to describe the trajectory paths in the out-of-plane direction, consider a small displacement z in the out-of-plane direction (z). The equation of motion in the z-direction is approximately given by

$$m\ddot{z} = -ez\left(\frac{\partial E_z}{\partial z}\right)_{z=0}. \tag{2.47}$$

From the divergence equation for the electric field around the equilibrium orbit

$$\frac{\partial}{\partial r}\left(R_0 E_r + rE_r\right) + \left(R_0 + r\right)\frac{\partial E_z}{\partial z} = 0. \tag{2.48}$$

Substituting in for E_r in terms of E_{R0}, N, R_0 and r, and using the fact that $1 << r/R_0$ and $N << r/R_0$, then the derivative of the electric field component in the z-direction with respect to z can be approximated to be

$$\left(\frac{\partial E_z}{\partial z}\right)_{z=0} \approx \frac{E_{RO}}{R_0}\left(N-1\right). \tag{2.49}$$

Putting this into the equation motion in the z-direction (Eq. 2.47), and expressing E_{R0} in terms of the velocity v_0 on the central curved axis, gives the following equation of motion

$$\ddot{z} + z\frac{v_0^2}{R_0^2}(N-1) = 0. \tag{2.50}$$

In terms of variations with respect to the path length, s, this equation becomes

$$z'' + \frac{z}{R_0^2}(N-1) = 0. \tag{2.51}$$

The solution to this equation takes the same form as before. If the analyzer entrance values of z and $\acute{z}$ are denoted by (z_1, $\acute{z}_1$) and their subsequent values at angle θ are (z_2, $\acute{z}_2$), then

$$\begin{bmatrix} z_2 \\ z_2' \end{bmatrix} = \begin{bmatrix} \cos Q_{ze}\theta & \dfrac{R_0}{Q_{ze}}\sin Q_{ze}\theta \\ -\dfrac{Q_{ze}}{R_0}\sin Q_{ze}\theta & \cos Q_{ze}\theta \end{bmatrix} \begin{bmatrix} z_1 \\ z_1' \end{bmatrix} \tag{2.52}$$

where $Q_{ze} = (N\text{-}1)^{1/2}$.

As expected, for the CDA, there is no focusing of the out-of-plane trajectories, since the electric field component in z is zero, $N = 1$ in the above equation predicts straight-line trajectories. On the other hand, $N = 2$ (SDA) provides exactly the same solution as the in-plane one (ie, $Q_{ze} = Q_{re}$ for $N = 2$), which means that out-of-plane trajectories will focus at the same point as in-planes ones. This important condition means that the 180° SDA provides an image of the source point, where rays in all directions, from a first-order optics point of view, focus at the same point at the analyzer exit. This condition is called stigmatic focusing, and it is an important advantage of the SDA over many other analyzers.

The 180° SDA is more popularly known as the Hemispherical Deflection Analyzer (HDA), and is one of the most widely used analyzers in surface science. The HDA is characterized by first-order focusing properties, and its relative base energy resolution is given by (Granneman and Van der Wiel, 1983)

$$\frac{\Delta E}{E} = \Delta\alpha^2 + \Delta\beta^2 \tag{2.53}$$

where $\Delta\alpha$ and $\Delta\beta$ represent in-plane and out-of-plane angular spreads respectively. This means at say ± 6° angular spreads, the relative base resolution is limited to be around 2%, compared to say 0.662% for the CMA. In practice, the HDA often operates with angular spreads of ±3°, where the relative base energy resolution is around 0.5%. The transmittance is also comparatively low, due to its small angular azimuthal angle collection range, 0.172% for ±3° around a 45° polar emission angle, about 50 times lower than that for the CMA. Figues 2-19(a) and 2-19(b) show simulations through the idealized HDA field distribution, showing that it has a full-width in-plane energy resolution of around 1% at an angular spread of ±6°.

There are several aspects of the HDA design that make up for its comparatively low energy resolution and transmittance. Its ability to pre-retard the incoming beam, keeping the input focal point at the entrance of the analyzer unchanged significantly improves the energy resolution. A series of lenses in a pre-analyzer decelerating column, as shown in Fig. 2-20(a), is able to achieve this, effectively lowering the pass energy through the analyzer by adjusting the potential V_P, and ensuring that the input focal point stays remains at the analyzer entrance plane by suitably biasing the potentials V_2, V_3, and V_4. The analyzer inner deflection plate is set to voltage V_P (retardation voltage), while the outer deflection plate is V_1. The input angles into the analyzer can also be varied electronically, for any given retarded pass energy, and there is an optimum magnification of the source at the analyzer entrance (Zouros and Benis, 2002; Benis and Zouros, 2008).

Figure 2-20(b) shows simulated ray tracing paths for both the deceleration focusing column and the analyzer. The voltages in this simulation are, V_P/E_0 = -0.576 V, V_1/E_0 = -0.91 V, V_2/E_0 = -0.78 V, V_3/E_0 = 0.8 V, V_4/E_0 = 0.54, corresponding to a deceleration factor (ratio of emission energy to pass energy) of 2. Note that in Fig. 2-20(b), electrons enter the HDA off-centre, this para-centric mode of operation compensates for the influence of fringing fields, which would otherwise have the effect of causing the output focal point to shift above the exit aperture/detector plane.

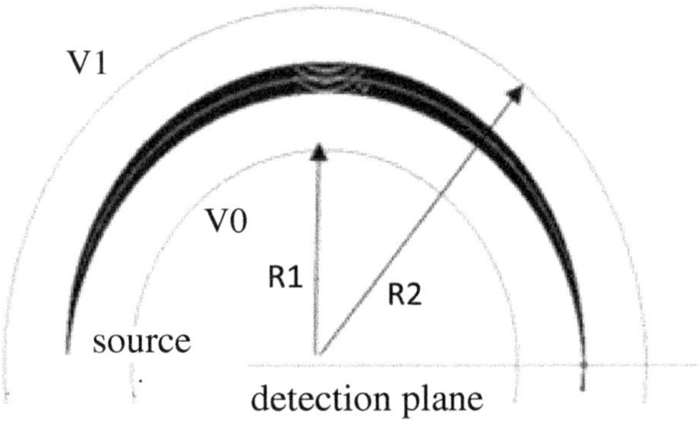

(a)

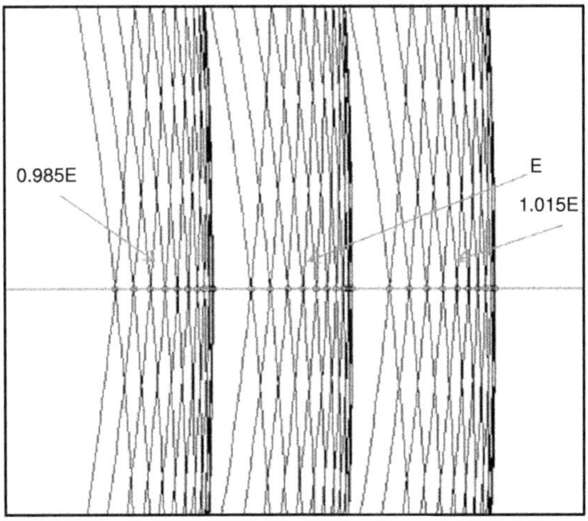

(b)

Fig. 2-19. Simulated ray paths through the idealized HDA: (a) Source to exit/detector plane. 21 trajectories in uniform steps over the angular range -3 to 3°, (b) Around the exit plane. 21 trajectories in uniform steps over the angular range -6 to 6° for energies ±1.5% around the pass energy E.

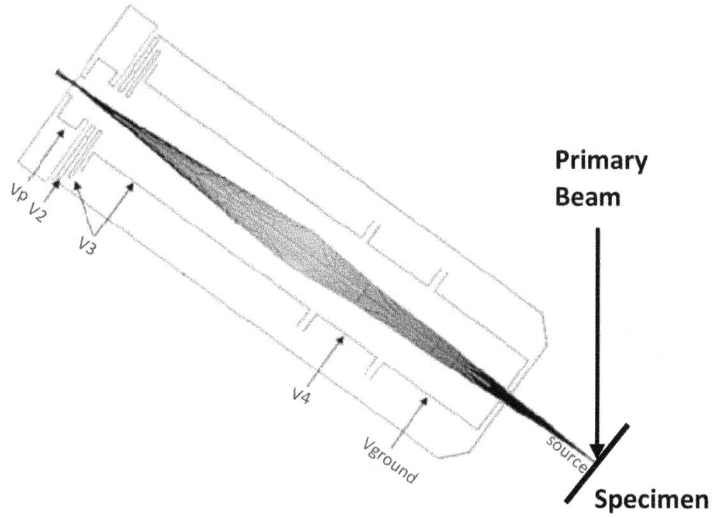

(a)

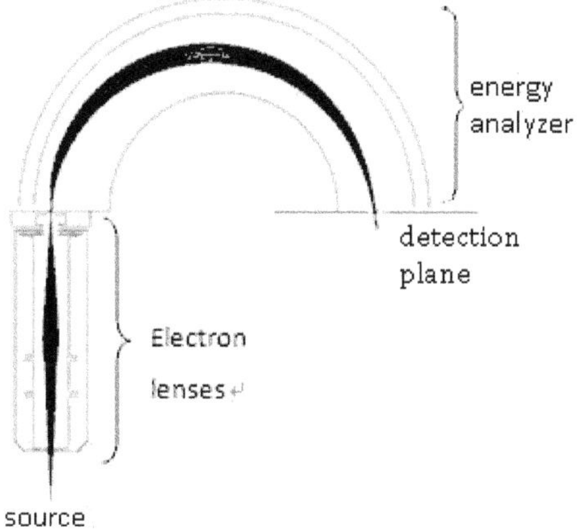

(b)

Fig. 2-20. HDA with input deceleration focusing column. V_P/E_0 = -0.576 V, V_1/E_0 = -0.91 V, V_2/E_0 = -0.78 V, V_3/E_0 = 0.8 V, V_4/E_0 = 0.54. (a) Direct ray tracing through input deceleration column with emission energy E_0. 21 trajectories plot over an angular range of -3 to 3° in uniform steps, (b) Direct ray tracing through both the deceleration column and analyzer. V_P and V_1 are the analyzer inner and out deflection plate potentials respectively.

A relative base energy resolution below 0.05% has been reported for the HDA (Zouros and Benis, 2002), both theoretically and experimentally. Unlike the CMA, the HDA resolution is not unduly compromised by imprecision on the specimen position. A single detector can also be placed at the analyzer exit plane, instead of an aperture slit, and the HDA can be operated in a multi-channel mode with a band-pass width of around ±5% of the pass energy.

In the present context of developing spectrometers for the SEM, the high energy resolution of the HDA is not a critical requirement, as it is for many applications in surface science. A simpler approach for SEMs is to use a 90° SDA, as shown in Fig. 2-21. Here direct ray tracing of electrons is plot for the source located outside the analyzer by a distance R_0 from the entrance deflector plate edge. First-order optics using Eqs. 2.45 and 2.52 predict that the exit focal plane also lies a distance R_0 from the exit deflector plate edge, where the source point, the origin O (center of circle defining the deflection plates), and exit focal point lie on a straight-line. This condition is known as Barber's rule. The dimensions of the zero volt outer casing electrode have been carefully selected so that the first-order optical properties of this analyzer are close to the ideal one predicted by Barber's rule. In general, the distances P and Q, as defined in Fig. 2-16(a) for the SDA are related to R_0 by

$$PQ = R_0^2 . \qquad (2.54)$$

For the 90° CDA, $P = Q$, and $\acute{r}_1 = -\acute{r}_2$, gives $P = Q = 0.35R_0$, indicating that the source and exit aperture/detector planes must be located closer to the analyzer than for the 90° SDA. The base energy resolution (full-width) for the idealized 90° SDA is $2\Delta\alpha^2$, which is around 0.55% for an angular spread of ±3° (Granneman and Van Der Wiel, 1983). As with the HDA, the 90° SDA has limited collection in the azimuthal angular direction, and its overall transmittance is 0.172% for ±3° around a 45° polar emission angle. An array of such analyzers can of course, in principle be used to provide better transmittance.

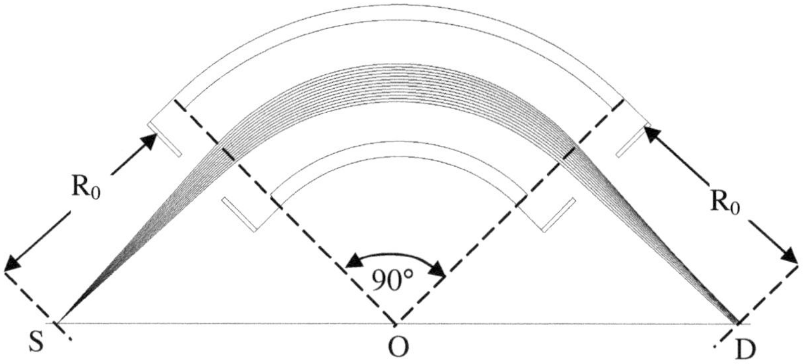

Fig. 2-21. 90° Spherical Sector (SDA). The input angle spread varies from -3° to 3° in uniform steps, in 1° steps.

2.4.4 *Magnetic deflector analyzers*

Magnetic deflection fields are commonly used for a variety of different applications in charged particle optics. In accelerator physics, they are required for guiding beams of electrons/ions and are often referred to as "bending magnets". The following summary is based upon the derivation given by Banford (1966c). Like electric deflection fields, magnetic deflection fields not only change the direction of charged particles, but they also have a focusing effect. Consider the action of a constant magnetic field created between two square iron plates whose sides measure $2R_0$ and top right-hand corner is labeled by the point O, as illustrated in Fig. 2-22. In this case, by suitably adjusting the strength of the magnetic field strength B, a beam of electrons leaving the point S situated a distance R_0 above them in the y direction, can be deflected through an angle of 90° and brought to a focus at point D, located a distance R_0 from the right-hand plate edge. Just as in the case for the 90° SDA, Barber's rule applies, where points S, O and D lie on a straight line. The trace-width w is dominated by first-order focusing, and the focal point distances from the deflector exit edge q_1 and q_2 correspond to input angular variations $-\Delta\alpha$ and $\Delta\alpha$ respectively.

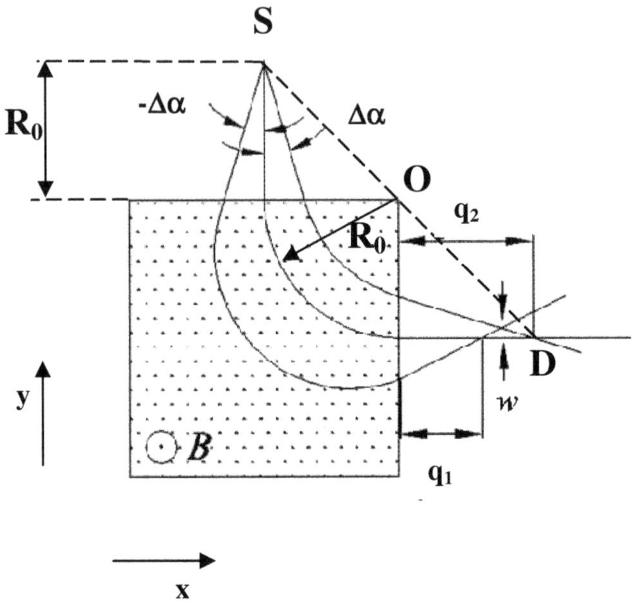

Fig. 2-22. Schematic diagram illustrating the bending and focusing action of a constant magnetic field **B**.

The equations of motion (non-relativistic) for an electron moving in the magnetic field is

$$m\ddot{x} = -e\dot{y}B$$
$$m\ddot{y} = e\dot{x}B. \tag{2.55}$$

We can recast these equations into the following form

$$\ddot{x} = -\dot{y}\omega$$
$$\ddot{y} = \dot{x}\omega \tag{2.56}$$

where ω is defined in terms of the magnetic field strength and electron charge to mass ratio by $\omega = B\eta$. The solution to these equations of motion has the following form

$$x = R\cos(\omega t + \varepsilon)$$
$$y = -R\sin(\omega t + \varepsilon) \tag{2.57}$$

where ε is a constant. If v_{x0} and v_{y0} are the initial velocities in the x and y directions respectively at time t = 0, then the magnitude of the initial velocity, v_0, is given by

$$v_0^2 = v_{x0}^2 + v_{y0}^2 = R^2\omega^2. \tag{2.58}$$

If the initial kinetic energy is E_0 (in electron volts), the initial velocity is given by

$$v_0^2 = 2\eta E_0. \tag{2.59}$$

On substituting in for ω and equating the above two equations of motion for the initial velocity, the radius of deflection R is found to be

$$R = \sqrt{\frac{2E_0}{\eta}}\frac{1}{B}. \tag{2.60}$$

For the simple type of magnetic deflector shown in Fig. 2-22, the first-order focusing properties, dispersion and spherical aberration of the trace-width can be derived analytically, since the trajectory ray path has an analytical solution. The trajectory path of the central ray lies on the arc of a circle centered at O with radius R_0. If the point O is defined to be the origin, x = 0, y = 0, then the entry point and exit points are located at ($-R_0$, 0) and, (0, $-R_0$) respectively. If now the entry point moves to a position ($R_0\tan(\Delta\alpha)$ $-R_0$, 0), caused by the input angle changing by $+\Delta\alpha$, then the center of the circle will now move to (x_c, y_c), where

$$x_c = -R_0 + R_0\tan\Delta\alpha + R_0\cos\Delta\alpha \approx R_0\left(\Delta\alpha - \frac{\Delta\alpha^2}{2}\right)$$
$$y_c = R_0\sin\Delta\alpha \approx R_0\Delta\alpha. \tag{2.61}$$

The approximation here assumes that $\Delta\alpha << 1$. The trajectory path will trace the arc of a circle of radius R_0 now centered at (x_c, y_c). The y-coordinate value and its slope at the exit point, y_2 and y_2' are given by

$$y_2 = y_c - \sqrt{R_0^2 - x_0^2} \approx R_0\Delta\alpha - R_0$$
$$y_2' = \frac{x_c}{y_2 - y_c} \approx -\Delta\alpha\left(1 - \frac{\Delta\alpha}{2}\right). \tag{2.62}$$

The exit point q_2 is then approximately given by

$$q_2 = \frac{y_2}{y_2'} \approx R_0\left(1 + \frac{\Delta\alpha}{2}\right). \tag{2.63}$$

Since $\Delta\alpha << 1$, the exit focal point is approximately independent of $\Delta\alpha$, and lies at R_0. The trace width w, can be estimated by taking the dependence on $\Delta\alpha$ into account, which is equal to the change in focal point position Δq_2

$$w = y_2'\Delta q_2 \approx \Delta\alpha\frac{R_0\Delta\alpha}{2} = \frac{R_0}{2}\Delta\alpha^2. \tag{2.64}$$

Indicating a second-order dependence on spherical aberration, where the aberration coefficient is $R_0/2$.

The energy dispersion is also straightforward. A change in energy, ΔE creates a relative change in radius, ΔR, and the trajectory path of the central ray is now centered at coordinates $(-\Delta R, 0)$ moving with a radius $R_0 - \Delta R$, the exit slope (at $x = 0$) is now $-\Delta R/R_0$, giving a dispersion of ΔR at point D. Since the radius depends on the square root of the initial kinetic energy, E_0, then the total trace width is given by

$$w = \frac{R_0}{2}\frac{\Delta E}{E_0} + \frac{R_0}{2}\Delta\alpha^2. \tag{2.65}$$

The relative base energy resolution, where the angular spread goes from $-\Delta\alpha$ to $\Delta\alpha$ is therefore $\Delta\alpha^2$, similar to the HDA electric deflector.

Figure 2-23 shows simulated trajectory ray paths through an idealized magnetic sector, where the magnetic field is a simple step function (constant value inside the analyzer, zero elsewhere), confirming that the full-width energy resolution is around $\Delta\alpha^2$ (approximately 0.25% for $\Delta\alpha = 50$ mrad).

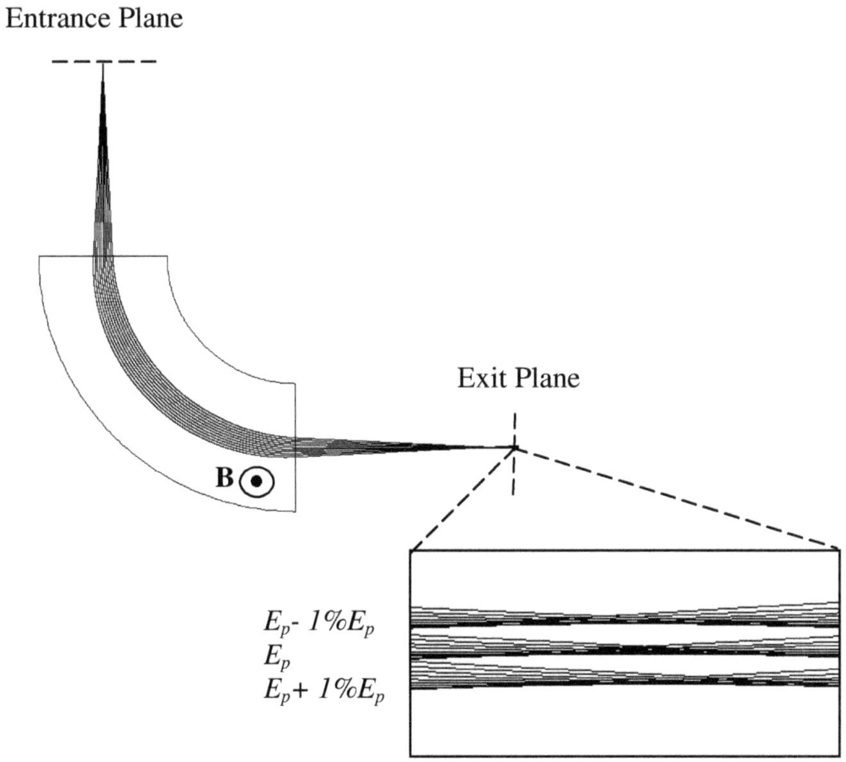

Fig. 2-23. Simulated trajectory paths through an idealized 90° magnetic sector for an input angular spread of -50 to 50 mrad in 10 mrad steps and initial energies of E_p - $1\%E_p$, E_p, and E_p+ $1\%E_p$.

For the more general case, which allows for variations in the *B*-field value, the trajectory path of the electron traces the arc of a circle centered at point *O*, as shown in Fig. 2-24, where the radius R_0 corresponds to a central curved axis. In order to derive first-order focusing properties for this situation, consider an orbit that deviates from the equilibrium orbit R_0, by a small distance *r*, as shown in Fig. 2-24.

The path length *s* on the central curved axis is directly proportional to the angle θ, where $s = \theta R_0$. Consider the situation where the component of the magnetic field in the z direction, *B*, varies as a function of $R_0 + r$, and is B_0 at $r = 0$. Assuming $r << R_0$

$$B(R_0 + r) = B_0 + r\left(\frac{\partial B}{\partial r}\right)_{r=0}. \tag{2.66}$$

If the parameter *n* is defined by

$$n = -\frac{R_0}{B_0}\left(\frac{\partial B}{\partial r}\right)_{r=0} \tag{2.67}$$

then this becomes

$$B(r + R_0) \approx B_0\left(1 - \frac{nr}{R_0}\right) \tag{2.68}$$

where B_0 is the magnetic field value on the central curved axis.

The equation of motion along the central curved axis in the radial direction is

$$\frac{mv_0^2}{R_0} = ev_0 B_0. \tag{2.69}$$

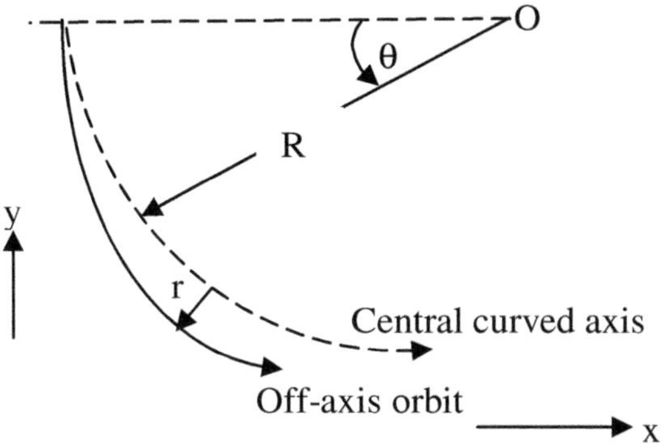

Fig. 2-24. Schematic illustration of off-axis ray path in a magnetic deflection field.

At a distance *r*, the equation of motion is

$$m\ddot{r} = \frac{mv_0^2}{(r+R_0)} - ev_0B_0\left(1 - \frac{nr}{R_0}\right). \tag{2.70}$$

Expanding this equation, using $(1+r/R_0)^{-1} \approx 1-r/R_0$ and using the equilibrium equation to eliminate B_0, gives the following equation of motion

$$\ddot{r} + \frac{v_0^2}{R_0^2}(1-n)r = 0. \tag{2.71}$$

In terms of derivatives with respect to the path length, *s*, where $s \approx v_0 t$, the in-plane paraxial equation for motion (near axis) is then given by

$$r'' + \frac{(1-n)}{R_0^2} r = 0. \tag{2.72}$$

Proceeding as before in the electric case, the solution can be expressed in matrix form in terms of the deflection angle θ, and the values of *r* and its derivative at the entrance of the deflector

$$\begin{bmatrix} r_2 \\ r_2' \end{bmatrix} = \begin{bmatrix} \cos Q_{rm}\theta & \dfrac{R_0}{Q_{rm}}\sin Q_{rm}\theta \\ -\dfrac{Q_{rm}}{R_0}\sin Q_{rm}\theta & \cos Q_{rm}\theta \end{bmatrix} \begin{bmatrix} r_1 \\ r_1' \end{bmatrix} \tag{2.73}$$

where $Q_{rm} = (1\text{-}n)^{1/2}$. For a uniform magnetic field, $n = 0$, this first-order in-plane solution is identical to the one for the electric SDA. It is interesting to note that the focusing action can be suppressed by $n = 1$, that is, electrons trace a circular path, without changing their angles relative to the curved axis. This type of action is useful for bending and transporting electrons, without disturbing their in-plane focal properties, and requires a distribution where the field strength is approximately inversely proportional to the radius

$$B(r+R_0) = B_0\left(1-\frac{r}{R_0}\right) \approx \frac{B_0}{R_0+r}. \tag{2.74}$$

For the equation of motion in the out-of-plane direction, if there is a small displacement z at the analyzer entrance, then the force is related to the component of the magnetic field in the radial direction. B_r, which on the curved axis, is zero. The equation of motion is approximately given by

$$m\ddot{z} = ev_0 z\left(\frac{\partial B_r}{\partial z}\right)_{z=0}. \tag{2.75}$$

Assuming that there is no current source in the central region of the magnetic deflector, curl $\boldsymbol{B} = 0$. In cylindrical coordinates this leads to

$$\left(\frac{\partial B_r}{\partial z}\right)_{z=0} = \left(\frac{\partial B_z}{\partial r}\right)_{r=0} = -\frac{nB_0}{R_0}. \tag{2.76}$$

Therefore, the equation of motion simplifies to

$$m\ddot{z} = -\frac{ev_0 znB_0}{R_0}. \tag{2.77}$$

Eliminating B_0 by using the equation of motion at the equilibrium orbit (Eq. 2.69), this equation becomes

$$\ddot{z} + nz\frac{v_0^2}{R_0^2} = 0. \tag{2.78}$$

Eliminating the double time derivative, the out-of-plane equation of motion is therefore given by

$$z'' + \frac{n}{R_0^2} z = 0. \tag{2.79}$$

This has the solution

$$\begin{bmatrix} z_2 \\ z_2' \end{bmatrix} = \begin{bmatrix} \cos Q_{zm}\theta & \frac{R_0}{Q_{zm}} \sin Q_{zm}\theta \\ -\frac{Q_{zm}}{R_0} \sin Q_{zm}\theta & \cos Q_{zm}\theta \end{bmatrix} \begin{bmatrix} z_1 \\ z_1' \end{bmatrix} \tag{2.80}$$

where $Q_{zm} = n^{1/2}$. For uniform field, $n = 0$, there is no out-of-plane focusing, as expected. However, for n = ½, the in-plane and out-of-plane first-order focal properties are identical, therefore stigmatic focusing is obtained at $Q_{zm} = Q_{rm} = 1/\sqrt{2}$. The magnetic field distribution for this condition needs to have the following form

$$B(r + R_0) = B_0\left(1 - \frac{r}{2R_0}\right). \tag{2.81}$$

This field consists of two parts, a main deflection field B_0, and a quadrupole like field term (field strength varies in proportion to the off-axis distance r). For the case where $n = 1$, $Q_{zm} = 1$, the case where in-plane focusing is suppressed, there will be out-of-plane focusing, at a deflection angle of 90° ($\cos Q_{zm}\theta = 0$).

There are many ways to superimpose a small quadrupole field distribution on to the main deflection field so that stigmatic focusing can be achieved. The most common way is to tilt the entrance/exit edges of the deflector, with respect to the electron beam direction. Figure 2-25 shows a schematic illustration of this for the exit edge of a square plate magnetic deflector. The exit edge in this case is tilted by an angle of γ with respect to the vertical direction. The fringe field, $B_{fn}(z)$ is the magnetic field normal to the exit edge, $B_{fn}(0) = 0$ at the mid-plane odd-symmetry where the magnetic potential is zero, but which is non-zero for out-of-plane trajectories, $z \neq 0$. $B_{fn}(z) < 0$ for $z > 0$, pointing out of the edge, and $B_{fn}(z) > 0$ for $z < 0$, point into the edge. For a vertical exit edge ($\gamma = 0$), $B_{fn}(z)$ is predominantly in the x-direction, parallel to the ray path of the electron, and therefore does not affect the motion of electrons. However, when the exit edge is tilted, an additional component of the fringe field in the y direction is created, $B_{fy}(z)$, and the fringe field component in the x-direction is modified to $B_{fx}(z)$ where

$$\begin{aligned} B_{fx} &= B_{fn} \cos \gamma \\ B_{fy} &= B_{fn} \sin \gamma. \end{aligned} \tag{2.82}$$

The component $B_{fy}(z)$ is perpendicular to the motion of the ray path (in the x-direction), and therefore a force in the out-of-plane (z-direction) is generated. The equation of motion governing the out-of-plane motion is

$$m\ddot{z} = -ev_0 B_{fn} \sin \gamma. \tag{2.83}$$

If we assume that the fringe fields are step-like and are confined to a distance d_f from the plate edge, then the scalar potential at a point z located well inside the plates, $\psi(z) = zB_0$, must fall to zero over the

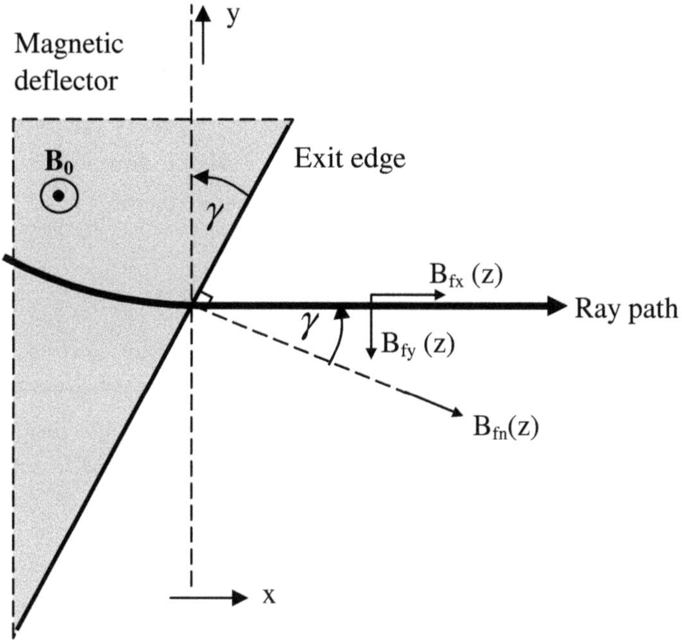

Fig. 2-25. Tilted exit edge magnetic deflector for stigmatic focusing.

distance d_f, where $B_{fn}(z) \approx \psi(z)/d_f = zB_0/d_f$. Now substituting for $B_{fn}(z)$ into the out-of-plane equation of motion, gives

$$m\ddot{z} = -ev_0 \frac{zB_0}{d_f} \sin\gamma. \tag{2.84}$$

On comparing this to Eq. 2.80, the out-of-plane equation of motion for a general deflection field distribution of $B_0(1- nr/R_0)$, then the equivalent field curvature parameter n is given by

$$n = -\frac{R_0 \sin\gamma}{d_f}. \tag{2.85}$$

For stigmatic focusing, we know that n = ½, so for a given field fringe distribution, a suitable angle can be found. Another helpful strategy in the design of magnetic deflector analyzers is to use curved edge exit/entrance edges, which can correct for the second-order spherical aberration, making it possible to achieve second-order focusing. The transmittance of magnetic deflector analyzers is relatively low, similar to the PPA, CDA or SDA, only a fraction of a percent, since collection in the out-of-plane direction is limited to only a few degrees or less.

2.4.5 *Wien filters*

Wien filters function by applying crossed electric and magnetic fields that are both perpendicular with respect to one another, and perpendicular with respect to the direction of motion. Figure 2-26 shows a rectangular Wien filter in which there is an electric field component, E_y and a magnetic field component B_z. A parallel beam of electrons travelling in the x-direction with a range of different velocities (kinetic energies) enter from the left-hand side. At the pass energy, E_P, the electric and magnetic forces on the beam of electrons are such that they cancel, and the beam travels in a straight line along a central ray.

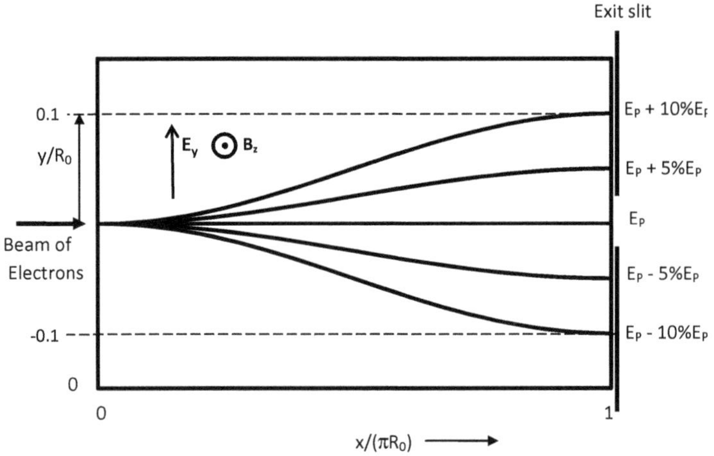

Fig. 2-26. Simple analytical solution of direct ray paths through a Wien filter. Rays enter the filter parallel to the x-axis and have a spread of different energies around the pass energy E_p.

The Wien filter is first and foremost a velocity filter, since cancellation of electric and magnetic forces occur at a given pass velocity, v_p, but since for a beam of electrons propagating in the x-direction, different initial velocities directly correspond to different initial energies, the filter action also results in energy selection. Electrons having energies different to the pass energy are deflected away from the central ray, in the y-direction. Figure 2-26 shows direct ray paths of electrons having energies of 5 and 10% above and below the pass energy. A slit placed at the Wien filter exit can therefore be used for energy filtering. The distances shown in Fig. 2-26 are shown relative to the radius around which the beam would travel if only the magnetic field were present, R_0, given by

$$R_0 = \frac{v_p}{\eta B_z} = \sqrt{\frac{2\phi_P}{\eta}}\frac{1}{B_z} \tag{2.86}$$

where ϕ_P is the beam voltage and η is the electron charge to mass ratio. In the y-direction, for an energy spread of 10% around the pass energy, the beam is deflected to a distance $0.1R_0$, while in the x-direction, the distance along the Wien filter is measured as a fraction of the filter length L_0, equal to πR_0, at this distance, each ray exits the filter parallel to the x-axis.

The direct ray paths shown in Fig. 2-26 were derived as simple analytical solutions to the following equations of motion

$$\begin{aligned} \ddot{x} &= -\dot{y}\omega \\ \ddot{y} &= -\eta E_y + \dot{x}\omega \\ \ddot{z} &= 0 \end{aligned} \tag{2.87}$$

where $\omega = \eta B_z$. E_y and B_z are assumed to be step functions that have constant value inside the filter and zero elsewhere. The pass velocity v_p, is found from the cancellation of electric and magnetic forces in the y-direction and is therefore given by

$$v_p = \frac{E_y}{B_z}. \tag{2.88}$$

It is straightforward to show that the general solution to the above equations of motion takes the following form

$$\begin{aligned} x &= \frac{\eta E_y t}{\omega} + \frac{c_1}{\omega} + R\cos(\omega t + \varepsilon) \\ y &= -\frac{\eta E_y}{\omega} + \frac{c_2}{\omega} + R\sin(\omega t + \varepsilon) \\ z &= c_3 t + c_4 \end{aligned} \tag{2.89}$$

where c_1, c_2, c_3, c_4, R, and ε are constants dependent on initial conditions at the filter entrance. It is clear that motion in the z-direction is determined only in terms of its initial conditions, which in this case, for simplicity, will not be discussed and assigned to be zero. If the initial velocities in the x and y directions at $t = 0$ are given by $v_{x0} = v_p$ and $v_{y0} = 0$ respectively (parallel input rays), then

$$\begin{aligned} v_{x0} &= v_p - R\omega\sin(\varepsilon) \\ v_{y0} &= R\omega\cos(\varepsilon) \\ \Rightarrow \varepsilon &= \frac{\pi}{2}, \; R = \frac{(v_p - v_0)}{\omega} = -\frac{\Delta v}{\omega} \end{aligned} \tag{2.90}$$

where $\Delta v = v_0 - v_p$, the change in velocity with respect to the pass velocity. From kinetic energy considerations, and using the above equations, relative changes in velocity are linked to relative changes in the beam voltage $\Delta\phi$ by

$$\frac{\Delta\phi}{\phi_0} = \frac{2\Delta v}{v_p} = \frac{2R}{R_0}. \tag{2.91}$$

Substituting this into the analytical solution for $y(t)$, stipulating that the initial positions at $t = 0$ are zero, and making the approximation that $t \approx x/v_p$, the ray path in the x-y plane is then approximately given by

$$y \approx -\frac{R_0}{2}\left(\frac{\Delta\phi}{\phi_0}\right)\left[\cos\left(\frac{\pi x}{L_0}\right)-1\right]. \tag{2.92}$$

This is the equation from which the direct ray paths shown in Fig. 2-26 were plot. It confirms that for $x = L_0$, all rays are parallel, irrespective of their initial energies and that dispersion in the y-coordinate varies linearly with relative changes in beam voltage (energy). The dispersion at the exit plane is given by

$$y \approx -R_0\left(\frac{\Delta\phi}{\phi_0}\right). \tag{2.93}$$

For the more general case where the incoming beam of electrons have initial angles in the x-y plane, as well as a spread in initial velocities, they are focused on the exit plane (for $x = L_0$), as indicated in Fig. 2-27(a), where numerically simulated trajectory paths were plot for initial angles ranging from, -50 to + 50 mrad in steps of 10 mrad, and energies of E_p - $5\%E_p$, E_p, and E_p + $5\%E_p$. In this case, E_y = 639.96 V/mm, B_z = 26.976 mT, ϕ_p = 1600 eV, and R_0 = 5 mm. Figure 2-27(b) shows trajectory ray paths around the exit plane, from which the full-width energy resolution is estimated to be around 0.79%.

An analytical approximation to the trajectory ray paths shown in Fig. 2-27 can be derived in a straightforward manner. Consider the situation of an electron having an angle α relative to the x-axis at the entrance of the filter (t = 0) with a velocity v_p (beam voltage ϕ_p), assuming that α is relatively small, expressions for R and ε can be obtained by equating velocity components

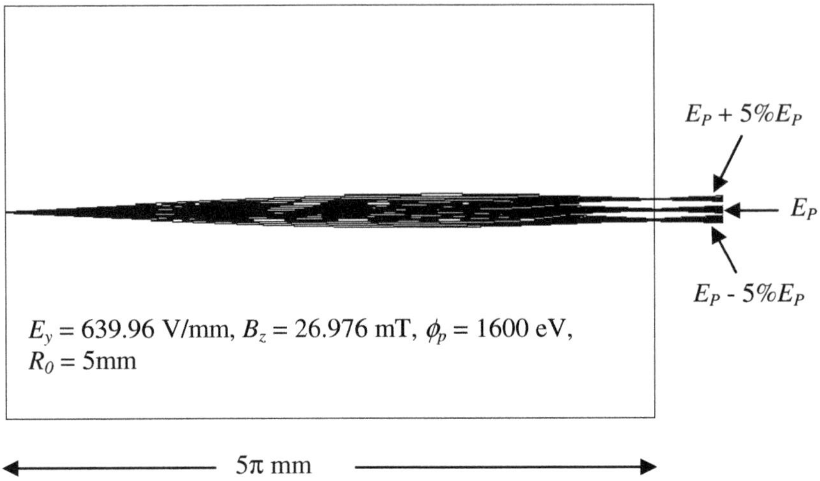

(a)

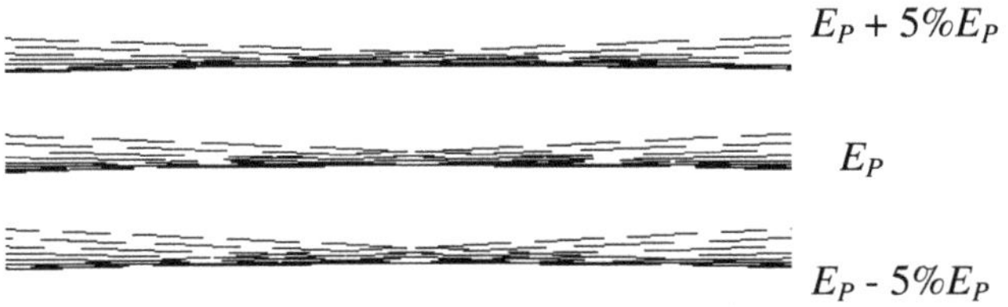

(b)

Fig. 2-27. Simulated trajectory paths through Wien filter for initial angles ranging from, -50 to + 50 mrad in steps of 10 mrad, and energies of E_p - 5%E_p, E_p, and E_p + 5%E: (a) Through entire filter, (b) Around exit plane.

$$\begin{aligned} v_p \cos\alpha &= v_p - R\omega \sin\varepsilon \\ v_p \sin\alpha &= R\omega\cos\varepsilon \\ \Rightarrow \varepsilon &\approx \frac{\alpha}{2},\ R \approx \alpha R_0. \end{aligned} \tag{2.94}$$

These expressions can be substituted into the analytically derived solutions $x(t)$ and $y(t)$ to give

$$x = v_p t - R\cos\varepsilon + R\cos(\omega t + \varepsilon)$$
$$y = -R\sin\varepsilon + R\sin(\omega t + \varepsilon). \tag{2.95}$$

Since $x(t)$ is still dominated by the first term, $v_p t$, motion in the x-y plane is now effectively a sine wave, with a zero point at the filter exit, confirming that the Wien filter has first-order focusing optics. Furthermore, an estimate of the full-width energy resolution of the filter can be made by putting in for $x = L_0$, and solving for t

$$t = \frac{1}{v_p}\left[L_0 + R\cos\varepsilon - R\cos(\omega t + \varepsilon)\right]. \tag{2.96}$$

This expression can be further simplified by noting that at $x = L_0$, $\cos(\omega t) \approx \cos(\pi)$, so $\cos(\omega t + \varepsilon) \approx -\cos(\varepsilon)$, giving

$$t \approx \frac{1}{v_p}\left[L_0 + 2R\cos\varepsilon\right]. \tag{2.97}$$

Using Eqs. 2.94 and 2.95, the angular dispersion on the exit plane due to the input angular spread is therefore

$$\omega t + \varepsilon \approx \frac{1}{R_0}\left[L_0 + 2R\cos\varepsilon\right] + \varepsilon \approx \pi + \frac{5}{2}\alpha$$
$$\Rightarrow y(x = L_0) \approx -3R_0\alpha^2. \tag{2.98}$$

Since the energy dispersion at the exit plane is given by $R_0\ (\Delta\phi/\phi_p)$, the full-width energy resolution is approximately given by

$$\frac{\Delta\phi}{\phi_p} \approx 3\alpha^2. \tag{2.99}$$

The analytically derived relative energy resolution prediction for ±50 mrad is 0.75%, close to the 0.79% obtained by the numerical ray tracing results depicted in Fig. 2-27.

If the Wien filter is to be put in the path of a focusing electron beam system, such as an electron microscope, subsequent cancellation of its angular/energy dispersive action can be achieved by using a second Wien filter which is excited in the opposite direction to the first one. This relies on the fact that the direction of the energy and angular dispersion changes with the sign of the electric/magnetic fields inside the filter. In this way, a pair of Wien filters can be used as a monochromator in an electron microscope, reducing the energy spread in the primary beam.

Another important property of the Wien filter is its ability to achieve stigmatic focusing, that is, act as a round lens, focusing both in-plane and out-of-plane electrons on to the same exit/image point. The simplest way to achieve this condition, is to apply an additional quadrupole electric field, E_2 (Tsuno, 1999), so that the electric field component (in the y-direction in this example) is modified to be

$$E_y = E_1 + 2E_2 y. \tag{2.100}$$

The strength of the electric quadrupole field is adjusted so that

$$\frac{E_2}{E_1} = \frac{1}{4R_0}. \tag{2.101}$$

In general, the stigmatic focusing condition is usually expressed in terms of a quadrupole electric and quadrupole magnetic field (Tsuno, 1999), but in practice, it is much easier to restrict it to the introduction of a single electric field component only. Figure 2-28 shows a schematic diagram for the cross-section of a Wien filter where the same electrodes can be used to generate an electric deflection field (by $\pm V_D$), a magnetic deflection field (by $\pm NI$), and an electric quadrupole field (by $\pm V_Q$).

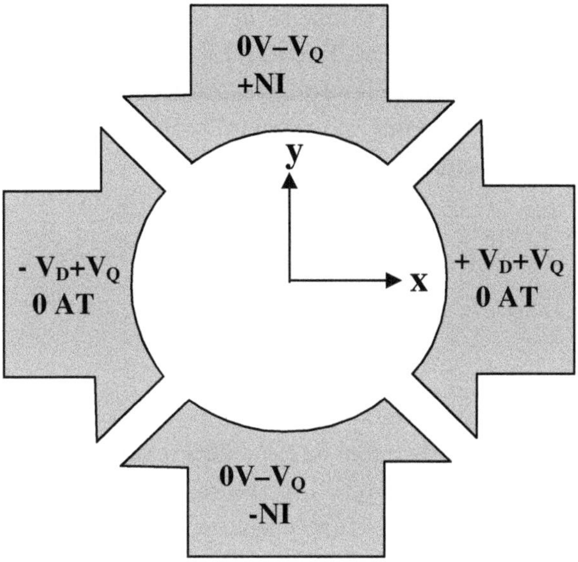

Fig. 2-28. Schematic diagram of the cross-section of a stigmatic focusing Wien filter. $\pm V_D$ for the electric deflection field, $\pm NI$ for magnetic deflection field , and $\pm V_Q$ for the electric quadrupole field.

2.4.6 *Magnetic collimation and time of flight spectrometers*

Time of flight spectrometers are widely used in accelerator physics and surface science, particularly for ion beam mass/energy selection. The technique is easier to apply to ions than electrons, since ions in general have orders of magnitude smaller charge-to-mass ratios than electrons, and therefore move much slower than electrons. Time of flight electron spectrometers have been used for photo-electron applications, where electron kinetic energies are relatively small, in the few eV range (Kruit and Read, 1983).

In the SEM, for electron energy time of flight spectrometers to be effective, there must be some way of minimizing the effect of initial angular spread on the transit time of scattered electrons, from specimen to detector, so that their flight times are mainly determined by initial kinetic energy. One way of doing this is to collimate them.

The collimation of secondary electrons in a sharply decreasing magnetic field was described in detail by Kruit and Read (1983). From Eq. 2.89, the general solution Wien Filter solution, putting $E_y = 0$, the trajectory path of an electron entering a region of constant magnetic field B_z (pointing in the z-direction) is given by

$$\begin{aligned} x &= R\sin(\omega t) \\ y &= -R\cos(\omega t) \\ z &= v_{z0}t \end{aligned} \tag{2.102}$$

where v_{z0} is the initial velocity in the z-direction and the angular frequency of rotation, ω, is equal to ηB_z. This trajectory path describes a helical path moving up the z-axis. For the case of a decreasing magnetic field strength along the z-axis, if the variation of the magnetic field along the optic axis is adiabatic, that is, if the field does not change significantly over one revolution of the electron's helical motion, then its angular momentum is conserved, and the path of an electron will tend to spiral around a flux line, as shown schematically in Fig. 2-29.

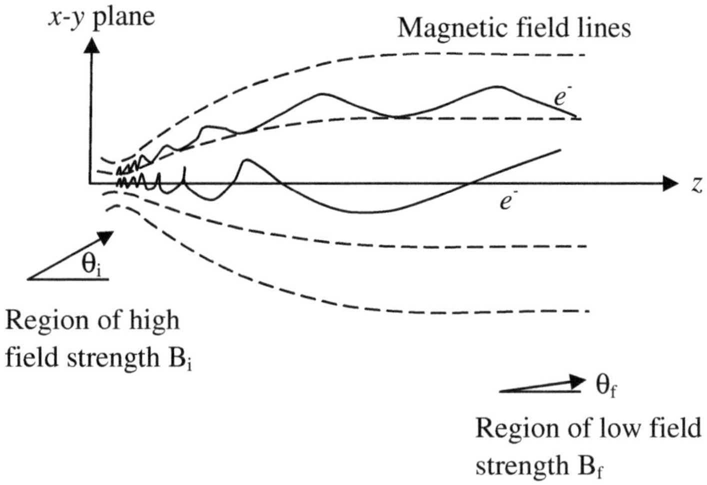

Fig. 2-29. Schematic diagram of electron trajectory paths through a sharply decreasing magnetic field.

If the electron initial emission angle and the field strength is given by θ_i and B_i, respectively, its angle θ_f at the region of low field strength B_f can be found from the following approximate equation:

$$\frac{\sin\theta_f}{\sin\theta_i} = \left(\frac{B_f}{B_i}\right)^{1/2}. \tag{2.103}$$

The radius r_f at the region of low field, compared to the initial radius r_i, is given by

$$\frac{r_f}{r_i} = \left(\frac{B_i}{B_f}\right)^{1/2}. \tag{2.104}$$

For an electron having an initial velocity v, traversing a distance d, its time of flight T is given approximately by

$$T = \frac{d}{v}\left(1 + \frac{B_f \sin^2\theta_i}{2B_i}\right). \tag{2.105}$$

If B_f is two or more magnitudes smaller than B_i, which is generally the case for in-lens objective lenses (since the emitted secondary electrons are allowed to spiral up far out of the lens field), then the secondary electrons at the detector plane will be nearly parallel. Their time of flight is largely determined by their initial energies, and they will tend to spread out in the radial direction. In practice, since the secondary electrons have a wide range of different energies, some of them will not have adiabatic behavior, particularly the higher energy ones. The collimating action of the objective lens is only valid for the low energy secondary electrons, ones with initial energies typically around a few eV.

Kruit and Read (1983) define a parameter χ for each point (z) along the optical axis that defines the degree to which an electron's behavior is adiabatic. This parameter is defined in terms of the change of the magnetic field on-axis ΔB during one period of rotation, relative to the absolute field value B at that point, and is given by

$$\chi = \frac{2\pi m v}{eB^2}\left|\frac{dB}{dz}\right|. \tag{2.106}$$

For adiabatic behavior to occur, $\chi \ll 1$. Substituting in for the electron mass and charge, and using the electron's kinetic energy, E,

$$\chi = 2.11\times10^{-5}\,\frac{\sqrt{E}}{B^2}\left|\frac{dB}{dz}\right|. \tag{2.107}$$

Chapter 3

In-lens Improvements

Figures 3-1(a)–3-1(e) illustrate some proposals to improve the performance of SEM objective lenses. Figure 3-1(a) depicts the conventional objective lens arrangement, where the specimen is placed in a field-free region below the lower lens pole-piece. Figure 3-1(b) shows the magnetic in-lens type of objective lens, where the specimen is placed in the lens gap region. The type of lens where the magnetic field extends beyond a single lens pole-piece, as shown in Figs. 3-1(c) and 3-1(d), is known as a semi-in-lens or single pole objective lens. Figure 3-1(e) depicts the retarding field lens, where the primary beam is slowed down just before it strikes the specimen. The specimen in this arrangement is often immersed in an electric field. Over the last decade or so, different SEM columns have been designed, utilizing the different objective lens designs shown in Figs. 3-1(b)–3-1(e). Each of these lens configurations will be examined in more detail in subsequent sections.

3.1 Magnetic immersion lenses

The flux lines and axial field distribution for a typical in-lens objective lens is shown in Fig. 3-2. The specimen is immersed in a strong magnetic field, which provides lower aberrations. The working distance, here, is the distance between the specimen surface and the lens upper pole piece. Simulated on-axis aberrations for this lens are presented in Table 3-1. The lens gap here is 4 mm.

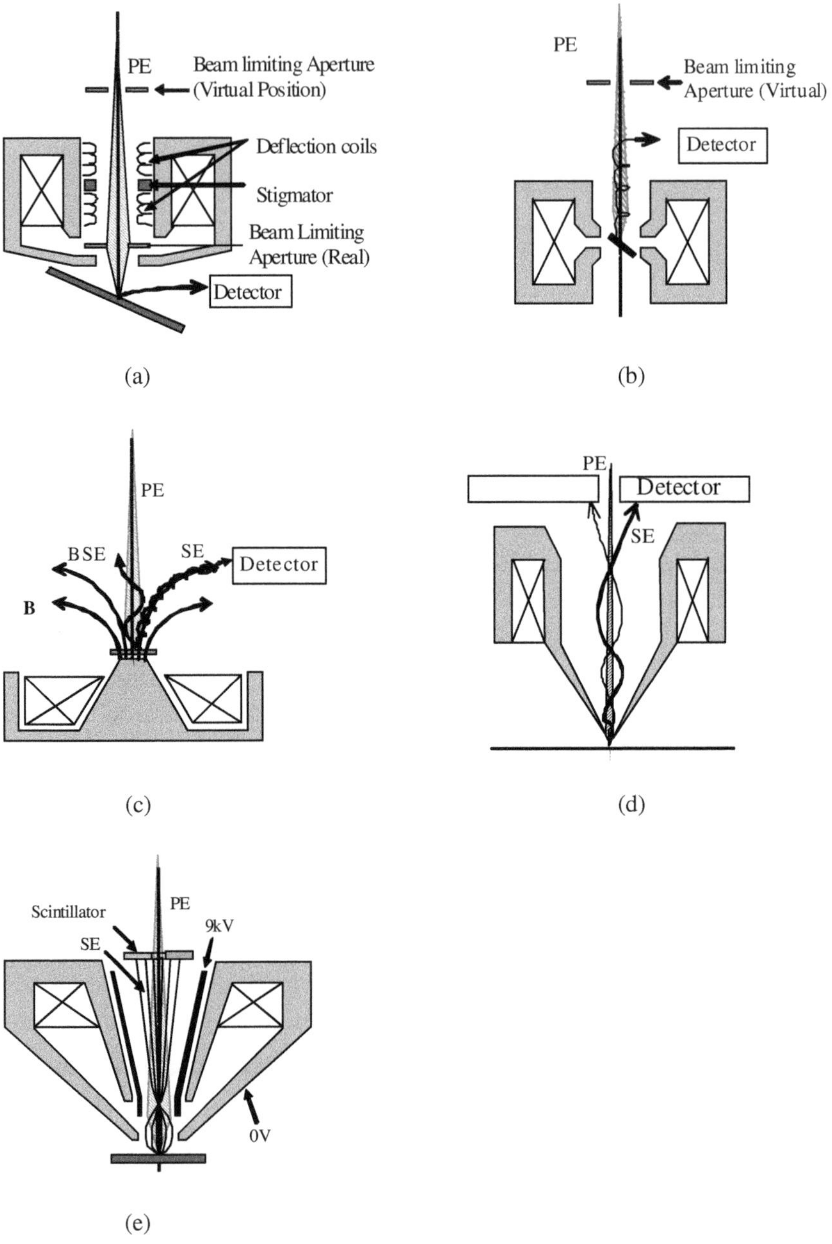

Fig. 3-1. Objective lens improvements: (a) Conventional lens, (b) In-lens, (c) Single pole lens below the specimen, (d) Single pole lens above the specimen, (e) Retarding field lens.

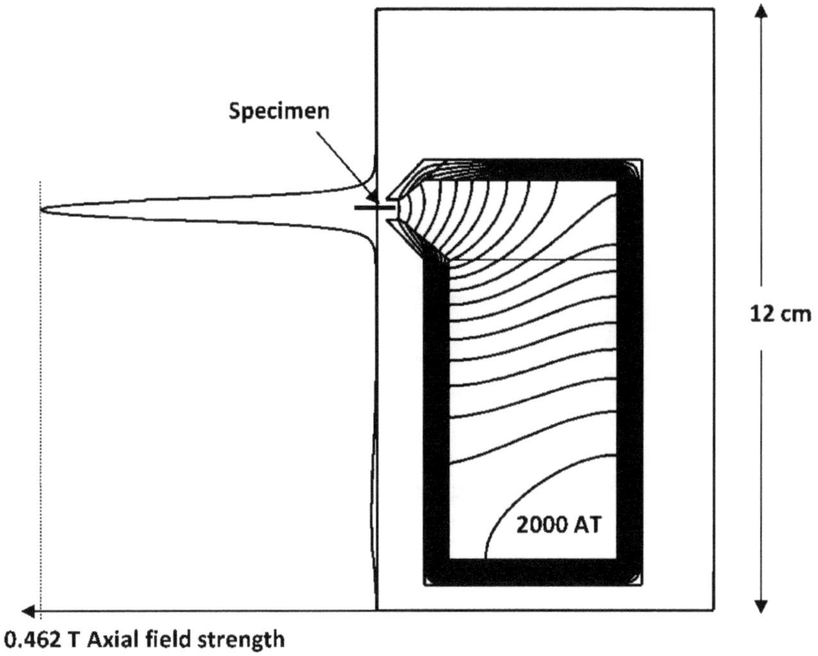

Fig. 3-2. Flux lines and axial field distribution of an in-lens objective lens.

Table 3-1. Simulated on-axis aberrations for a magnetic in-lens objective lens.

Beam Energy (keV)	Working Distance (mm)	f (mm)	C_s (mm)	C_c (mm)	Excitation (AT)
1	0.5	0.935	0.36	0.6	1892
1	1	1.1	0.546	0.747	1306
20	2	1.51	0.937	1.07	3374

From Table 3-1 it is clear that f, C_s, and C_c are around an order of magnitude smaller than for conventional lenses. Graphs of the probe diameter as a function of probe current are shown in Figs. 3-3(a) and 3-3(b) for a tungsten and field emission source respectively. They show that the spatial resolution of the in-lens objective lens is typically a factor of 3 to 4 times better than that of a conventional lens (see Fig. 1-14,

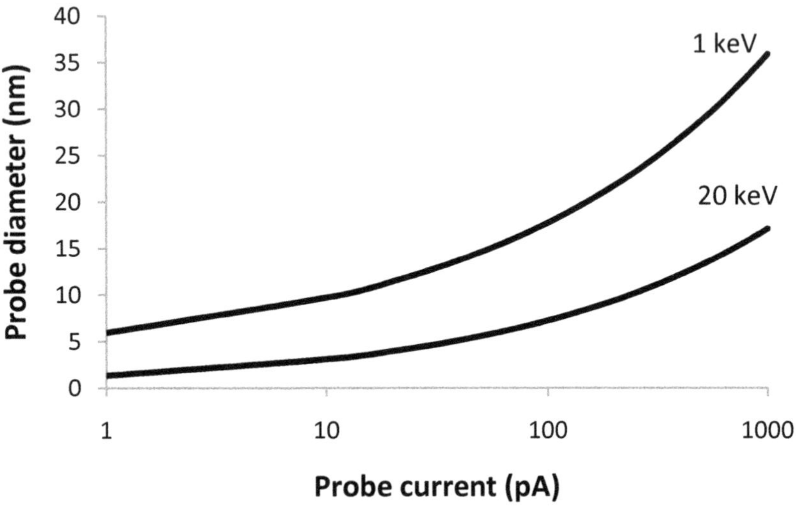

(a)

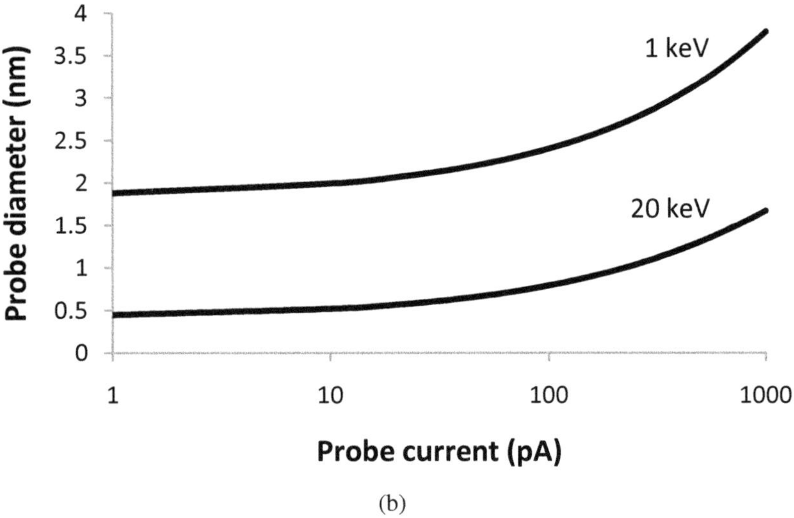

(b)

Fig. 3-3. Predicted probe diameter as a function of probe current for an in-lens objective lens: (a) Tungsten gun (f=0.935 mm, C_s=0.36 mm, C_c=0.6 mm, β=2×10^4 Acm^{-2}sr^{-1} at 1 keV, f=1.51 mm, C_s=0.937 mm, C_c=1.07 mm, β=2×10^5 Acm^{-2}sr^{-1} at 20 keV, ΔE= 1.5 eV), (b) Field emission gun (f=0.935 mm, C_s=0.36 mm, C_c=0.6 mm, β=10^7 Acm^{-2}sr^{-1} at 1 keV f=1.51 mm, C_s=0.937 mm, C_c=1.07 mm, β=10^8 Acm^{-2}sr^{-1} at 20 keV, ΔE= 0.4 eV).

Chapter 1). One drawback of most in-lens systems is that they only allow the use of relatively small specimen sizes. The specimen height must typically be less than 3 mm, while the width is restricted to be smaller than 6 mm. The specimen holder has a side-entry arrangement, similar to that used in transmission electron microscopes.

Collection of secondary electrons in this lens takes advantage of the fact that SEs are collimated in a strongly decreasing magnetic field, an effect described at the end of the last chapter, and are best collected by a detector above the lens. If a conventional E-T scintillator detector is used, as is often the case, it creates asymmetric electric fields that cause astigmatism and misalignment of the primary beam. This problem is particularly noticeable at low voltages (Pawley, 1990). An overhead rotationally symmetric detector with a small hole in it to allow the primary beam to pass overcomes this problem. A micro-channel plate or round scintillator detector may be used for this purpose. The trajectory paths of electrons having initial energies of 3 eV, emitted of a wide range of initial angles (up to 90°) are shown in Fig. 3-4(a). These electrons spiral up tracing out a helical path, and are collimated in the vertical direction.

The degree of electron collimation can be estimated for each point along the optical axis (z) by the adiabatic parameter χ, given by Eq. 2.107. For a significant amount of collimation to occur, χ should be much less than unity. Consider an immersion lens that has a magnetic field value of 0.5 T at the specimen, and falls to around 1 mT at the lens exit bore, located a distance of 3 mm way. Furthermore, let us assume that the rate at which the magnetic field drops beyond this point is small, say 1 mT/mm. For a 1 eV electron, χ is approximately 0.014 at the point of emission and around 21.1 at the lens exit, indicating that most the collimation occurs close to the specimen, while at the lens exit, electrons will tend to spread out.

For backscattered electrons, there will be a wide degree of angular dispersion. From first-order optics considerations alone, their paths

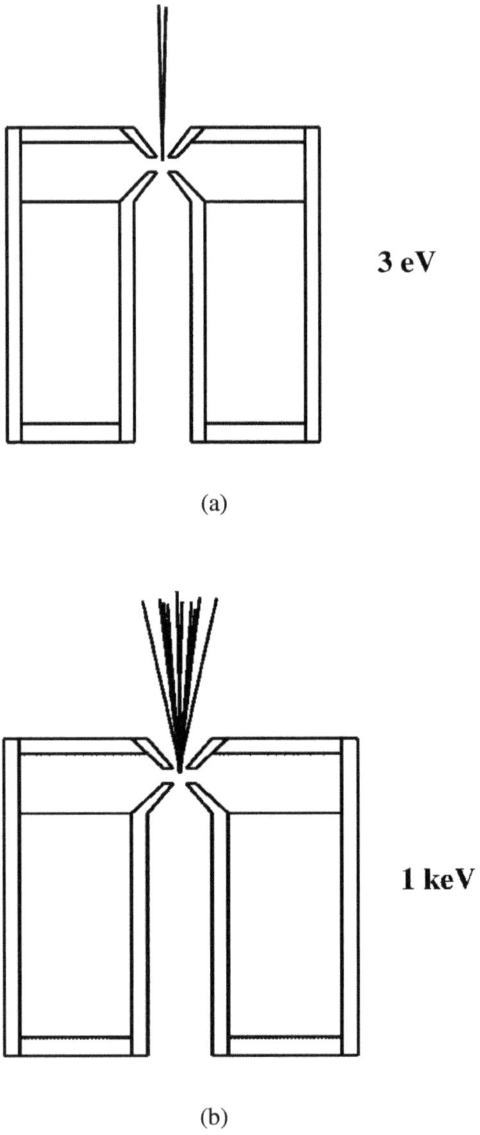

Fig. 3-4. Simulated electron trajectory paths for the in-lens objective lens over a wide range of emission angles: (a) 3 eV secondary electrons, (b) 1 keV backscattered electrons.

should all be parallel to the axis, or close to parallel, reproducing the path of the incoming primary beam. However, backscattered electrons traverse highly nonlinear regions of the lens field, and are projected back through the lens over a range of different angles. The transport efficiency of backscattered electrons, just like that of the secondary electrons, is predicted to be high. Figure 3-4(b) shows the trajectory paths for backscattered electrons at an energy 1 keV emitted over a wide range of different angles. Both for backscattered electrons and secondary electrons, a small fraction will be lost through the hole at the center of the detector, which allows the primary beam to pass.

There have been several variations of the in-lens proposal, which for the most part have been concerned in overcoming its restriction on specimen size. Figure 3-5 shows the flux lines of an in-lens system where the specimen is placed on a lower flat iron pole-piece. This type of lens was proposed by Nakagawa *et al.* (1991). The lower plate is physically separated from the rest of the magnetic circuit, but is of course, magnetically coupled to it. There are, in effect, two gaps of the same

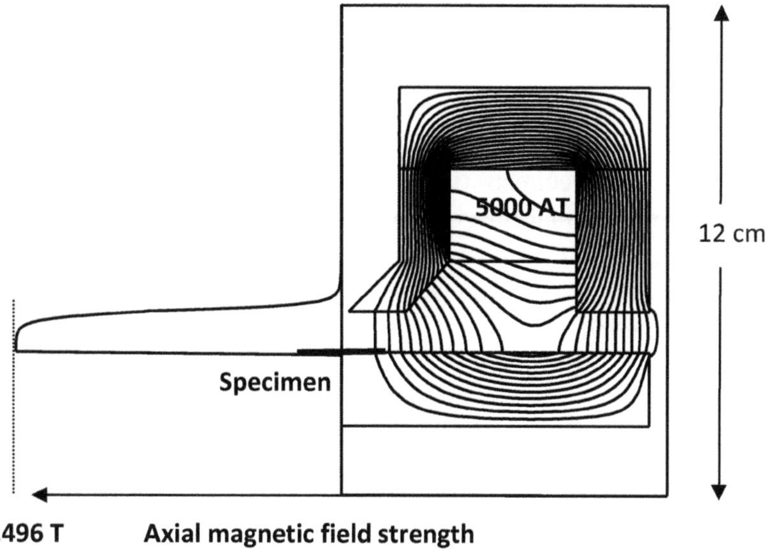

Fig. 3-5. Flux lines for a double gap flat-plate in-lens objective lens.

size in the magnetic circuit, and in this way large specimens can be used. Nakagawa proposed its use for large integrated wafer specimens. Dinnis (1996) proposed a similar lens for making voltage contrast measurements. The maximum allowable specimen height is now determined by the gap size. The lens shown in Fig. 3-5 has a 1 cm gap, which allows for specimen heights more than a factor of three times bigger than most in-lens systems. As the gap size increases, the resolution of the lens for a given working distance falls. Also, additional lens excitation current (coil excitation) must compensate for the larger gap size and additional gap in the magnetic circuit. Table 3-2 tabulates the on-axis aberration coefficients calculated for this lens under similar conditions that were used for the previous in-lens. Comparing Table 3-1 to Table 3-2 shows that the on-axis aberration coefficients are comparable to the previous in-lens design but, as expected, the flat-plate in-lens objective lens requires much higher levels of coil excitation to drive it, around a factor of three times more.

Another variation on the in-lens design was given by Mizuno *et al.* (1994), who combined the objective lens with the specimen chamber. Figure 3-6 shows the simulated flux lines for such a lens. Like the double gap flat-plate in-lens objective lens, the in-specimen chamber lens can be used for large specimens. However, the in-specimen chamber lens does not require an extra gap in the magnetic circuit, and is more efficient in terms of coil excitation strength. In addition, the in-specimen chamber lens allows for specimen tilt, which in many cases provides extra information in the final image. Both the double gap flat-plate in-lens, as well as the in-specimen chamber lens, are suited to the observation of large integrated circuit wafers, and in fact were designed with this application in mind.

Table 3-2. Simulated on-axis aberrations for a magnetic flat-plate in-lens objective lens.

Beam Energy (keV)	Working Distance (mm)	f (mm)	C_s (mm)	C_c (mm)	Excitation (AT)
1	0.5	0.992	0.378	0.643	4750
1	1	1.15	0.562	0.786	3345
20	2	1.56	0.934	1.11	8845

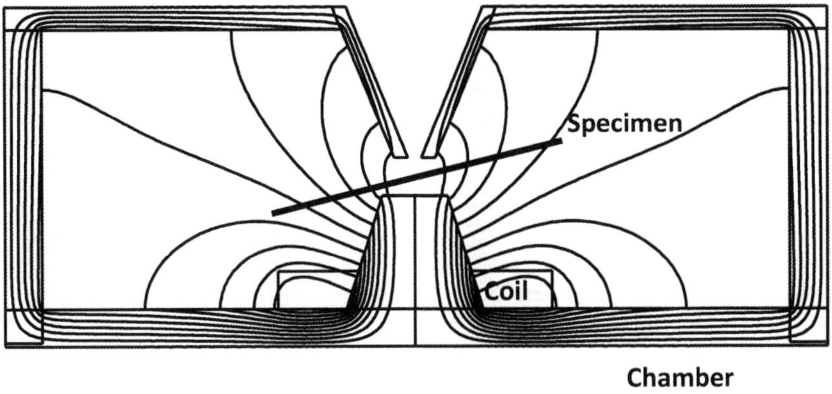

Fig. 3-6. Flux lines for an in-specimen chamber objective lens.

3.2 Magnetic semi-in-lens designs

In-lens arrangements inevitably place some restriction on the specimen size, since they operate by putting the specimen in the gap of a magnetic circuit. There is another type of magnetic immersion lens arrangement, sometimes called a single pole-piece lens or semi-in-lens objective lens, which overcomes this limitation. Here the magnetic field around a single pole-piece tip effectively determines the optics of the lens, and so it does not place any restrictions on the specimen size. The specimen is immersed in the magnetic field that protrudes beyond a single pole-piece tip. The flux lines for a semi-in-lens arrangement are shown in Fig. 3-7. This type of lens was proposed by Shao and Lin (1989), and was designed to observe large integrated circuit wafer specimens. Shao and Lin proposed a scintillator/photomultiplier detector arrangement situated within the lens bore, where the secondary electrons are deflected in the transverse direction. But this proposal is unnecessarily complicated. It is easier to allow the secondary electrons to spiral up through the lens bore and either detect them at the top of the lens with a micro-channel plate, or deflect them to one side and use a conventional E-T scintillator/photomultiplier arrangement, as is usually done for in-lens objective lenses.

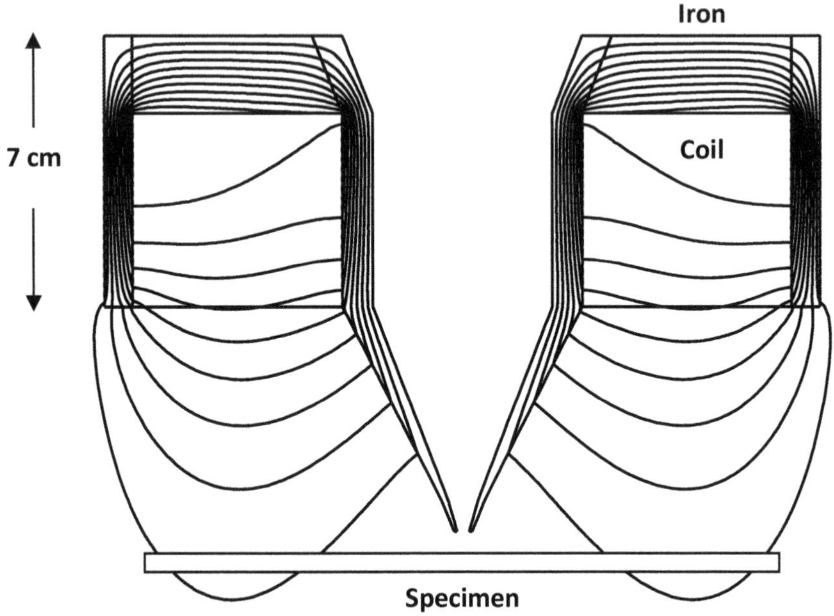

Fig. 3-7. Flux lines for a semi-in-lens objective lens placed above the specimen.

The peak value of the axial field for a 4 mm diameter bore is 0.177 Tesla (coil excitation of 3,000 AT), and it lies around 2 mm below the single pole tip. The axial field distribution and flux lines for these conditions are shown in Fig. 3-8(a). An important point to recognize about this semi-in-lens arrangement is that it is well suited for low voltages but unsuitable for high primary beam voltages. Figures 3-8(b) and 3-8(c) show that as the lens excitation strength is increased to 11,700 AT and 20,430 AT, respectively, the lens suffers from severe saturation effects. An extra peak in the axial field distribution at an excitation strength of 20,430 AT causes a primary beam travelling at an energy of 20 keV to focus well before the main axial field peak, in effect, degrading the lens performance. Some of the optical properties calculated for the lens at a primary beam voltage of 1 kV are shown in Table 3-3. The working distance, here, is defined to be the distance from the surface of the specimen to the single pole-piece tip.

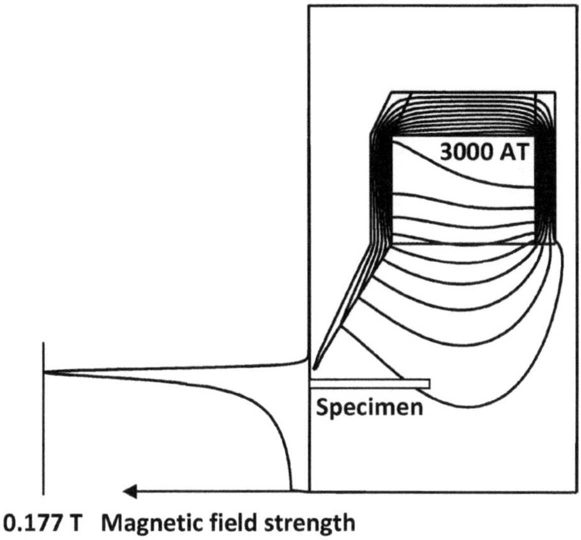

(a)

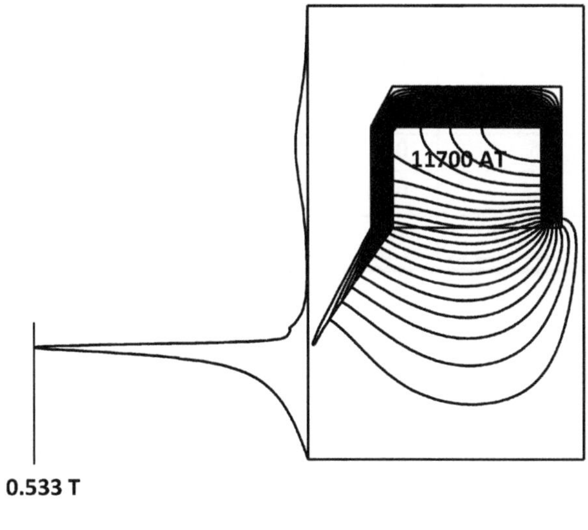

(b)

Fig. 3-8. Flux lines and axial field distribution for the semi-in-lens objective lens at different excitation strengths: (a) 3,000 AT, (b) 11,700 AT, (c) 20,430 AT.

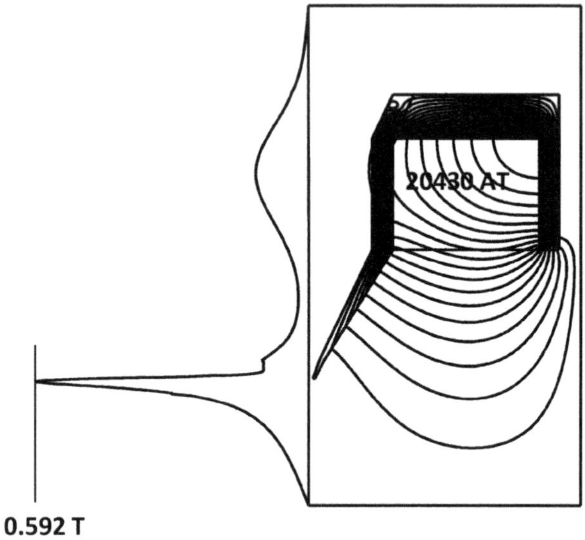

(c)

Table 3-3. Simulated on-axis aberrations for a semi-in-lens objective lens placed above the specimen.

Beam Energy (keV)	Working Distance (mm)	f (mm)	C_s (mm)	C_c (mm)	Excitation (AT)
1	0.5	0.979	0.375	0.632	6510
1	1	0.656	0.499	1.11	4533
1	2	1.56	0.963	1.13	2634

Table 3-3 shows that the semi-in-lens system has aberration coefficients similar to those found for in-lens arrangements, but requires much higher coil excitation strengths. This also makes it unsuitable for high primary beam voltage operation. Wasted magnetostatic energy in this lens is contained in the relatively large break in the magnetic circuit. Figures 3-9(a) and 3-9(b) present the probe diameter/current dependence for different excitation strengths at a primary beam voltage of 1 kV. They illustrate the significant improvement that can be obtained by operating at shorter working distances and higher excitation strengths.

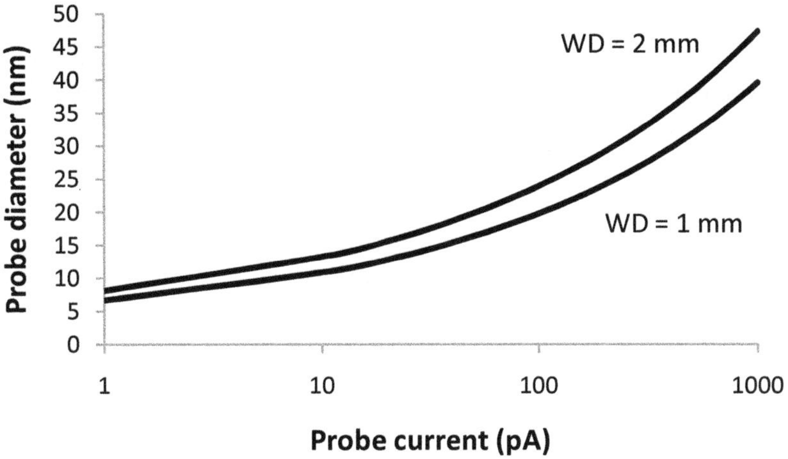

(a)

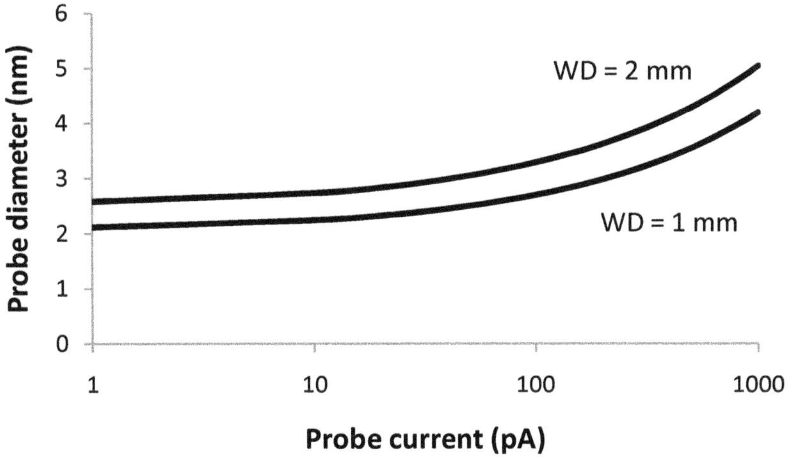

(b)

Fig. 3-9. Predicted probe diameter as a function of probe current for the semi-in-lens objective lens at a primary beam energy of 1 keV (f=1.12 mm, C_s=0.507 mm, C_c=0.759 mm, WD=1 mm, f=1.56 mm, C_s=0.963 mm, C_c=1.13 mm, WD= 2 mm): (a) Tungsten gun (ΔE=1.5 eV, $\beta=2\times10^4$ Acm^{-2}sr^{-1}), (b) Field emission gun (ΔE=0.4 eV, $\beta=10^7$ Acm^{-2}sr^{-1})

Figures 3-10(a) and 3-10(b) illustrate the trajectory paths for 3 eV secondary electrons and 1 keV backscattered electrons emitted from a specimen placed 1 mm below the single pole-piece tip of the semi-in lens layout. The secondary electrons spread out a bit in the radial direction as they spiral up through the lens bore, but their collection efficiency is expected to be high. The backscattered electron collection efficiency is also expected to be high, but the wider angle ones will be lost, since they inevitably collide with the inner walls of the lens bore. It should be noted that if a working distance of greater than 2 mm is used, then the secondary collection efficiency for low energies will significantly drop since they initially pass through an increasing magnetic field gradient. The solution to this problem is to apply a moderate size electric field in the vertical direction, extracting the secondary electrons through the single pole-piece tip. This can be achieved by applying a small positive voltage to a cylindrical electrode, which is placed inside the lens bore. Another problem for lenses of this type, particularly for those that are relatively long, is magnetic stray fields within the lens bore. These fields may cause low energy secondary electrons to spread out and collide with the inner walls of the lens bore. To circumvent this happening, a low current magnetic solenoid coil can be placed inside the lens bore. Tsuno *et al.* (1996) present simulations of electron trajectory paths illustrating these type of effects.

Like the in-lens arrangement, the semi-in-lens objective lens can be integrated with the specimen chamber. Figures 3-11(a) and 3-11(b) show a schematic drawing of this semi-in-lens-chamber layout and its corresponding flux lines and axial field distribution. This type of scheme is suitable for the inspection of large integrated circuit wafers and allows for specimen tilt, but unlike its in-lens counterpart, it works best for low primary beam voltages (typically, under 3 kV). It is useful where a large amount of space below the lens is required, such as the case for thick specimens (> 10 mm). For an excitation strength of 3,000 AT, the peak field strength is 0.15 Tesla, which is able to focus a 1 keV primary beam at a distance of 2 mm below the single pole-piece tip. For these conditions, $f = 1.86$ mm, $C_s = 0.911$ mm, and $C_c = 1.28$ mm.

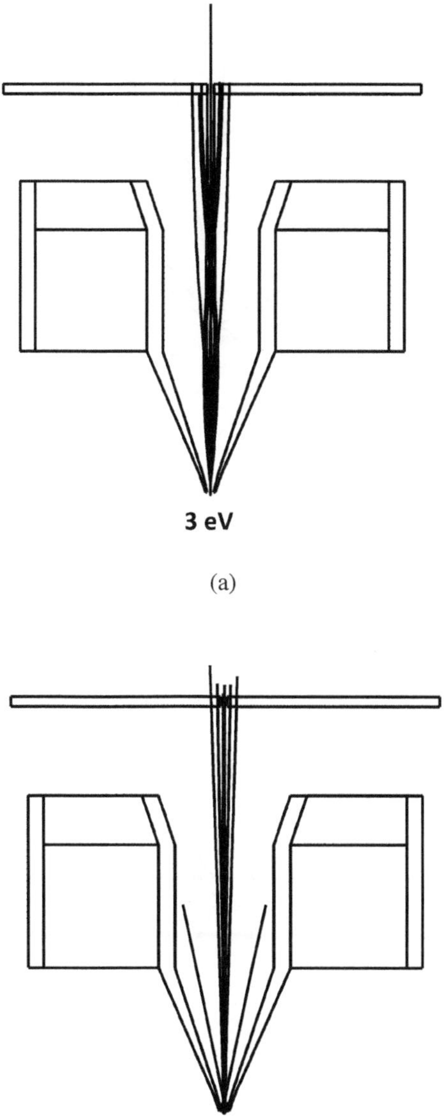

(a)

(b)

Fig. 3-10. Simulated trajectory paths in the semi-in-lens objective lens over a wide range of emission angles: (a) 3 eV secondary electrons, (b) 1 keV backscattered electrons.

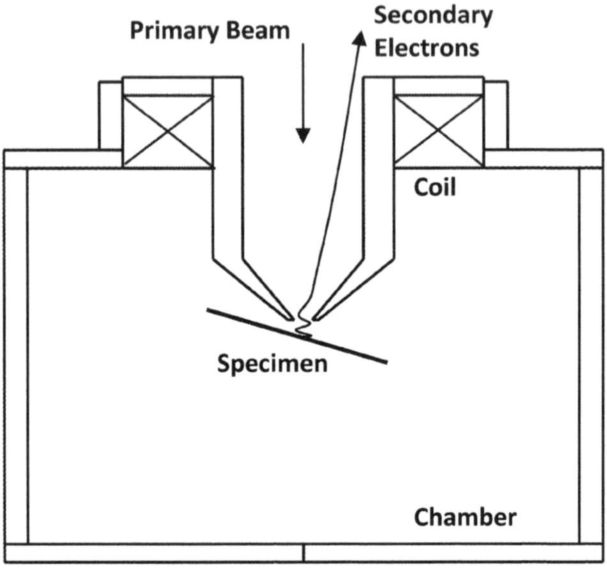

(a)

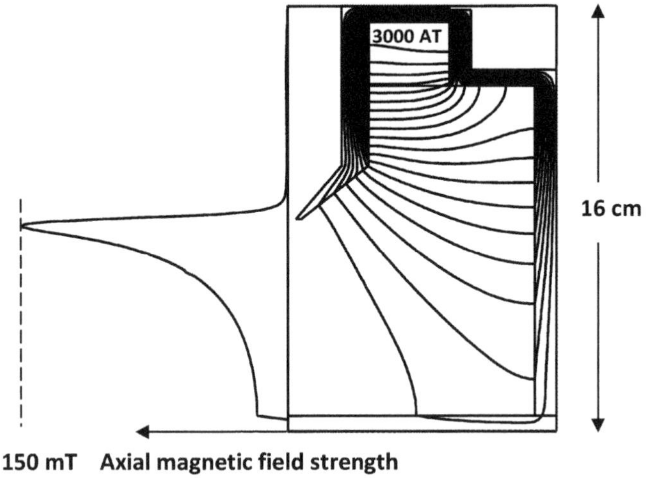

(b)

Fig. 3-11. Semi-in-chamber objective lens: (a) Schematic drawing, (b) Flux lines and axial field distribution.

The axial field distribution and flux lines for another possible single pole-piece lens are shown in Fig. 3-12. This type of lens was presented by Mulvey (1982) and was, in fact, the first type of single pole-piece lens to be proposed. Here the specimen is placed above the single pole-piece tip. The working distance, measured from the pole-tip to the surface of the specimen, must naturally be greater than the specimen thickness. This type of single pole-piece arrangement is well suited for TEM applications, where the primary beam travels through the specimen and down the lens bore. In this case, the bore of the lens needs to be progressively widened (not shown in the Fig. 3-12), and is generally known as the snorkel lens. For the SEM, there is no reason for having a lens bore hole, since the primary beam does not pass through the specimen. The conical shape of the single pole-piece allows for specimen tilt. Figure 3-13(a) shows the simulated axial field distribution for a 1 mm hole in the lens bore. The lens tip has a 2 mm diameter, producing a peak field of 0.26 Tesla for an excitation strength of 900 AT. If there is no hole in the lens bore, the field strength at the pole-piece tip is 0.34 Tesla. At a working distance of 2 mm, with or without a hole in the lens bore, the magnetic field strength is around 0.11 Tesla.

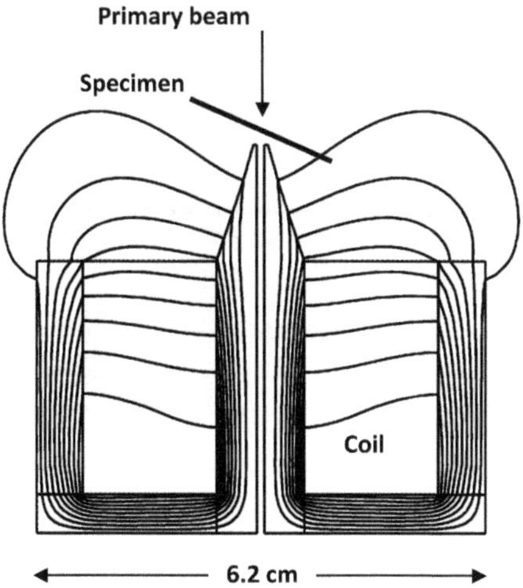

Fig. 3-12. Single pole lens below the specimen.

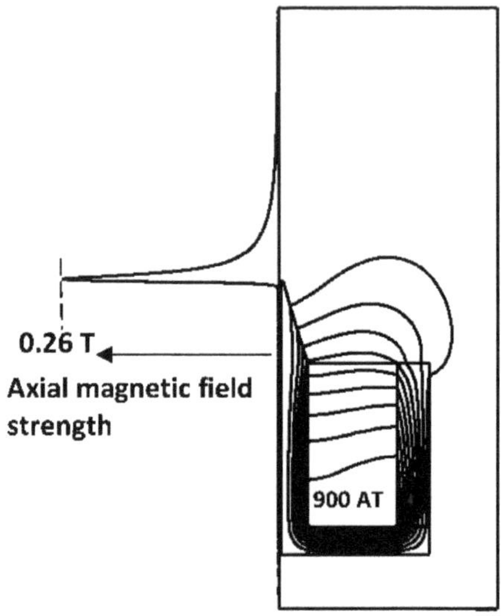

(a)

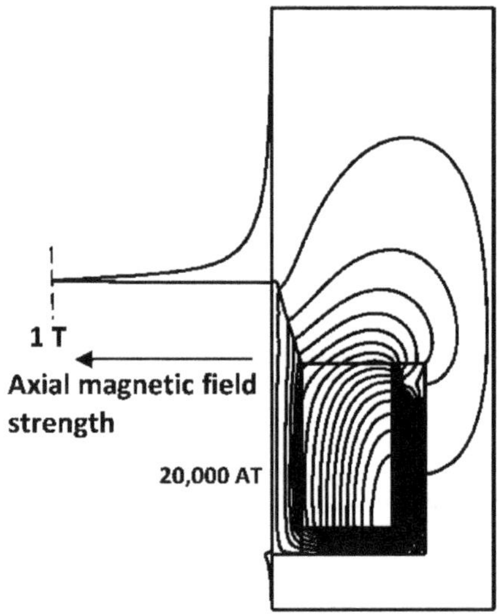

(b)

Fig. 3-13. Flux lines and axial field distribution for a single pole lens below the specimen at different lens coil excitation strengths: (a) 900 AT (with central hole), (b) 20000 AT (no central hole).

A single pole-piece lens placed below the specimen does not give as good resolution at low primary beam voltages as the previous case where it was placed above the specimen. This is because the axial field strength falls off much more slowly beyond the tip than within it. However, for high primary beam voltages, where the field close to the tip is more important, it performs better. Saturation effects, although they exist, are not as severe as they are for the previous single pole-piece lens above the specimen. Figure 3-13(b) shows the axial field distribution and flux lines for an excitation strength of 20,000 AT. The peak field strength at the tip surface is 1 Tesla, without a hole in the lens bore. This field strength is sufficient to focus a 20 keV beam 1 mm above the lens pole piece tip. Such operating conditions are, of course, only possible for thin specimens. Some of the on-axis aberration characteristics for this lens are summarized in Table 3-4 and given in Figs. 3-14(a) and 3-14(b).

Table 3-4 shows that for all the beam energies considered (1, 5, 20 keV), the chromatic aberration coefficient is a factor of around 2 to 3 times smaller than its value for conventional lenses, while the spherical aberration is improved by over an order of magnitude. Since it is chromatic aberration that dominates at low primary beam voltages and spherical aberration that prevails at high voltages, then these simulation results predict that the single pole-piece lens below the specimen will perform better at high primary beam voltages. But this advantage can only be gained where the specimen is thin (typically less than 1 mm). Figures 3-14(a) and 3-14(b) show that at high primary beam voltages, the probe diameter of the lens is comparable to that predicted for in-lens objective lenses. The relatively high excitation strengths required under these conditions will necessitate the use of an efficient lens cooling system. Note that for this type of lens, it is necessary to increase the excitation strength in order to increase the working distance. At high beam voltages, lens heating constraints may cause problems. The advantage of using this type of lens is that it frees up valuable space immediately above the specimen, providing more room for detectors and attachments in the specimen chamber.

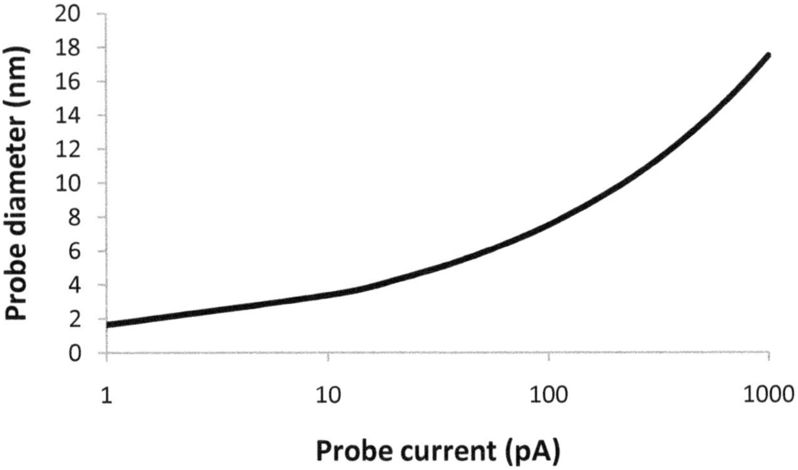

(a)

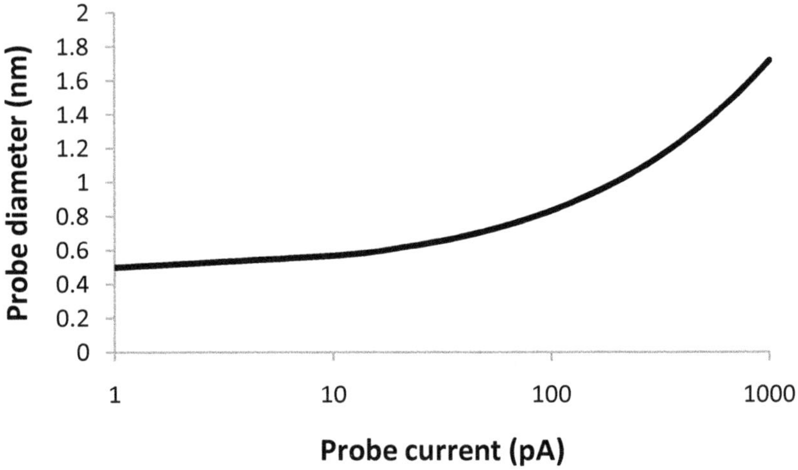

(b)

Fig. 3-14. Predicted probe diameter as a function of probe current for the single pole lens below the specimen (f=5.22 mm, C_s=0.992 mm, C_c=2.81 mm, 20 keV, WD=1 mm): (a) Tungsten gun (ΔE=1.5 eV, β=2.10^5 Acm^{-2}sr^{-1}), (b) Field emission gun (ΔE=0.4 eV, β=10^8 Acm^{-2}sr^{-1}).

Table 3-4. Simulated on-axis aberrations of a semi-in-lens objective lens placed below the specimen.

Beam Energy (keV)	Working Distance (mm)	f (mm)	C_s (mm)	C_c (mm)	Excitation (AT)
1	1.08	3.78	0.665	1.95	900
1	2.4	4.92	1.03	2.71	1200
1	3.41	5.5	1.25	3.12	1500
5	1.68	4.73	0.943	2.57	5000
20	1	5.39	1.06	2.9	25000

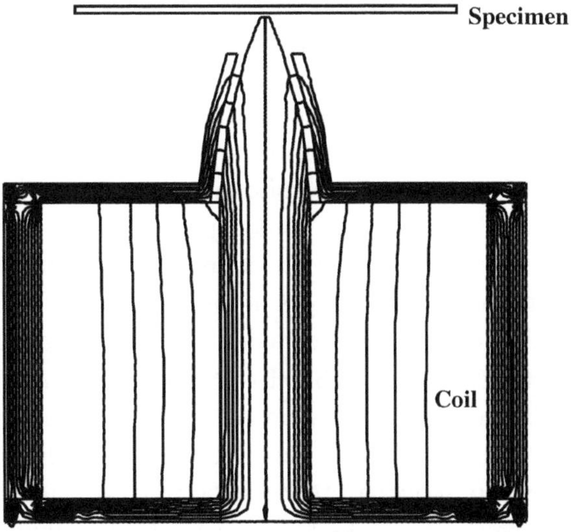

Fig. 3-15. Flux lines of an improved single-pole lens below the specimen, (Khursheed, 1996).

There have been some variations of the Mulvey single pole below-the-specimen lens design. Khursheed introduced a design which involved covering up the top part of the coil with iron, providing a return path for the magnetic flux just below the tip Khursheed (1996). Simulated flux lines for this lens are given in Fig. 3-15. Khursheed predicted that such an arrangement improves the axial aberrations by around 30%.

Tsai and Crewe (1998) introduced a "gapless magnetic objective lens", which like Khursheed's proposal, covers up the coil with iron, providing a return path for the flux. Their lens design involves mounting a conical magnetic shroud on to the top of the lens, which shorts the lens tip to the rest of the magnetic circuit. The magnetic material at the lens tip is driven into saturation, and so effectively forms a magnetic gap in the circuit around the tip. Tsai and Crewe's proposal is in fact equivalent to Khursheed's proposal, but has a smaller gap in the magnetic circuit around the pole tip. In both cases, the predicted aberrations are around 30% better than the original single pole lens proposed by Mulvey.

3.3 Electric retarding field lenses

Another high resolution SEM objective lens used in recent years is the retarding field lens. It is based on the concept of using a high primary beam energy through most of the column, and then slowing it down just before it reaches the specimen. This type of lens is particularly well suited to achieving high resolution at low landing energies. Its main advantage is that low beam voltages at the specimen are obtained, while the column and gun are operated under high beam voltage conditions. The overall performance of the SEM is then greatly improved, since the brightness of the gun is typically an order of magnitude better at high beam voltages than it is for low beam voltages, and the influence from stray fields is greatly reduced. At the same time, by having low landing energies at the specimen, all the advantages of operating at low primary beam voltages are obtained, such as elimination of charging, low radiation damage to the specimen, a reduced beam/specimen interaction volume, and a higher secondary emission yield.

There have been many different retarding field lens designs. For a more complete survey of them, the reader is referred to Mullerova and Lenc (1992) and Frank and Mullerova (1999). Retarding field lenses can be divided into the "low energy" type, where landing energies at the specimen surface are typically 1 keV, or the "very low energy" type, where landing energies are lower than 50 eV. At tens of eV, the electron wavelength becomes comparable to inter-atomic distances, and this makes possible the imaging of various diffraction contrast dependent parameters, such as surface crystallinity, ultra-thin specimen layers, and magnetic microstructure. Mullerova and Lenc (1992) and Frank and Mullerova (1999) describe work mainly on the very low energy regime. The following summary describes low voltage SEMs, rather than very low voltage SEMs.

Consider the idealized retarding lens shown in Fig. 3-16(a). It retards the primary beam at energy eV_p down to a landing energy of $e(V_p\text{-}V_s)$ at the specimen. The specimen is placed a distance W below an electrode with which it shares the same voltage, $-V_s$. A distance L separates this shielding electrode from an electrode further up the column at 0 V. Notice that there are grids at the centre of each electrode so that the potential is assumed to vary linearly in each region.

Let k be the ratio of the energy drop across the lens

$$k = \frac{V_p}{V_s}. \tag{3.1}$$

For the case of no electric field at the specimen (Fig. 3-16(a)), Frank and Mullerova (1999) refer to it is an "immersion lens", in order to distinguish it from what they call a cathode lens, where there is only one 0 V electrode above the specimen, as shown in Fig. 3-16(b). However, this terminology is misleading, since an immersion lens must "immerse" the specimen in a field. It is perhaps clearer to name this latter lens a full immersion retarding field lens, since in practice, most immersion retarding field lenses will lie between the two extremes of having a field free specimen to one which has a strong electric field at its surface (up to 5 kV/mm or more).

For an incoming parallel primary beam, in the idealized field free specimen case, Frank and Mullerova report the chromatic and spherical aberration coefficients to be approximately given by

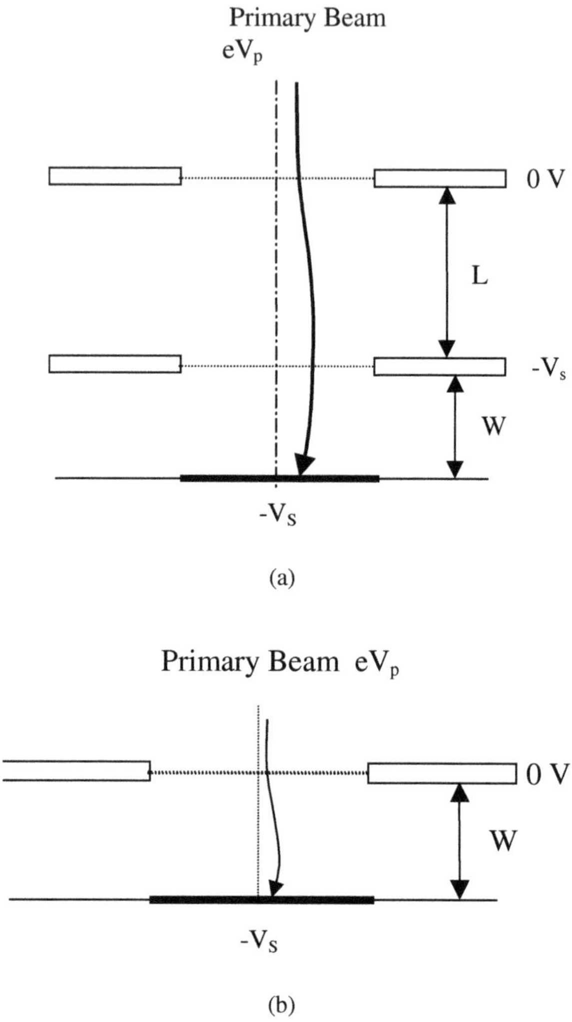

Fig. 3-16. Retarding field lenses: (a) Idealized field free layout, (b) Full immersion.

$$C_s \approx C_c \approx \left[\frac{1}{2}W + L\left(k^{1/2}+1\right)^{-2}\right]\left(1-k^{-1}\right). \tag{3.2}$$

Similarly, in the full immersion retarding field case, for $k >> 1$,

$$C_s \approx C_c \approx \frac{W}{k} = W\frac{V_s}{V_p} = \frac{V_s}{E_s} \rightarrow \tag{3.3}$$

where E_s is the electric field strength at the specimen surface. Clearly in the specimen field free case, the chromatic and spherical aberration coefficients can at best approach the value of $W/2$. In the full immersion case, their limiting values will be determined by how large an electric field strength can be tolerated at the specimen surface. If the breakdown electric field value is taken as 10 kV/mm, then at a landing beam voltage of 1 kV, C_c and C_s will have the limiting values of 0.1 mm.

A purely electrostatic retarding field lens was proposed by Zach (1989), which was later followed by a combined electric/magnetic one (Frosien *et al.*, 1989). Figure 3-17 shows a schematic diagram for a compound electric/magnetic retarding field lens. Here a liner tube, sometimes called the "booster tube", leaves the gun assembly and ends a distance of around 10 mm from the specimen surface. The booster electrode is connected to a high voltage power supply, which typically elevates it to a voltage of 9 kV. The specimen is shielded from the booster tube by an additional electrode at ground potential, creating a weak to moderate electric field strength at the specimen surface (50 to 500 V/mm). The magnetic axial field peaks at around the same position where the booster electrode ends, and the specimen is immersed in a weak magnetic field (10 mT). Figures 3-18(a) and 3-18(b) show the axial field distributions, equipotential lines, and flux lines for this compound retarding field lens. In the compound lens presented by Frosien *et al.* (1989), the magnetic field region was designed to lie within the lens assembly. But a magnetic field at the specimen is considered here, since it improves the lens optics. The secondary and backscattered electrons travel at high energies back through the lens (higher than 9 keV), and are collected by a round scintillator detector placed above the lens. The on-axis aberration performance calculated for this lens at low primary beam voltages is summarized in Table 3-5.

The simulation results in Table 3-5 predict that the retarding field lens can provide chromatic aberration coefficients that range from 1 to 2.5 mm over a working distance of 1 to 4 mm. This is over one order of magnitude better than its corresponding values for conventional lenses. The spherical aberration coefficient values are also relatively low (1.75 to 4.32 mm), although not as low as those obtained by in-lens magnetic lenses. If there is no leakage magnetic field at the specimen, as in the original proposal by Frosien *et al.* (1989), the spherical aberration is much higher, typically between 5 to 8 mm. A crucial parameter here is

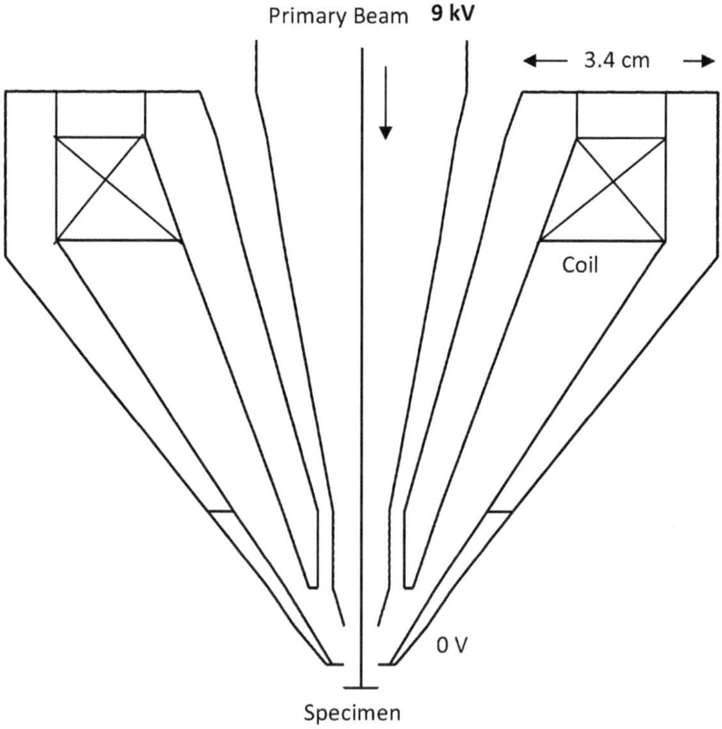

Fig. 3-17. Retarding field objective lens layout.

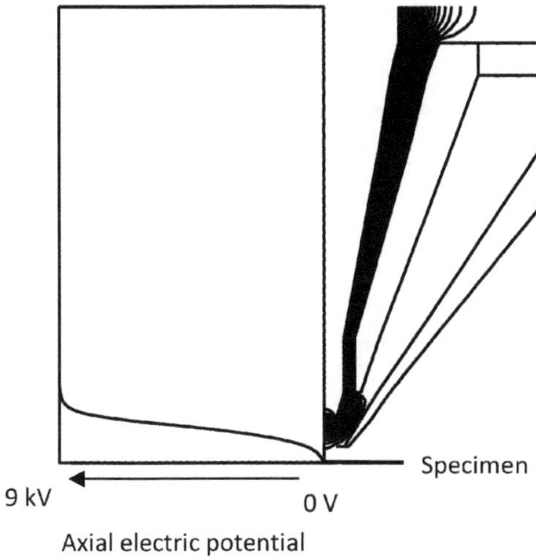

(a)

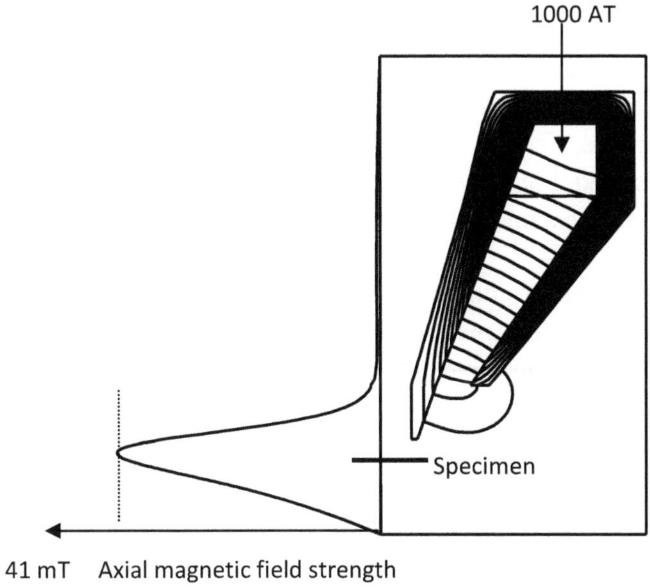

(b)

Fig. 3-18. Simulated field distributions for the retarding field objective lens: (a) Equipotential lines and axial potential distribution, (b) Magnetic mlux lines and axial magnetic field distribution.

Table 3-5. Simulated on-axis aberrations of a compound retarding field lens.

Beam Energy (keV)	Working Distance (mm)	Electric Field Strength at Specimen (V/mm)	f (mm)	C_s (mm)	C_c (mm)
1	1	68.2	4.50	1.75	1.16
1	2	118.9	4.72	2.25	1.48
1	3	218.9	4.92	3.08	1.91
1	4	406.4	5.17	4.32	2.43
0.5	1	68.2	3.34	1.0	0.706
0.5	2	118.9	3.48	1.41	0.976
0.5	3	218.9	3.57	2.08	1.34
0.5	4	406.4	3.72	3.19	1.83

the electric field strength at the specimen surface. As the working distance decreases, it increases. The presence of strong electric fields at the specimen surface is usually undesirable. This is because for dielectric specimens, they greatly increase the surface charging effect (by reducing the number of secondary electrons that return to the specimen surface), which in turn creates strong transverse electric fields that distort and blur the final image. For most specimens, however, weak to moderate fields can be tolerated. For purely metal specimens, the allowable electric field strength at the specimen can be strong (>1 kV/mm).

Low landing energies down to 200 eV have been observed with the type of retarding field lens presented in Fig. 3-17 (Frosien *et al.*, 1995). The simulation results in Table 3-5 show that the aberration coefficients for a landing energy of 500 eV are also small. This means that the final probe size increase, due to the primary beam energy being lowered, is not as great as it is for other high resolution lenses. Figure 3-19(a) shows the probe diameter against probe current dependence at the primary beam energies of 1 keV and 500 eV at a working distance of 2 mm for a field emission gun. The probe size for the primary beam landing energy of 500 eV is only around 1.3 times greater than its value at 1 keV landing energy. Figure 3-19(b) shows the probe diameter versus current dependence for the working distances of 1 mm and 3 mm at a

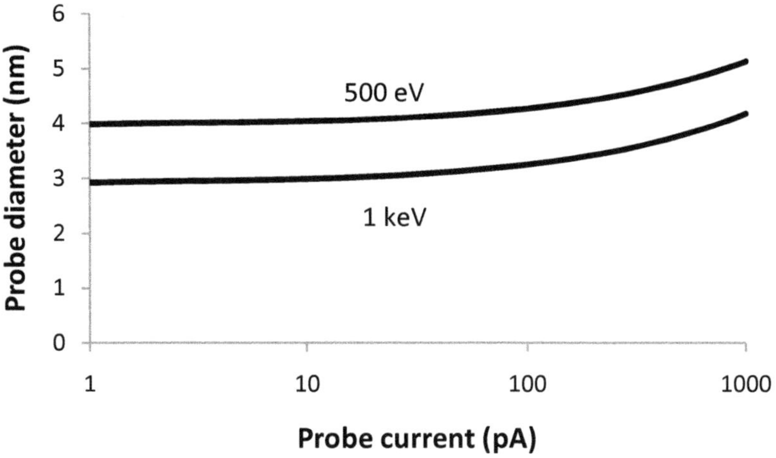

(a)

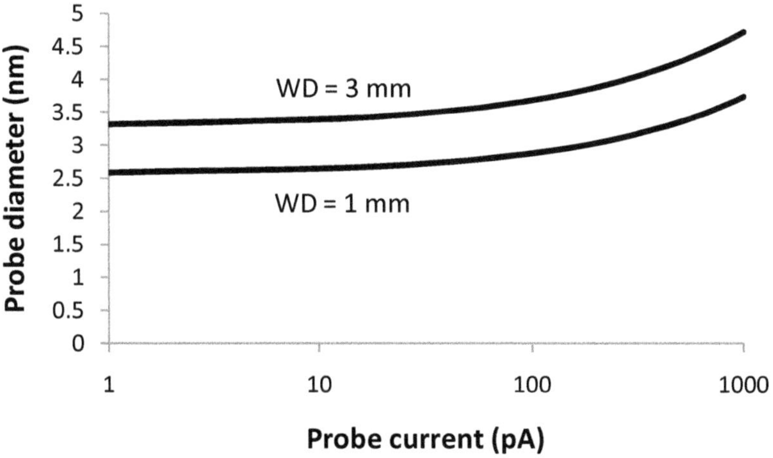

(b)

Fig. 3-19. Predicted probe diameter as a function of probe current for the retarding field lens (field emission gun): (a) Working distance of 2 mm, (β=5×10^7, ΔE=0.4 eV, 1 keV, f=4.72, C_s=2.25, C_c=1.48 mm, 500 eV, f=3.48, C_s=1.41, C_c=0.976 mm), (b) Primary beam voltage of 1 kV, (β=5×10^7, ΔE=0.04 eV, WD=1 mm, f=4.5, C_s=1.75, C_c=1.16 mm, WD=3 mm, f=4.92, C_s=3.08, C_c=1.91 mm).

primary beam landing energy of 1 keV. A 3 to 4 nm probe diameter is predicted for the compound retarding lens.

Figures 3-20(a) and 3-20(b) show electron trajectory paths of 3 eV secondary electrons and 1 keV backscattered electrons for a working distance of 4 mm over a wide range of emission angles. Except for the wide-angle backscattered electrons, all electrons travel back through the lens bore and will be collected by a detector placed at the top of the lens. Since a substantial number of these electrons lie close to the optic axis (within 2 mm), the collection efficiency of all scattered electrons will depend on the diameter of the detector primary beam passing hole. A diameter of greater than 4 mm may reduce the overall collection efficiency by over 50%.

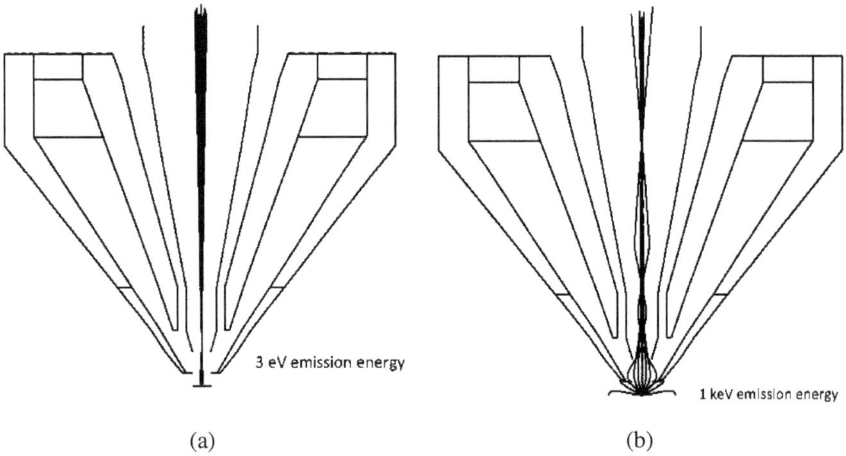

Fig. 3-20. Simulated trajectory paths in the retarding field lens over a wide range of emission angles: (a) 3 eV secondary electrons, (b) 1 keV backscattered electrons.

3.4 Mixed field in-lens designs

In certain applications such as voltage contrast, strong electric field strengths at the specimen are desirable (Khursheed *et al.*, 1991; Dubbledam, 1993). In these situations, the compound retarding field lens can be designed to immerse the specimen with strong magnetic and electric field strengths. Such a design has been presented by Beck *et al.*

(1995). Figure 3-21 shows a schematic drawing of a lens layout based on Beck *et al.'s* design. Figures 3-22(a) and 3-22(b) depict its respective electric and magnetic field distributions. Here, the specimen is immersed in a magnetic field strength of 60 mT and an electric field strength of 1 kV/mm. The specimen is located 6 mm below the booster electrode. This relatively large working distance allows for power connections to be made to a wafer or integrated circuit specimen. The improved aberrations for this lens are given in Table 3-6, and the probe diameter versus current dependence at the primary beam landing energies of 1 keV and 500 eV for a field emission gun SEM is shown in Fig. 3-23.

These results illustrate that a significant improvement in the probe resolution is predicted for a compound immersion retarding field lens where high electric field strengths at the specimen can be applied. The probe diameter is predicted to be less than 2.5 nm at 1 keV, which rivals the performance of magnetic in-lens objective lenses. Further improvements can, of course, be made by increasing the electric field strength at the specimen (decreasing the working distance). For voltage contrast applications, field strengths up to 5 kV/mm have been used (Beha and Clauberg, 1993; Khursheed *et al.*, 1991).

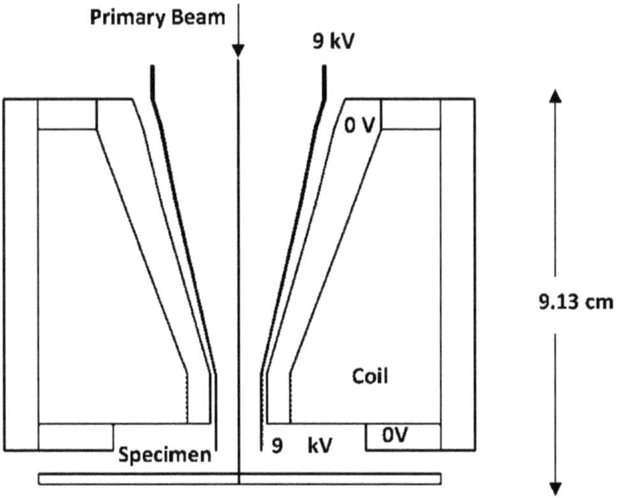

Fig. 3-21. Compound field immersion retarding field lens layout.

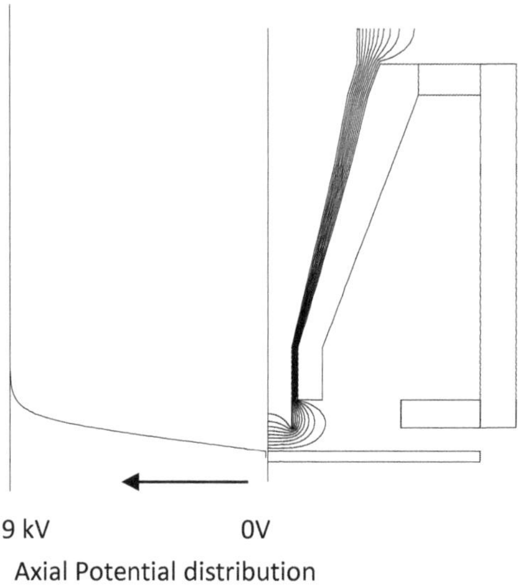

(a)

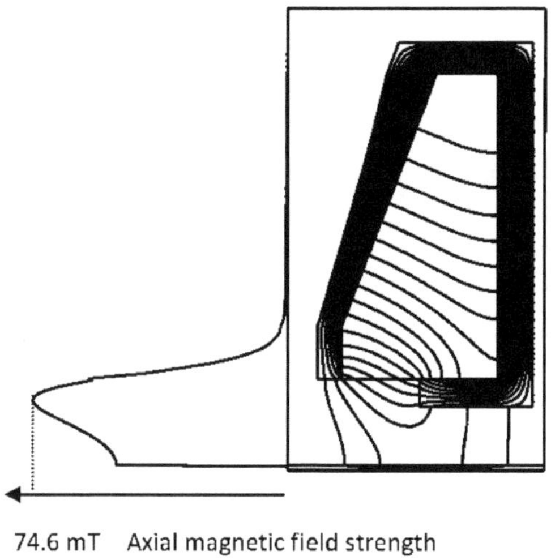

(b)

Fig. 3-22. Simulated field distributions for the compound immersion retarding field lens: (a) Equipotential lines and axial potential distribution, (b) Magnetic flux lines and axial magnetic field distribution.

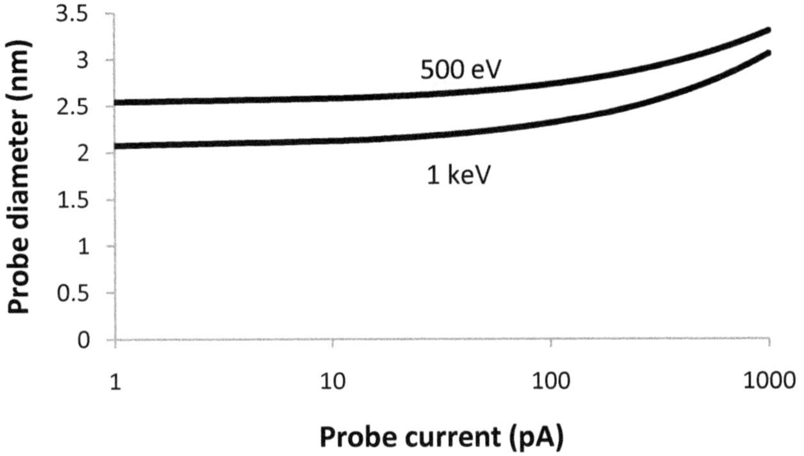

Fig. 3-23. Predicted probe diameter as a function of probe current for the compound field immersion retarding field lens (1 keV, f=3.75 mm, C_s=1.0 mm, C_c=0.742 mm, β=5×10^7, ΔE=0.4 eV, 500 eV, f=2.8 mm, C_s=0.513 mm, C_c=0.399 mm, β=5×10^7, ΔE=0.4 eV).

Table 3-6. Simulated on-axis aberrations of a compound immersion retarding field lens placed above the specimen.

Beam landing Energy (keV)	Specimen Electric Field (V/mm)	Working Distance (mm)	f (mm)	C_s (mm)	C_c (mm)
1	1054	6.	3.7	1.0	0.74
0.5	1054	6.	2.8	0.51	0.4

Figures 3-24(a)–3-24(c) show the trajectory paths of electrons leaving the specimen having initial energies of 3 eV, 500 eV, and 1 keV. Due to the presence of the sharply decreasing magnetic field gradient at the specimen, the 3 eV secondary electrons are collimated, and many of them will be lost through the detector primary beam-passing hole. Some means of deflecting these electrons off-axis is required. Figures 3-24(b) and 3-24(c), however, indicate that there is wide angular dispersion for the backscattered electrons.

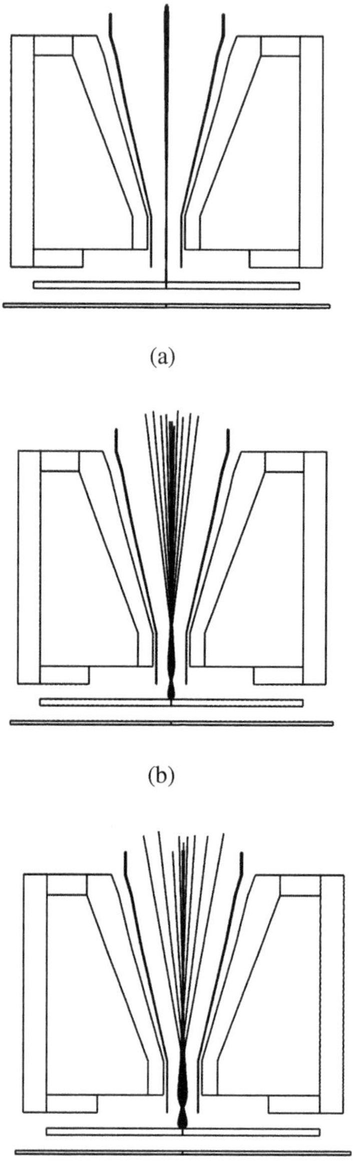

Fig. 3-24. Simulated trajectory paths in the compound immersion retarding field lens over a wide range of emission angles: (a) 3 eV secondary electrons, (b) 500 eV backscattered electrons, (c) 1 keV backscattered electrons.

In order to improve SE detection in the compound field immersion retarding lens, Meisburger *et al.* (1992) proposed the use of two matched Wien filters, one in the upper half of the column, and another in the lower half of the column, placed above the specimen and objective lens, as shown in Fig. 3.25. The Wien filters are excited in opposite directions, in order to cancel their angular/energy dispersion properties, and minimize aberrations on the primary beam, in the manner already described in Chapter 2.5. The lower Wien filter is used to deflect secondary electrons off-axis, and so improve their collection efficiency.

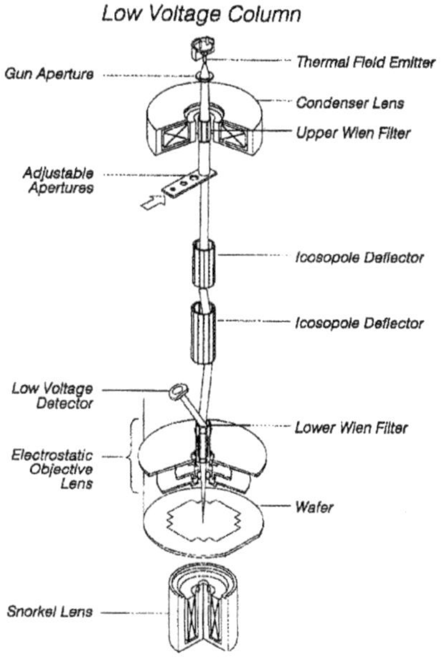

Fig. 3-25. Schematic layout of the dual Wein filter low voltage SEM column using a mixed field immersion lens proposed by Meisburger *et al.* (1992).

If the electric field in the lower Wien filter is in the x-direction, and the magnetic field is in the y-direction, and assuming that the primary beam velocity in the z-direction is many orders of magnitude greater than its other components, the Wein filter condition expressed in terms of the primary beam voltage U_p is

$$E_x = \left(2\eta U_p\right)^{1/2} B_y \tag{3.4}$$

where η is the electron charge to mass ratio. Let the magnetic scalar potentials on the pair of magnetic deflector plates be given by -*NI* and *NI* (the excitation strength), and the voltages on the electric deflector plates be given by $-U_d$ and U_d. Assuming that the plate separation distance for each set of deflector plates is the same, then the potential U_d that must be applied to the electric plates for a given magnetic deflector excitation *NI* is

$$U_d = 0.745 NI \sqrt{U_p}. \tag{3.5}$$

This equation indicates how the voltage required to drive the electrostatic field inside the Wien filter must change with the primary beam voltage. Figure 3-26 shows the result of using 1 cm long deflectors with a plate to plate separation distance of 2 cm, designed for a primary beam energy of 10 keV through the Wien filter (giving a landing energy of 1 keV at the specimen). An excitation strength of ±20 AT is applied to the magnetic deflector plates, while a voltage of ±1490 V is applied to the electric deflector plates. The secondary electrons leave the specimen with an energy of 1 eV and are emitted over a wide range of angles. They have an energy of 9,001 eV as they pass through the Wien filter. The beam of secondary electrons is deflected 26.56° from the optic axis. The Wien filter can in this way significantly improve the collection efficiency of secondary electrons. Knell and Plies (1998, 1999, and 2000) further developed mixed electric/magnetic field semi-in-lens designs. They proposed a Magnetic-electrostatic detector objective lens (MEDOL) design, which like the Beck *et al.* lens, uses pole-pieces/electrodes above the specimen, however, in the Knell and Plies MEDOL design, the magnetic gap is defined more radially, where both inner and outer pole pieces start from approximately the same height above the specimen.

Another type of compound immersion retarding field lens was presented by Mullerova and Lenc (1992). They use a single pole magnetic lens placed beneath the specimen, and a simple parallel-plate electric immersion lens above it. They reported high resolution for very low

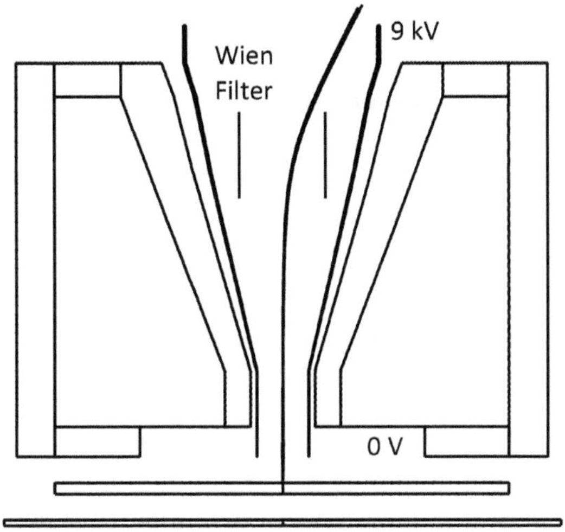

Fig. 3-26. Simulated trajectory paths of 1 eV secondary electrons over a wide range of emission angles in the compound immersion retarding field lens with a Wien filter 1 cm long, 2 cm plate separation, ±20 AT on magnetic plates, and ±1,490 volts on electric plates.

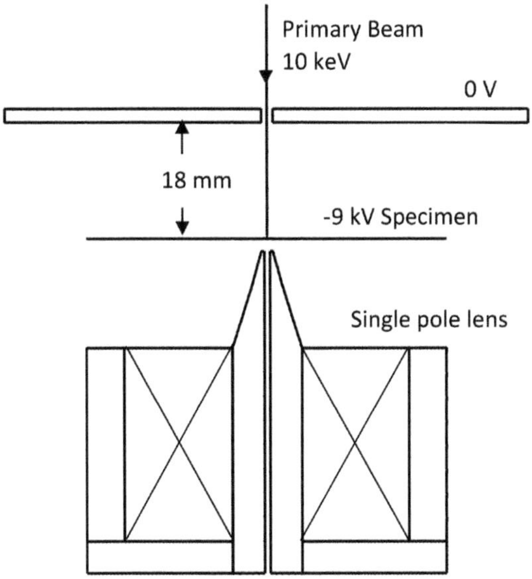

Fig. 3-27. A single pole compound field immersion retarding field lens.

beam landing energies at the specimen, typically 25 nm at a landing energy of 10 eV. A schematic drawing for a lens based on this principle is shown in Fig. 3-27. The primary beam energy down the column is high, around 10 keV. The primary beam retarding part of this arrangement is achieved by biasing the specimen to a high negative voltage. A 10 keV primary beam lands on the surface of a -9 kV specimen with an energy of 1 keV. Here, it is assumed that the single pole lens can also be biased to the same negative voltage that is applied to the specimen. If the specimen is placed 2 mm above the single pole lens tip, the magnetic field strength at its surface is around 0.11 T for a coil excitation strength of 900 AT, while the electric field strength is around 0.5 kV/mm.

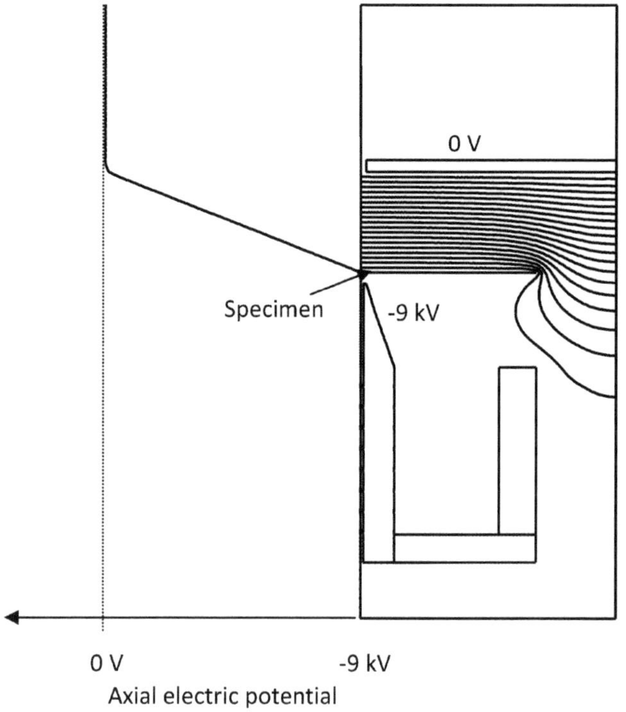

Fig. 3-28. Equipotential lines and axial potential distribution for the single pole compound field immersion retarding field lens.

The electric equipotential lines and axial potential distribution for this lens are shown in Fig. 3-28. As with the previous compound field immersion retarding field lens, a strong retarding electric field exists just above the specimen. Simulated on-axis aberration coefficients for this lens are given in Table 3-7, while its probe diameter-current dependence for a field emission gun is shown in Figure 3-29.

Table 3-7. Simulated on-axis aberrations of a compound field immersion retarding field lens with the magnetic circuit situated below the specimen.

Beam Energy in column (keV)	Landing Energy (keV)	Electric Field Strength at Specimen (V/mm)	f (mm)	C_s (mm)	C_c (mm)
6	1	280	2.06	0.546	0.945
5.5	0.5	280	1.57	0.411	0.646
10	1	506	1.65	0.431	0.688
9.5	0.5	506	1.22	0.3	0.441

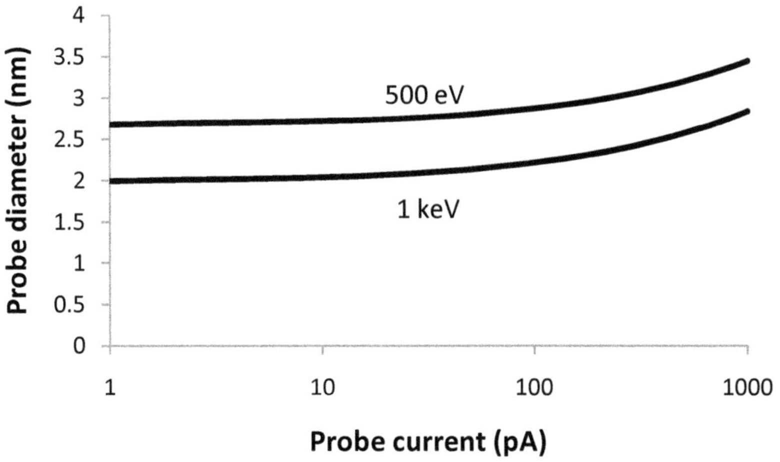

Fig. 3-29. Predicted probe diameter as a function of probe current for the single pole compound immersion retarding field lens. Field emission gun, 1 keV, f=1.65, C_s=0.431, C_c=0.688 mm, β=5×10^7, ΔE=0.4 eV, 500 eV, f=1.22, C_s=0.3, C_c=0.441 mm, β=5×10^7, ΔE=0.4 eV, WD=2 mm (magnetic), WD=18 mm (electric).

The on-axis aberrations are similar to the previous compound immersion retarding field lens arrangement, and are comparable to those obtained for magnetic in-lens objective lenses. They show how the retarding field arrangement overcomes the previous shortcomings of the magnetic single pole lens (placed below the specimen). The beam now moves through the slowly falling magnetic axial field part of the distribution relatively quickly (at a high energy), and is focused mainly by the part of the axial field distribution that rises up sharply, close to the single pole tip. However, this improved performance can only be obtained for thin specimens (< 2mm).

The trajectory paths for secondary electrons leaving the specimen at initial energies of 3 eV emitted over a wide range of angles are shown in Fig. 3-30(a). As with the previous compound immersion retarding field lens, the sharply decreasing magnetic field gradient at the specimen collimates them, and most of the secondary electrons are likely to be lost through the primary beam-passing hole in the detector. A means of deflecting them off-axis is required. The backscattered trajectory paths for 500 eV and 1 keV, shown in Figs. 3-30(b) and 3-30(c) respectively, are more widely dispersed.

A Wien filter can be used to deflect the secondary electrons off-axis, so that a normal E-T detector can be used to detect them. Figure 3-31 shows the result of using 1 cm long deflectors with a plate to plate separation distance of 2 cm, designed for a primary beam energy of 10 keV through the Wien filter (giving a landing energy of 1 keV at the specimen). An excitation strength of ± 20 AT is applied to the magnetic deflector plates while, a voltage of $\pm 1{,}490$ V is applied to the electric deflector plates. The secondary electrons leave the specimen with an energy of 1 eV and are emitted over a wide range of angles. The trajectory paths for these secondary electrons show that they are deflected by around 20° from the optic axis.

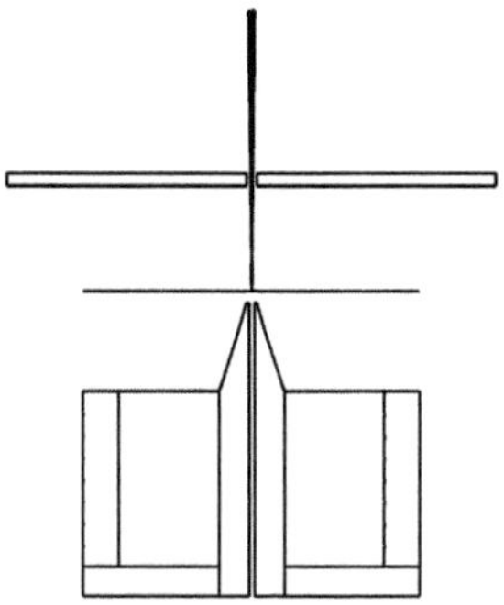

(a)

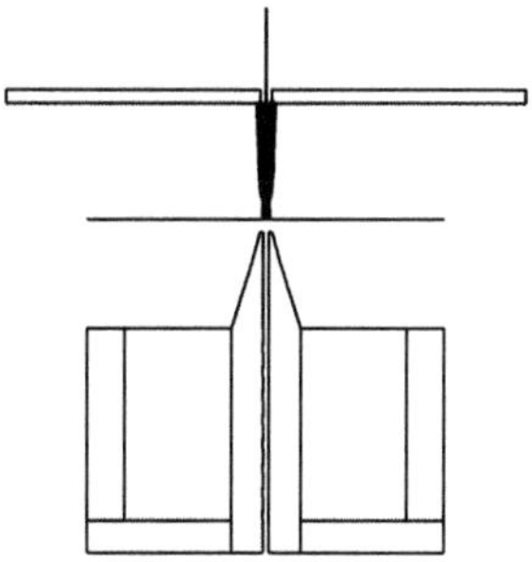

(b)

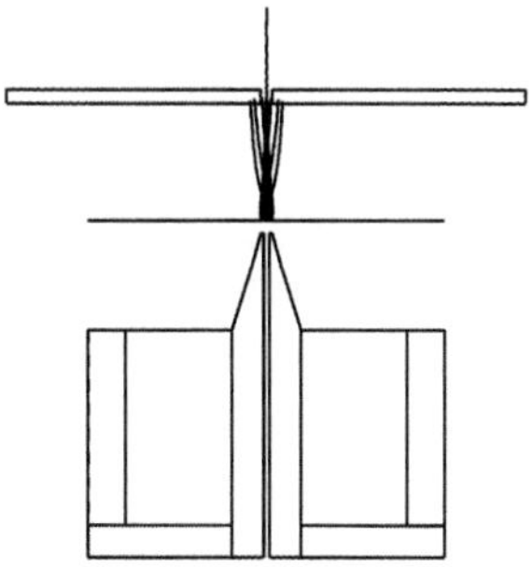

(c)

Fig. 3-30. Simulated trajectory paths for the single pole compound immersion retarding field lens over a wide range of emission angles: (a) 3 eV secondary electrons, (b) 500 eV backscattered electrons, (c) 1 keV backscattered electrons.

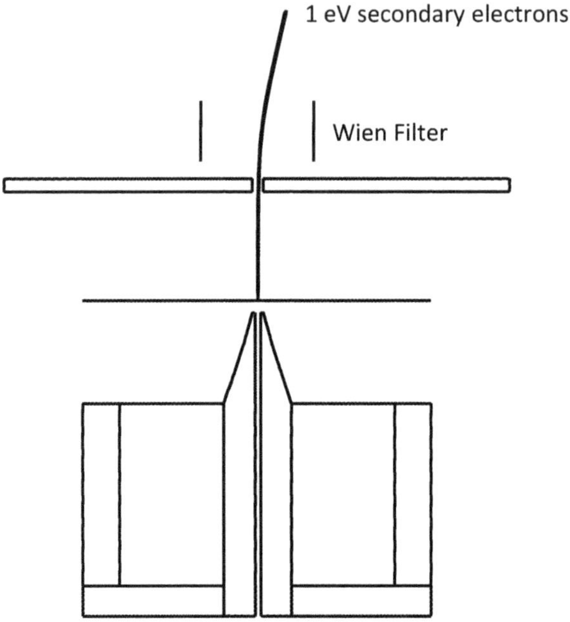

Fig. 3-31. Simulated trajectory paths of 1 eV secondary electrons over a wide range of emission angles in the single pole compound immersion retarding field lens with a Wien filter 1 cm long, 2 cm plate separation, ±20 AT on magnetic plates, and ±1,490 volts on electric plates.

3.5 Selected in-lens image examples

An indication of how much retarding electric field/magnetic immersion lenses can improve the image resolution of a conventional SEM can be obtained through the use of simple high resolution lens attachments for the SEM based upon permanent magnet technology (Khursheed *et al.*, 2001; Khursheed and Karuppiah, 2002; Khursheed, 2002a). These attachments are compact enough to be placed on the specimen stage, below the final lens, as illustrated for the add-on lens example shown in Fig. 3-32. The specimen is placed within the lens attachment, and the SEM is operated as normal, using its existing BSE and SE detectors. The add-on lens immerses the specimen in a strong magnetic field and is

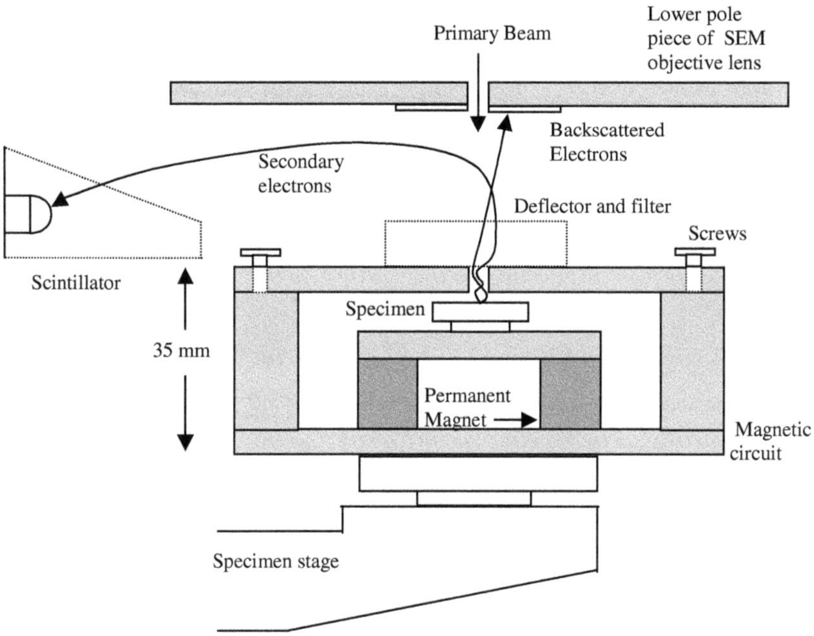

Fig. 3-32. Schematic layout for an add-on lens attachment inside the specimen chamber of a conventional SEM (Khursheed, 2002).

able to focus the primary beam over a distance of several millimeters, effectively becoming the final lens for the SEM. The existing SEM objective lens is used for fine focusing only. The primary beam enters the lens attachment through a top-plate hole having a diameter of 4 mm or less, and secondary electrons are strongly collimated as they travel up back through this hole, before being deflected and collected by the SE detector in the normal way. Small motors within the lens attachment, or attached to its exterior, provide specimen movement in x, y and z directions. Coarse focusing is provided by movement in the z-direction. These types of add-on lens attachments can provide focal lengths well below 3 mm, typically an order of magnitude smaller than conventional SEM objective lenses. The specimen can be biased in order to retard the primary beam inside the lens attachment.

Figure 3-33(a) shows flux lines from a simulated field distribution of an add-on magnetic immersion lens attachment, while Fig. 3-33(b) compares the simulated axial magnetic field distribution with an experimentally measured one. Simulations predict that for a 1 keV primary beam, this add-on lens has a focal length of 1.24 mm, C_C = 0.788 mm and C_S = 0.537 mm, around an order of magnitude lower than their corresponding values for conventional SEM objective lenses.

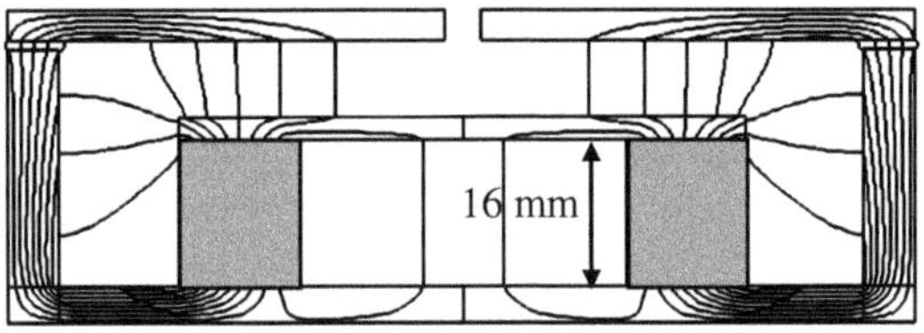

(a)

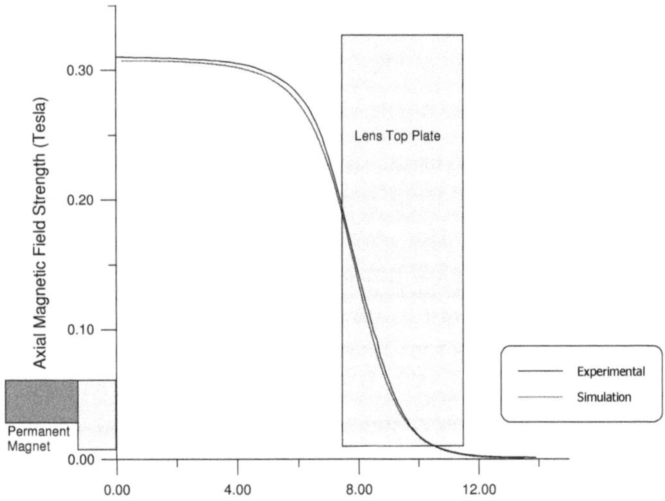

(b)

Fig. 3-33. Field distribution for a magnetic immersion add-on lens attachment: (a) Simulated flux lines, (b) Comparison of simulated axial field strength with experimentally measured values.

Figure 3-34 presents experimental SE images with and without the add-on lens attachment obtained on a JEOL 5600 tungsten gun SEM for primary beam voltages of 1 kV and 3 kV. A gold-on-carbon test specimen is used. The image demagnification is 100,000 and the add-on lens attachment improves the image resolution by over a factor of 4. Similar results for a field emission SEM are shown in Fig. 3-35. In this

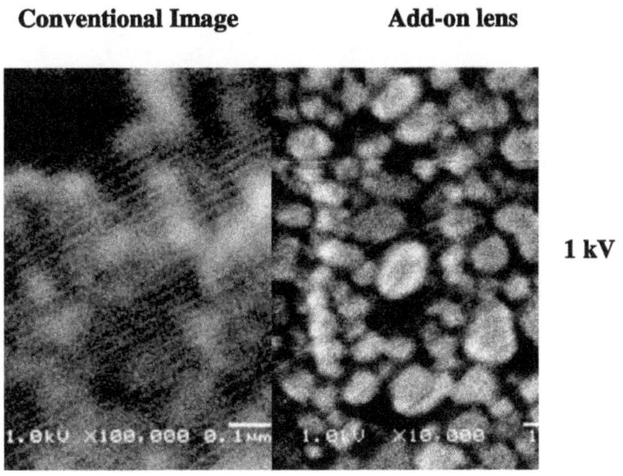

(a)

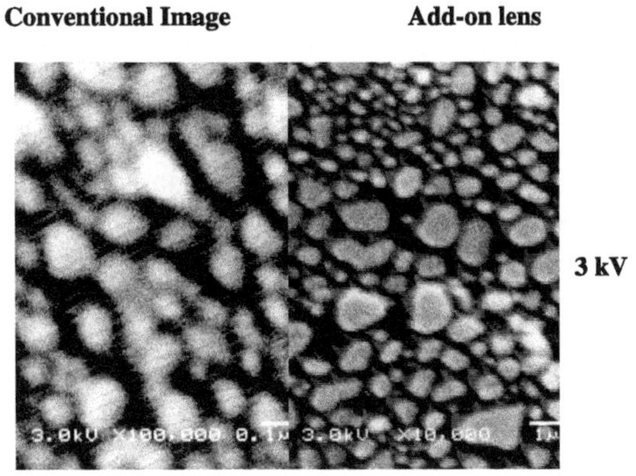

(b)

Fig. 3-34. Experimental images of a gold-on-carbon test specimen inside a JEOL 5600 tungsten gun SEM with and without the add-on lens. The image demagnification is 100,000: (a) Primary beam voltage of 1 kV, (b) Primary beam voltage of 3 kV.

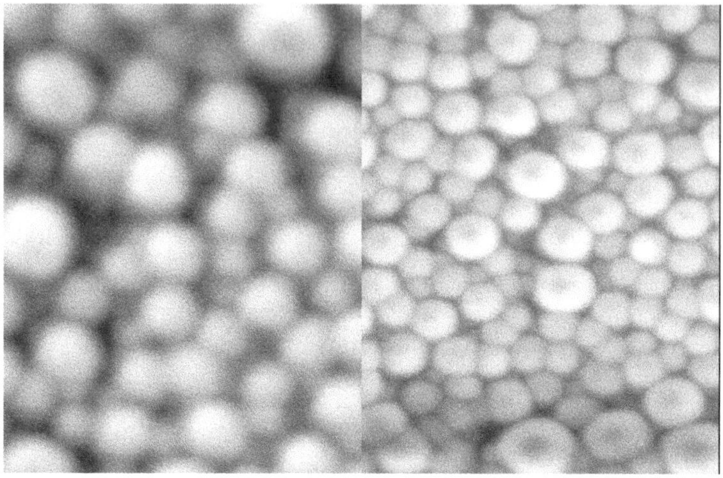

Conventional lens **Add-on lens**

Fig. 3-35. Experimental images of a tin-on-carbon test specimen inside a Philips XL30 FEG SEM with and without the add-on lens for a 1.1 kV primary beam. The image demagnification is 100,000.

case a Philips XL30 FEG SEM is used to image a tin-on-carbon test specimen with a demagnification of 100,000. The primary beam voltage is 1.1 kV, and an estimated image resolution of less than 2 nm is obtained with the add-on lens.

Figure 3-36 shows the layout design and simulated field distributions for a mixed field add-on lens attachment. In this case, the specimen is biased to negative voltages, retarding the primary beam so that it has low landing energies, in addition to the specimen being immersed in a magnetic field. The permanent magnet height is 5 mm, and is sufficient to focus a 6 keV beam on to a specimen biased to -5 kV, giving a landing energy of 1 keV. Simulations predict that this mixed field add-on lens has a focal length of 0.718 mm, $C_C = 0.251$ mm and $C_S = 0.235$ mm, less than half their corresponding values for the magnetic immersion lens attachment (magnetic field alone) at the same landing energy.

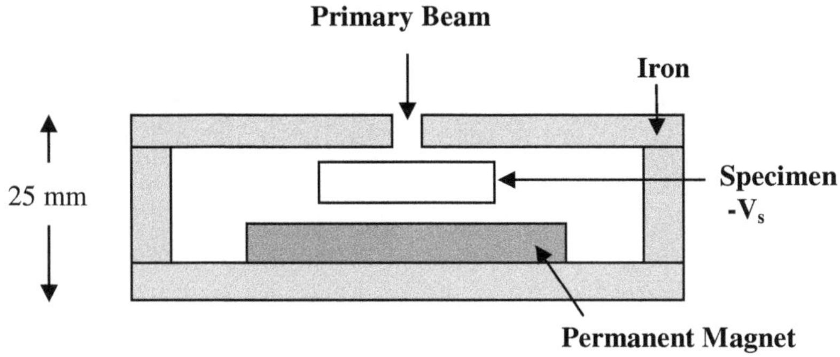

(a)

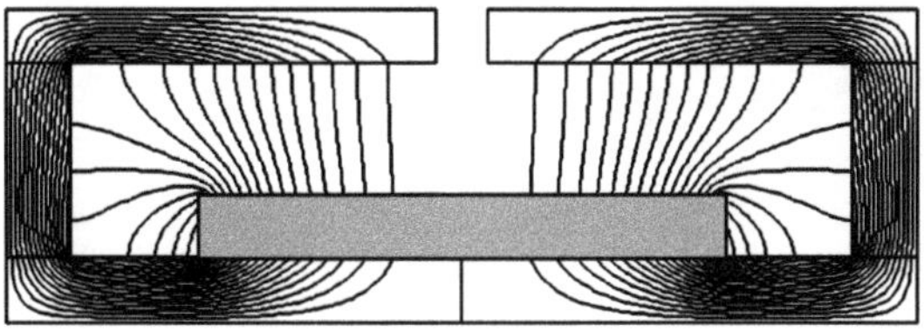

(b)

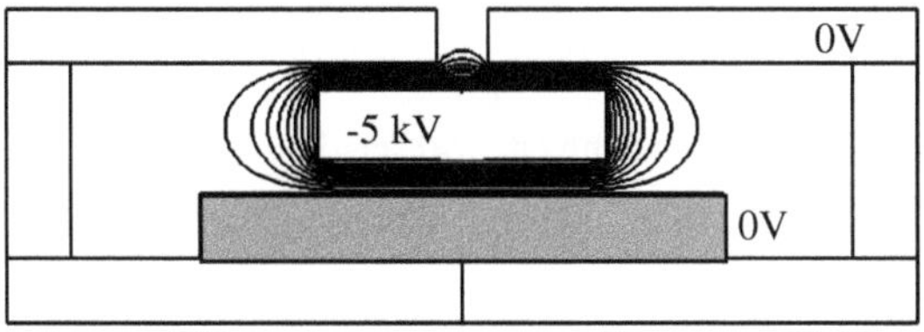

(c)

Fig. 3-36. Schematic diagram of a mixed field immersion add-on lens attachment: (a) Conductor/electrode layout, (b) Flux lines for simulated magnetic field distribution to focus a 6 keV primary beam on to the specimen, (c) Equipotential lines for simulated electrostatic field distribution.

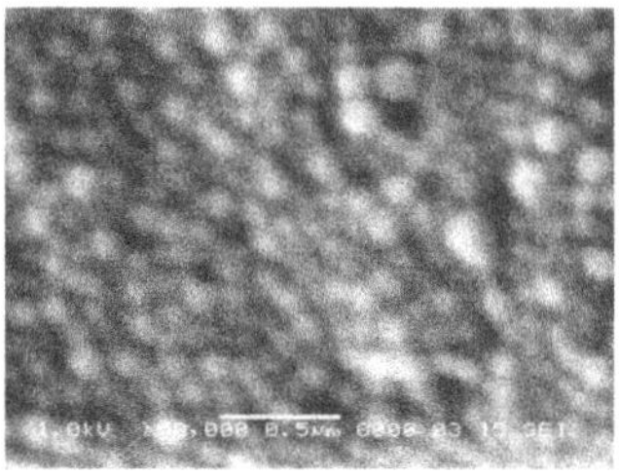

(a)

(b)

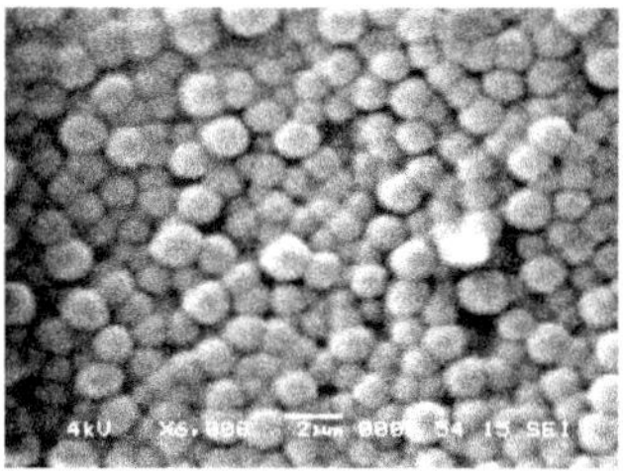

(c)

Fig. 3-37. Experimental images of a tin-on carbon test specimen using a mixed field add-on immersion lens attachment inside a JEOL 5600 tungsten gun SEM: (a) Conventional objective lens at 1 kV primary beam voltage (1 kV landing voltage) for a demagnification of 100,000. The working distance is 3 mm, (b) Mixed field immersion lens attachment with a landing voltage of 1 kV (7 kV beam and -6 kV specimen), (c) Mixed field immersion lens attachment with landing voltage of 600 V (4 kV beam and -3.4 kV specimen).

Figure 3-37 shows experimental images taken from a tin-on-carbon test sample inside a tungsten gun JEOL 5600 SEM, with and without the mixed field add-on lens attachment. Figure 3-37(a) shows that at a beam voltage of 1 kV without the add-on lens attachment (landing voltage of 1 kV), no detail in specimen surface can be observed (the demagnification is 100,000). On the other hand, Fig. 3-37(b) shows that at the same landing voltage with the add-on lens attachment, deposited tin particles are clearly visible. Figure 3-37(c) shows that even at a landing voltage of 600 V, tin particles are clearly visible. By taking a line-scan in Fig. 3-37(c), the image resolution was measured to be 3.6 nm (30 to 80% of the signal height across the edge of a tin particle), see Khursheed and

Karuppiah (2002) for more details. Conventional tungsten gun SEMs are usually not able to function at such low landing energies, and even at high beam energies, their image resolution is generally above 3 nm. These kinds of results illustrate the practical advantage of using mixed field immersion lenses over conventional SEM objective lenses.

Chapter 4

Sub-nanometer Probe Diameters

The in-lens designs described in the last chapter provide significantly better image resolution than conventional objective lenses, typically below 4 nm at low landing energies (< 1 keV). Further improvement in the image resolution can be achieved in a variety of different ways.

4.1 Monochromators and immersion objective lenses

To achieve aberration limited probe diameters of less than 1 nm at the specimen with a landing energy of 1 keV, Zach and Haider (1995) estimate that the spherical aberration coefficient (C_s) of an objective lens needs to be below 200 μm, while the chromatic aberration coefficient (C_c) should be smaller than 100 μm (for 0.5 eV energy spread). One way to relax this constraint is to filter the energy spread present in the primary beam with an energy monochromator. Although a cold field emission gun can in principle provide low energy spreads (down to 0.1 eV in some instances), it requires the use of a very high vacuum level, and suffers from instability in its output current. The reliable, widely used Schottky field emission source, together with a monochromator and immersion objective lens, is a more practical alternative, and as will be shown in the following pages, is able to provide sub-nanometer probe diameters at low primary beam landing energies (≤ 1 keV).

There have been several proposals to filter the energy spread in the primary beam of the SEM through the use of a monochromator. One of the simplest approaches, proposed by Henstra and Chmelik (2006), illustrated in Fig. 4-1, is to use the energy dispersive properties of a

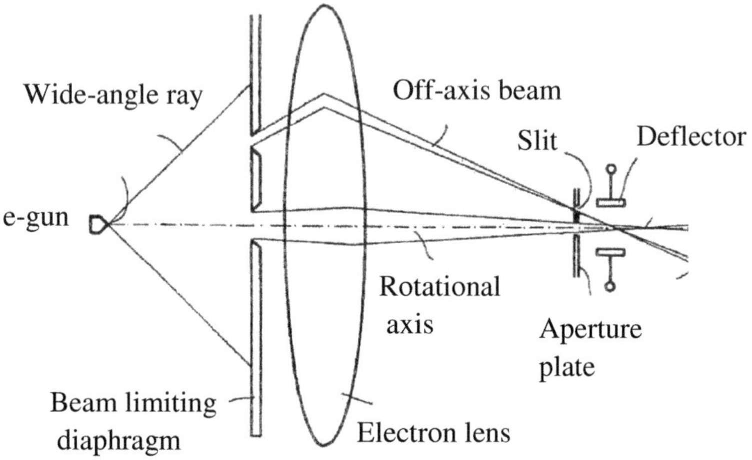

(a)

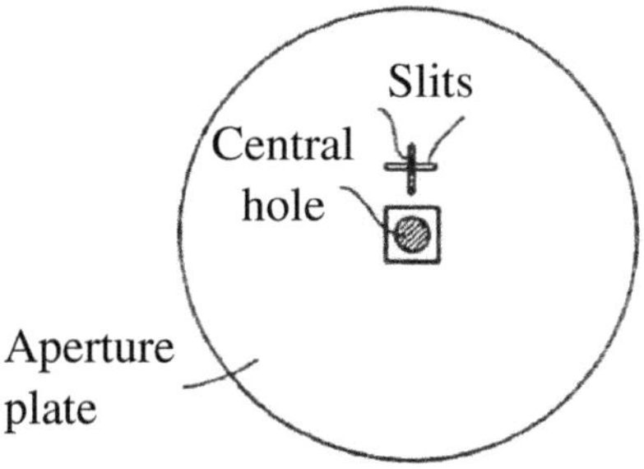

(b)

Fig. 4-1. Principle of a lens aberration energy monochromator: (a) Layout diagram, (b) Location of slits and hole in the aperture plate.

round electron lens placed immediately after the electron gun unit. Before entering the lens, beam limiting holes/slits define an on-axis and off-axis (wide-angle) beam. The electron lens then subsequently brings the off-axis beam to a focus on an aperture plate which has eccentric (off-axis) slits, effectively energy filtering it. A deflector unit then subsequently brings the off-axis beam on to the optical axis. When no energy filtering is required, the deflector is switched off, and central holes in the beam limiting diaphragm and aperture plate define the propagation of a round beam down the optical axis of the SEM. The deflector is therefore able to switch between two different modes of operation, an unfiltered high current beam (conventional case), and a low current energy filtered beam (incorporation of a monochromator).

There are a variety of other monochromator proposals for the SEM, some are based upon the use of magnetic sector deflectors, others rely on Wien filters, and yet others depend on electric sectors combined with a magnetic sector deflectors, see Ose *et al.* (2008) for more details. A detailed review of primary beam monochromators proposed for the SEM lies beyond the scope of this book. In the present context, it will be assumed that the energy spread within the primary beam can be filtered to be 0.15 eV, a factor of 2 to 3 times smaller than the energy spread present on Schottky field emission sources (0.4 eV).

The objective lens improvements described so far fall into three broad categories: those that immerse the specimen in a magnetic field, those that immerse the specimen in an electric field, and those that immerse the specimen in a mixed electric-magnetic field combination. These different configurations are depicted in Figs. 4-2(a)–4-2(c) in terms of the kind of axial field distributions that they generate. The electric retarding field lens is deliberately referred to here as an "electric immersion lens", in order to simplify the comparison, although strictly speaking, an electric field present at the specimen surface does not immerse it in the same way as a magnetic field immerses a non-magnetic specimen, however, for the purpose of this discussion, such subtleties will be overlooked. In addition, although the retarding field lens is referred to here as an electric immersion lens, it does employ an auxiliary focusing lens that is usually magnetic. Unlike the mixed field immersion lens depicted in Fig. 4-2(c) however, that also employs overlapping magnetic and electric fields, the focusing magnetic field of the electric retarding field lens does not immerse the specimen in a magnetic field, which is to say, it has zero magnetic field strength at the specimen surface.

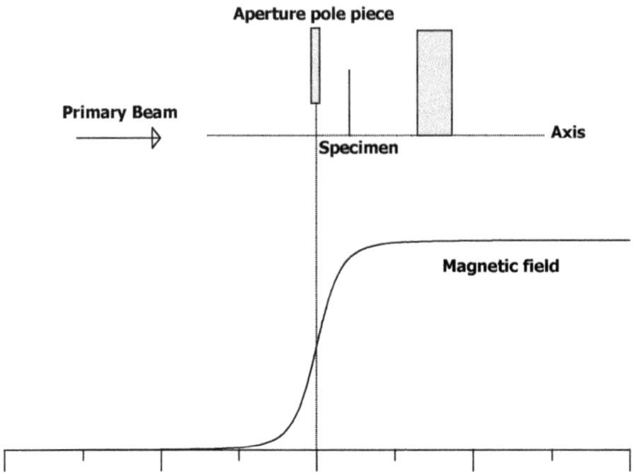

(a)

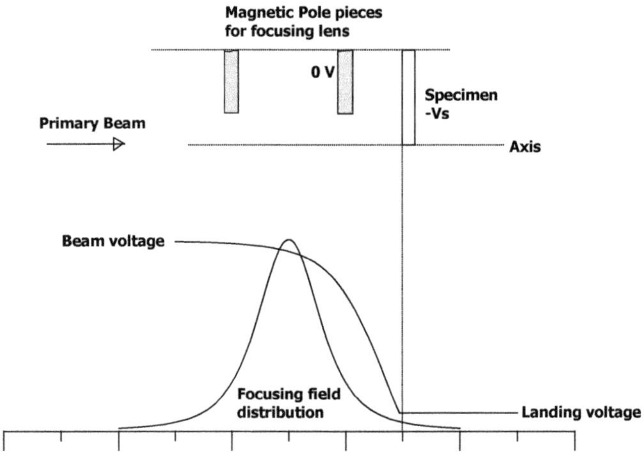

(b)

Fig. 4-2. Axial field distributions for immersion objective lenses: (a) Pure magnetic field, (b) Retarding electric field, (c) Mixed field combination.

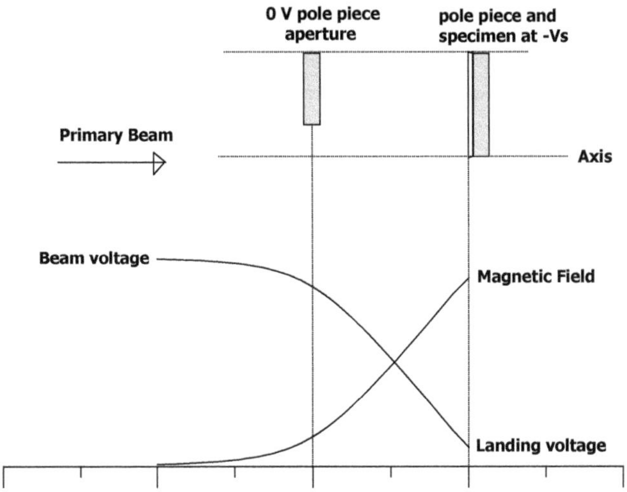

(c)

In the following work, it is assumed that the resolution of immersion lenses is limited by vacuum electrical field breakdown (10 kV/mm) and magnetic saturation (around 1 Tesla for iron). It should also be stated from the outset that only on-axis aberration limitations of immersion lenses are considered. Other effects that are important at high resolution such as the field-of-view, projected source size, the primary beam/ specimen interaction volume, vibrations at the specimen and surface charging, are not taken into account.

An aperture lens model assumes that all electrodes in the immersion lens are infinitely thin, as shown in Fig. 4-3, and that the field strength reaches a constant value on either side of the aperture. Here, the radius of the aperture is given by R, the aperture plate, biased to V_1, is located a distance W from the specimen, which is biased to V_2. The field strength on the right hand side of the lens is $E = -(V_2-V_1)/W$, and if a field-free region is assumed on the left-hand side, then the potential along the z-axis, $V(z)$, is given by (Hawkes and Kasper, 1989a)

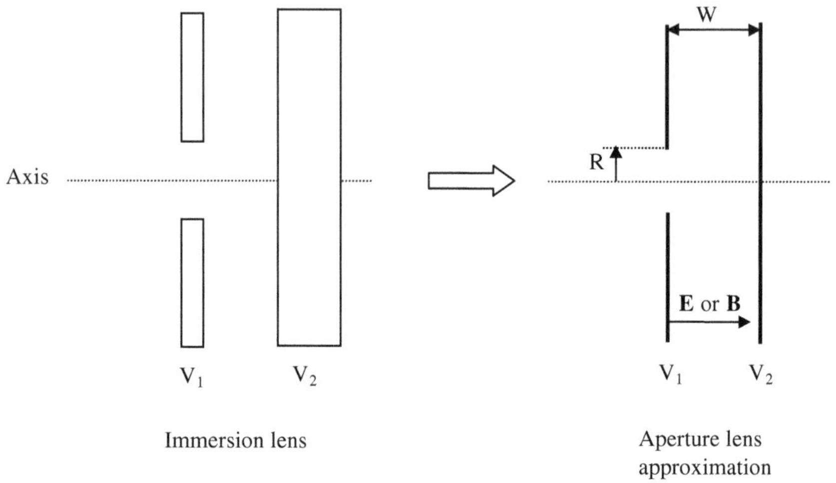

Fig. 4-3. The aperture lens approximation.

$$V(z) = V_1 - \frac{E}{2} z - \frac{E}{\pi} R \left(1 + \frac{z}{R} \tan^{-1}(z/R) \right). \tag{4.1}$$

Note that as the aperture radius *R*, tends to zero, the potential distribution is linear in *z*, as expected. Khursheed (2002b) shows that this aperture axial potential distribution is close to a finite element potential solution that takes into account the aperture electrode thickness (3 mm thick) for an aperture hole radius of 1 mm, and a working distance of 1 mm.

The aperture lens model has the merit of providing a smooth axial field distribution that can easily be differentiated, facilitating the accurate calculation of on-axis aberration coefficients. In the following estimates of the final probe size, the root-sum formula by Barth and Kruit (1996) is used to combine spherical, chromatic and diffraction aberrations together. For the retarding field lens case, a focusing magnetic field *B*(*z*) is assumed, which is approximated by the following Glaser field distribution

$$B(z) = \frac{B_0}{\left[1+(z/a)^2\right]^2} \tag{4.2}$$

where B_0 is the peak of the distribution, and the parameter a controls how sharply the distribution falls on either side of the peak value, effectively controlling the width of the distribution.

For a given set of aberrations and source energy spread, the final aperture radius is varied systematically so that the probe diameter is a minimum. For the purposes of this study, it is assumed that the final aperture can be varied in this way. A comparison of aberration predictions based upon the aperture lens model for the magnetic, retarding field and mixed field immersion lenses is given in Table 4-1 for a fixed landing energy of 1 keV.

Table 4-1. Simulation of aberrations for aperture lenses at magnetic saturation and electrical breakdown limits for a landing energy of 1 keV.

Aperture Lens Model at 1 keV Primary Landing Energy	Magnetic Lens (1 Tesla)	Electric Retarding Field Lens (10 kV/mm)	Mixed Field Lens
f (μm)	257	860	163
C_s (μm)	168	713	50.45
C_c (μm)	187	144	58.2
d_p (nm) (Root-sum method)	0.87	1.2	0.625

For the magnetic immersion lens, the focal length, f, C_s, and C_c are predicted to be, 257, 168 and 187 μm respectively, giving a probe diameter of 0.87 nm. The lens bore diameter is varied so that the calculated probe diameter is a minimum. An optimum lens bore size exists because while chromatic aberration steadily decreases as the lens bore radius becomes smaller, spherical aberration becomes large once the radius is small relative to the working distance. As the magnetic field distribution approaches a step function, the point at which the primary beam focuses does not change significantly. The working distance in this case is therefore allowed to vary. The optimum lens bore radius is 0.25 mm and the 1 keV primary beam focuses at a point 0.3 mm below the aperture lens electrode (with an axial peak field strength of 1 Tesla). In practice, the required lens bore size to achieve this condition will not be so small, since the zero plate thickness in the aperture lens model underestimates how sharply the field distribution rises. Finite element simulations show that a 0.5 mm radius hole in a 3 mm thick lens aperture plate provides similar aberrations. Lens bore diameters down to 0.5 mm can be manufactured with standard methods. Although the 0.3 mm working distance for the magnetic immersion lens is considerably smaller than that used for conventional lenses, it is easily achievable with motorized specimen height movement.

In the case of the electric retarding field lens the strength of the focusing magnetic field is adjusted to provide a focal point 1 mm below the aperture lens electrode. An 11 keV incoming primary beam is retarded down to a landing energy of 1 keV by biasing the specimen to –10 kV. For a Glaser magnetic field distribution which has $a = 1$ mm, the position that minimized the aberration coefficients was found to be approximately 1 mm to the left of the centre of the electric field lens distribution, as shown in Fig. 4.2(b). Note that although the magnetic and electric field distributions overlap, the specimen is still free of a magnetic field (it is still an electric immersion lens). The probe diameter is found to be a minimum for an aperture radius of around 1 mm. The on-axis aberrations for this radius are calculated to be: $f = 860$ μm, $C_s = 713$ μm, and $C_c = 144$ μm, giving a probe diameter of 1.2 nm. This probe diameter is around 38% higher than that predicted for the immersion magnetic lens.

For the mixed field immersion lens, a 10 kV/mm electric field strength is maintained on the specimen side of the aperture, while the strength of the magnetic field is adjusted to focus the primary beam on to the specimen, located at a working distance of 1mm. For simplicity, the radius of the lens aperture electrode is specified to be 1 mm, the same as for the electric retarding field lens. The aberration coefficients are found to be, C_s = 50.45 μm and C_c = 58.2 μm with f = 163 μm, giving a probe diameter of 0.625 nm. This probe diameter is around 2 times smaller than the retarding field electric lens, and around 1.4 times smaller than the immersion magnetic lens.

Figures 4-4(a)–4-4(d) show simulation predictions for how C_s and C_c vary with landing energy at electric field strengths of 2, 5 and 10 kV/mm and a peak magnetic field of 1 Tesla. The electric field strengths of 2 and 5 kV/mm are of particular interest here, since 10 kV/mm is rarely used in practice. Designing a lens to operate close to the vacuum electrical breakdown value is not easy. The magnitude of the electrical field may be below the breakdown limit in some parts of the lens, while in other parts, typically around corners, it may exceed the limit. Unlike magnetic saturation, electrical breakdown is not a self-limiting process.

The simulation results shown in Figs. 4-4(a)–4-4(d) indicate that mixed immersion lenses are predicted to have significantly lower aberration coefficients than pure field immersion lenses (for all the energies examined). The electric retarding field lens has comparable or better aberration coefficients than the magnetic immersion lens at low landing energies: at an electric field strength of 10 kV/mm, the spherical aberration coefficient is better for landing energies below 200 eV, while for the chromatic aberration coefficient, it is better for landing energies below 1,500 eV. However, this advantage is dramatically reduced as the electric field strength is decreased.

An important result of these simulations is that for the mixed field immersion lens operating at the moderate electric field strength of 5 kV/mm, C_s is less than 100 μm and C_c is less than 84 μm, satisfying Zach and Haider's criteria for 1 nm probe resolution at a landing energy of 1 keV.

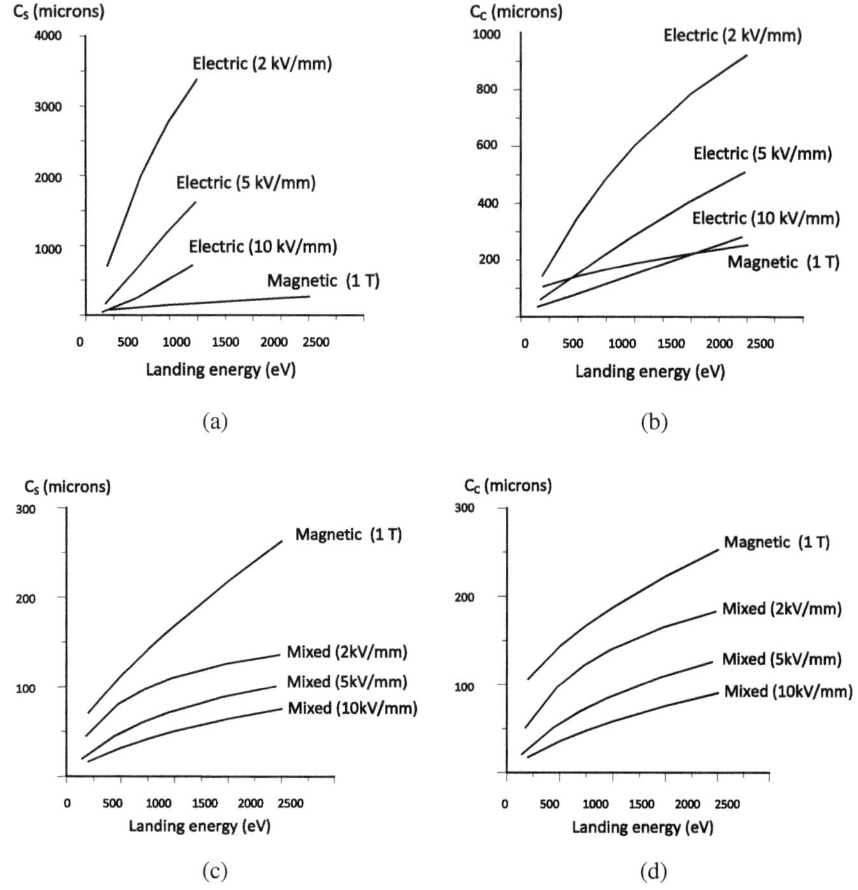

Fig. 4-4. Simulated immersion lens aberration coefficients as a function of primary beam landing energy: (a) Spherical aberration for immersion magnetic and electric retarding field lenses, (b) Chromatic aberration for immersion magnetic and electric retarding field lenses, (c) Spherical aberration for immersion magnetic and mixed field lenses, (d) Chromatic aberration for immersion magnetic and mixed field lenses.

Figure 4.5 shows how the simulated aberration limited spot diameter varies as a function of final semi-angle (ΔE = 0.4 eV). The minimum value of 0.742 nm, lies at a semi-angle of 30.24 mrad. To the left-hand side of this value the aberration probe size is dominated by diffraction, while to the right-hand side of the graph, the probe size is dominated by

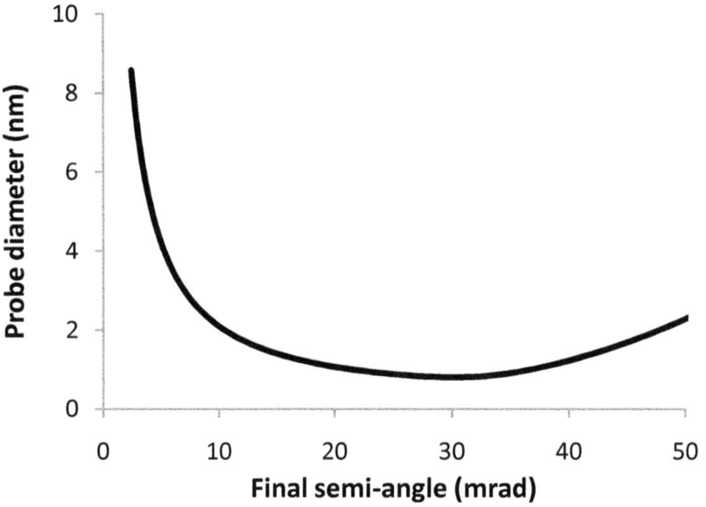

Fig. 4-5. Simulated aberration limited probe diameter for the 5 kV/mm mixed immersion lens as a function of final semi-angle (0.4 eV spread, 1 keV landing energy).

spherical and chromatic aberration (mostly spherical aberration in this case). The simulation results highlight an important characteristic of immersion lenses in general. The optimum semi-angle (30.24 mrad) is significantly larger than its value for conventional SEM objective lenses, which typically lies below 5 mrad. In the case of the immersion lens, the high optimum semi-angle value comes from its relatively short focal length (0.205 mm in this case). A short focal length necessitates the use of a small aperture. The aperture radius for these conditions is approximately 6.2 μm. In general, to achieve high resolution with immersion lenses, the aperture diameter should be kept below say 20 μm. This means that although the semi-angle at the specimen is relatively high for immersion lenses, it is not translated into high probe currents.

Another point about immersion lenses in general which emerges from these simulations is that since the final-semi-angle at the specimen is relatively large, around 30 mrad, the depth of focus at high resolution (probe sizes that approach 1 nm), will be relatively small. At a probe resolution of say 1 nm, the depth of focus (resolution divided by final semi-angle), will be of the order of only 33 nm. This is typically two-orders of magnitude smaller than for conventional lenses (several μm) which operate at higher probe diameters (> 5 nm) and have lower final semi-angles (< 5 mrad). This low depth of focus means that it is much more difficult to find the optimum focusing conditions for immersion lenses than for conventional lenses in practice.

Figures 4-6(a)–4-6(c) show the simulation results for the aberration limited probe diameter as the landing energy is varied ($\Delta E = 0.15$ eV). For an electric field strength of 10 kV/mm and landing energies below 300 eV, electric retarding field lenses are predicted to provide smaller aberration limited probe diameters than magnetic immersion lenses, while above 300 eV, magnetic immersion lenses are expected to perform better. Mixed field immersion lenses are consistently predicted to provide smaller probe diameters than their pure-field counterparts, ranging from a 1.5 to 2 factor of improvement in the probe diameter. At a landing energy of 200 eV and a maximum electric field strength of 10 kV/mm, a mixed field lens is predicted to provide on-axis aberrations as low as $C_s = 16.21$ μm and $C_c = 17.41$ μm with a focal length of $f = 84$ μm, giving an aberration limited probe size of 0.89 nm.

It is also apparent that as the electrical field strength decreases, the landing energy at which the magnetic immersion lens and electric retarding field lens are predicted to give a comparable aberration limited probe size also drops. At an electric field strength of 5 kV/mm, the crossing point drops to just under 200 eV, while for 2 kV/mm, the crossing point falls to less than 100 eV (not shown due to reasons of scale). These simulation results indicate that for most practical situations, where an electric field strength of 5 kV/mm or less is used, the electric retarding field lens is only advantageous for very low landing energies (less than 200 eV). The best form of immersion lens in all cases is the mixed field immersion lens.

Mixed field immersion lenses, together with a way of reducing the energy spread in the beam, with say the use of a monochromator, are

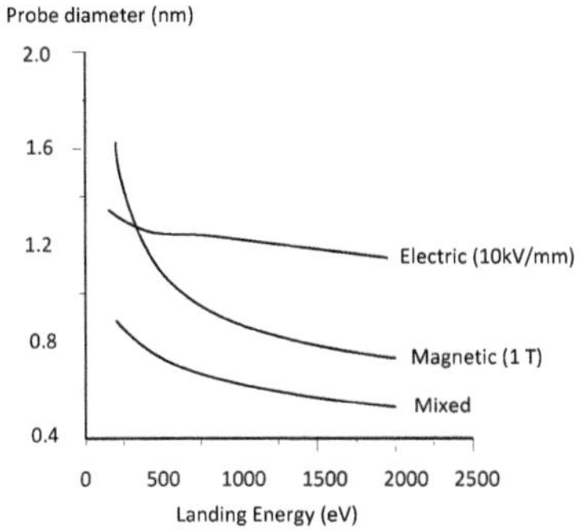

(a)

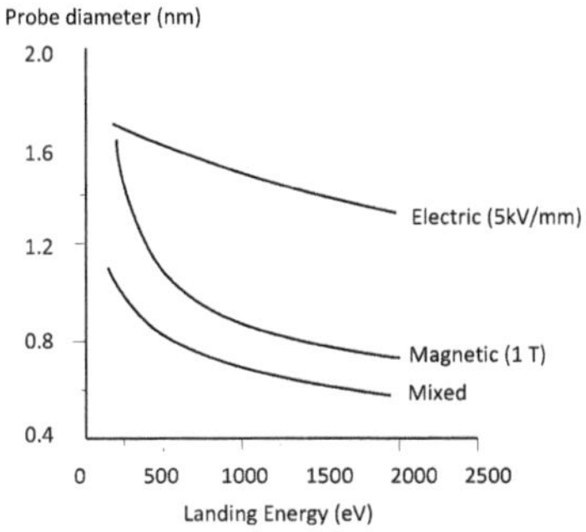

(b)

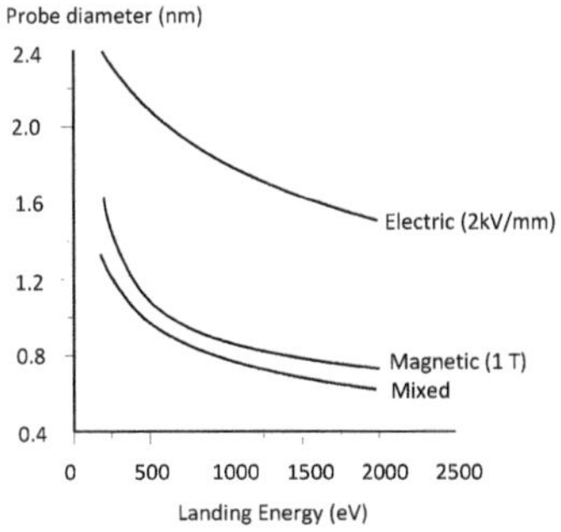

(c)

Fig. 4-6. Simulated probe diameters as a function of landing energy for immersion lenses ($\Delta E = 0.15$ eV): (a) Electric field strength of 10 kV/mm, (b) Electric field strength of 5 kV/mm, (c) Electric field strength of 2 kV/mm.

predicted to be able to provide sub-nanometer probe primary beam resolutions at the specimen.

Figure 4-7 shows two sub-nanometer resolution images taken from the FEI Magellan 400L SEM FEI at a beam landing energy of 1 keV. This SEM incorporates a monochromator of the type proposed by Henstra and Chmelik (2006), together with a magnetic immersion lens. For the type of specially prepared specimens shown in Fig. 4-7, TEM/STEM like resolution can be obtained.

The disadvantages of low landing energy monochromator-immersion objective lens SEMs are that, the monorchomator reduces the primary beam current, the specimen is exposed to high electric/magnetic field strengths, and the depth of focus is relatively small.

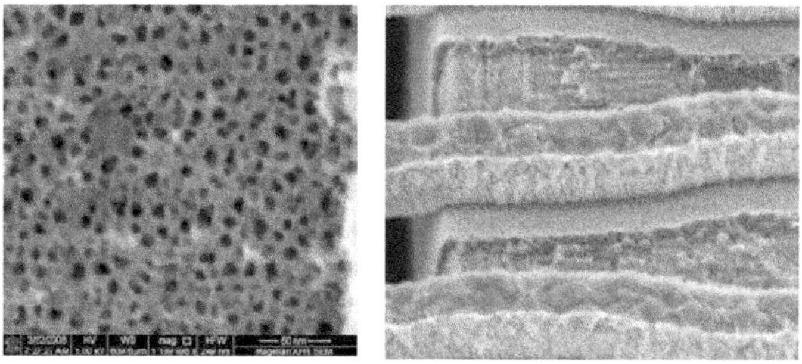

(a) (b)

Fig. 4-7. Sub-nanometer resolution SEM images using the FEI Magellen 400 L SEM using a monochromator together with an immersion lens for a beam landing energy of 1 keV (courtesy of FEI Inc.): (a) Deprocessed sample, stripped back to the poly-silicon level is magnified to 1,200,000 times, (b) 40° tilted view of deprocessed sample at 300,000 times demagnification.

4.2 Aberration correctors

The subject of aberration correction for electron microscopes comes from Scherzer (1936), who was the first to demonstrate that for round lens focusing systems that have continuous, source-free, static (time invariant) field distributions, on-axis chromatic and spherical aberration coefficients, C_C, and C_S are both inherently positive (also derived in Appendices 1.2 and 1.3 of this book), meaning, that their values can only be minimized, not made zero. Scherzer (1947) then went on to explore possible ways to achieve aberration correction by abandoning some of the assumptions behind his paraxial round lens theory. Some ways of circumventing Scherzer's theorem include, using three dimensional (non-rotationally symmetric) field distributions, use of electron mirrors, introduction of field discontinuities, regions of space-charge on-axis, and rapidly time-varying fields. Reviews of these attempts at aberration correction have been made by Septier (1966) and Hawkes and Kasper (1989b).

An aberration corrector proposal for low voltage SEMs was made by Zach (1989). Zach and Haider (1995) later carried out work on a quadrupole/octupole aberration corrector proposal, which was further developed by Uno *et al.* (2005). Zach estimated that his aberration corrector was theoretically capable of achieving around 1 nm probe resolution at low landing energies (0.5 to 1 keV) in the SEM. The aberration correction unit is placed above the objective lens and scanning/stigmator units, as shown in Fig. 4-8, and functions by altering the optics of the primary beam in such a way that it compensates for the effect of axial chromatic and geometric aberrations of the objective lens in the final image.

The principle of chromatic aberration correction based upon the use of three dimensional fields is illustrated in Fig. 4-9. In round lens systems, electrons with a higher energy than the primary beam energy E_P, $E_P + \Delta E$, focus beyond the Gaussian plane, while those with energies below it, $E_P - \Delta E$, focus before the Gaussian plane. In order to compensate for this effect, a chromatic aberration unit creates energy dispersion of the primary beam so that electrons of different energies enter the objective lens at different angles, making it appear that more energetic electrons come from a point further from the objective lens than the less energetic electrons. More energetic electrons inside the corrector unit experience greater focusing than less energetic ones.

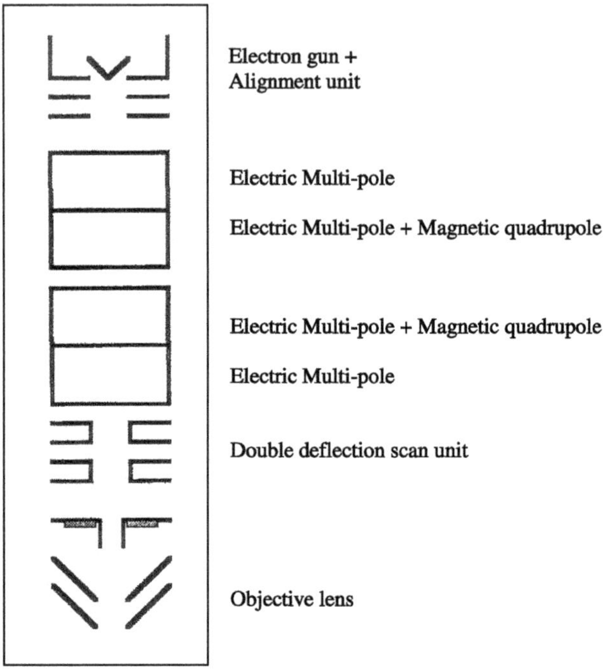

Fig. 4-8. Aberration corrector layout for the SEM as proposed by Zach and Haider (1995).

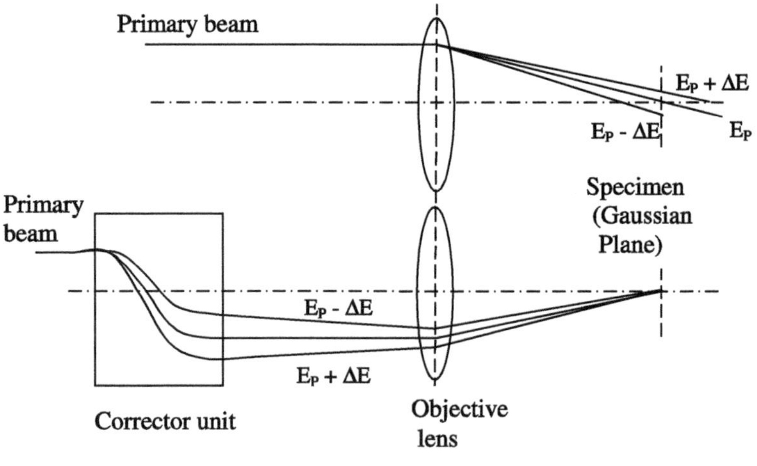

Fig. 4-9. Schematic illustration of chromatic aberration correction.

Figure 4-10(a) depicts the kind of superimposed three dimensional field distributions that can achieve chromatic aberration correction. These fields are generated from electric and magnetic quadrupoles rotated 45° with respect to one another. The quadrupoles are excited in such a way, so that the magnetic force acting on an electron passing through it (at the primary beam energy E_P), is balanced by a corresponding electric force. However, electrons having energies different from the primary beam energy will experience a force, either towards the axis, or away from it, depending on whether the energy is greater or less than the E_P, and depending on where the electron is located in the transverse plane (x-y plane). If the magnetic force is given by F_m and the electric force, by F_e, then for more energetic electrons, $E_P + \Delta E$, $|F_m| > |F_e|$, while for less energetic electrons, $E_P - \Delta E$, $|F_m| < |F_e|$.

Figure 4-10(b) shows an electron at a point above the centre point O, in this case, more energetic electrons will diverge with respect to electrons at E_P, while less energetic electrons will converge with respect to electrons at E_P. In the case of a point on the left hand-side of the centre point O, as shown in Fig. 4-10(c), more energetic electrons will converge with respect to electrons at E_P, while less energetic electrons will diverge with respect to electrons at E_P. In order to achieve a stronger focusing action for the more energetic electrons, the primary beam must be projected on to the mixed quadrupole unit having a line focus, where it has extension in the x-direction, and ideally, no extension in the y-direction. A similar mixed field quadrupole unit, rotated by 90° is then required to correct for chromatic aberration in the y-direction, into which a line focus in y is projected. It is relatively straightforward to show that the strength of force is proportional to ΔE (Rose, 2009), and such mixed field quadrupole field distributions will therefore correct for chromatic aberrations.

Figure 4-11 shows a four quadrupole chromatic aberration corrector based upon the use of the principles just outlined. The unit consists of two outer electric quadrupoles, Φ_{q1} and Φ_{q4}, and two inner mixed field quadrupole units, $\Phi_{q2} + \psi_{q1}$, and $\Phi_{q3} + \psi_{q2}$. The excitations on all four

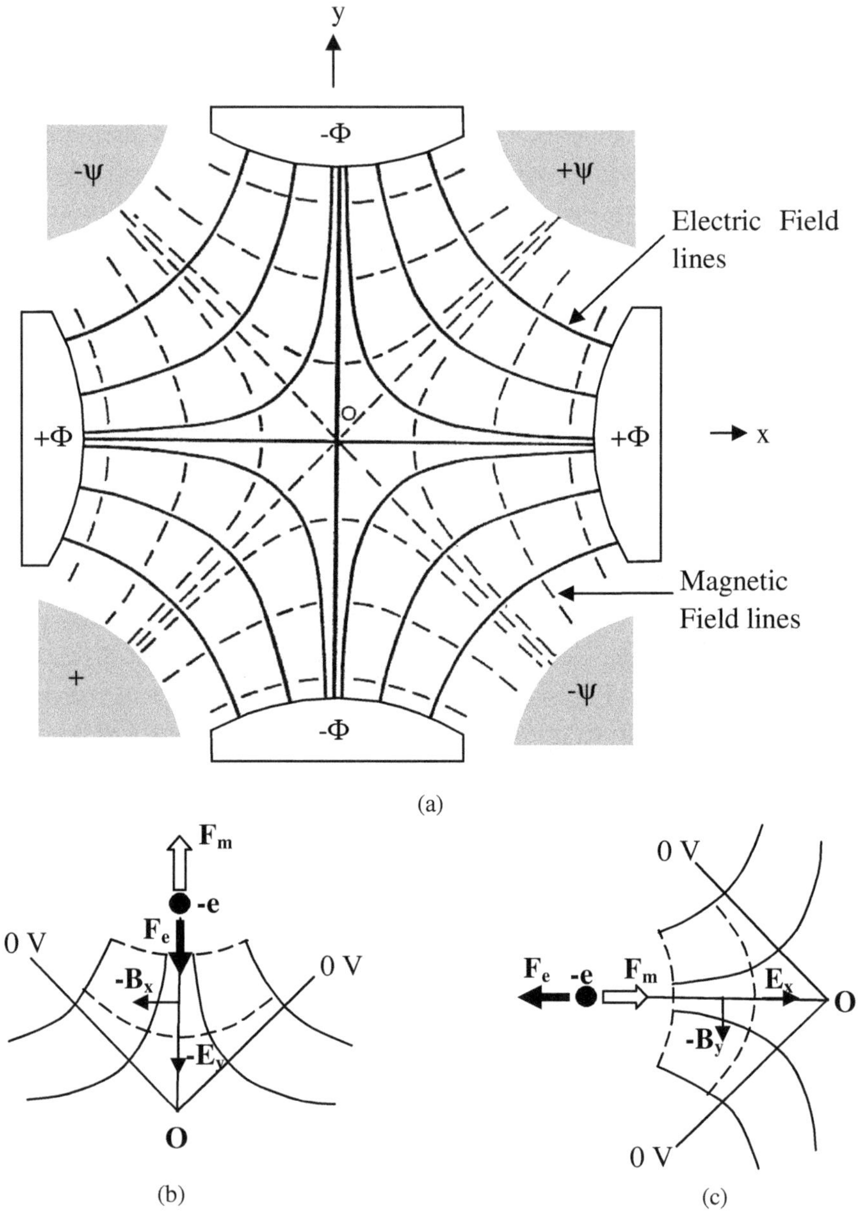

Fig. 4-10. Focusing action of superimposed magnetic and electric quadrupoles: (a) Form of superimposed field lines, (b) Forces acting on electron above the centre point O. F_m denotes magnetic force while F_e denotes electric force, (c) Forces acting on electron to the left hand side of centre point O.

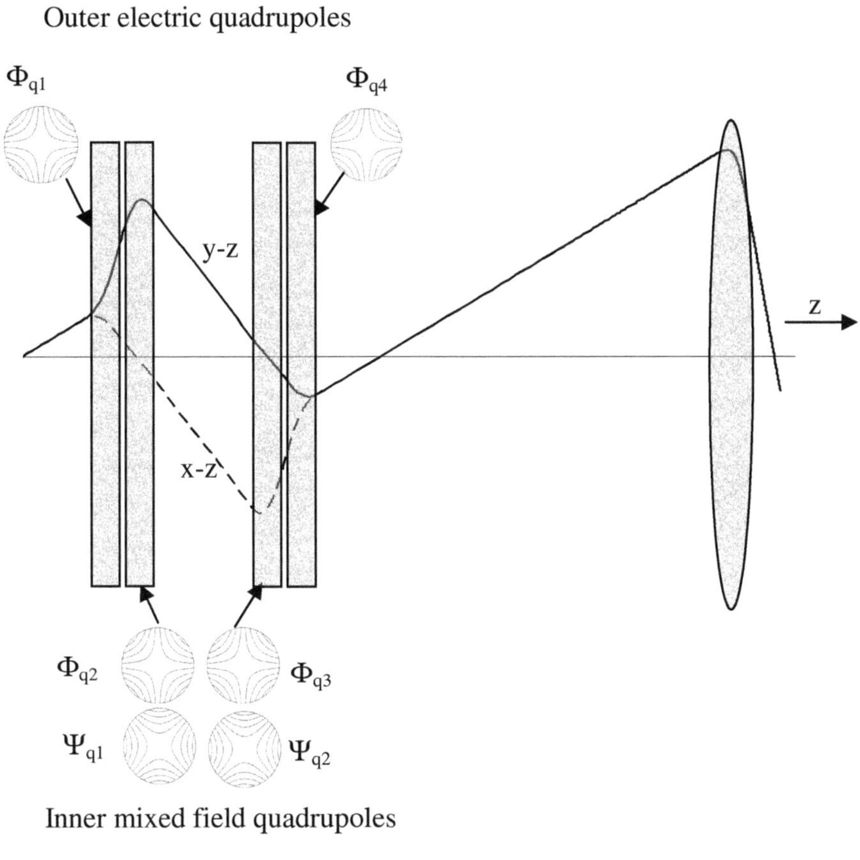

Fig. 4-11. Schematic layout of a four quadrupole chromatic aberration corrector unit with simulated first-order ray paths in the *x-z* and *y-z* planes.

electric quadropoles are typically set to achieve the first-order *x-z* and *y-z* ray trajectory paths shown in Fig. 4-11, where a *y-z* line focus is projected into the first inner mixed field quadrupole, $\Phi_{q2} + \psi_{q1}$, and *x-z* line focus is projected into the second mixed field quadrupole, $\Phi_{q3} + \psi_{q2}$. The excitation on the inner quadrupoles are then fine-tuned to correct for the chromatic aberrations of the whole system: aberrations from the quadrupoles of the corrector as well as objective lens aberration.

Figure 4-12 shows simulated ray paths in the *x-z* and *y-z* planes through both the corrector and objective lens in three-dimensional perspective. Note how the cross-sectional shape of the beam changes as it progresses through the corrector, starting as a round beam, it is projected to be a line focus in the *y-z* plane, then a line focus in the *x-z* plane, and subsequently brought back into being a round beam as it exits the corrector exit unit.

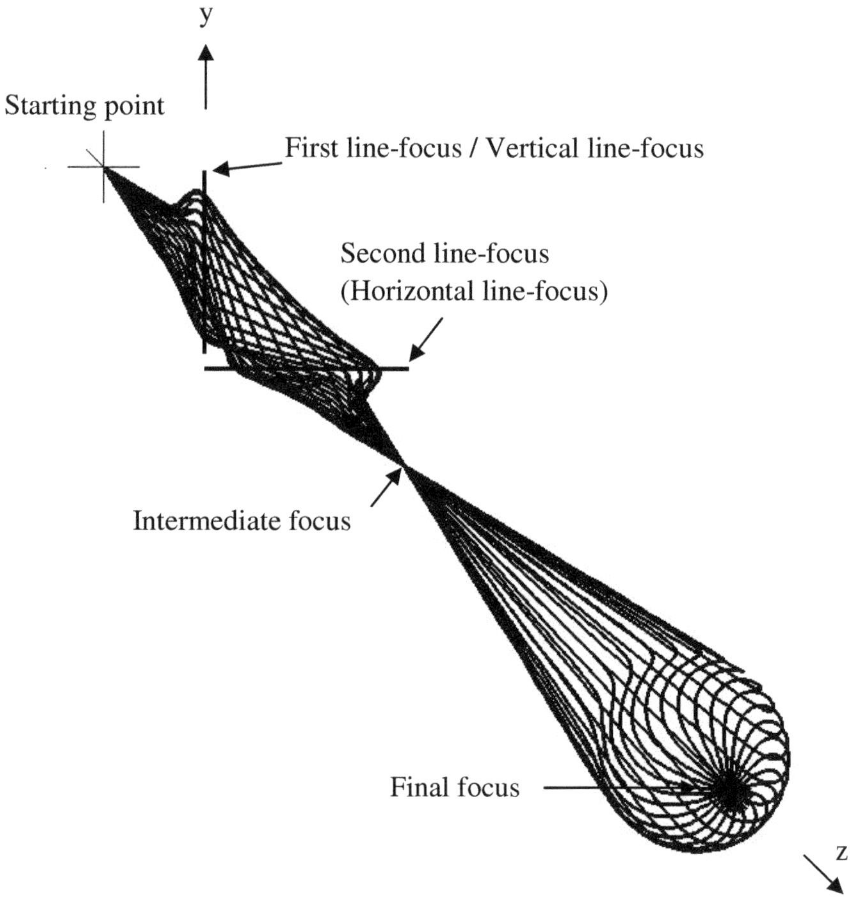

Fig. 4-12. Simulated first-order paraxial ray paths through a quadrupole chromatic aberration corrector with objective lens in three-dimensional perspective.

The principle of spherical aberration correction is illustrated in Fig. 4-13. The normal objective lens provides a greater focusing action on wide-angle trajectories than narrow-angles ones, which results in wider-angle trajectories being focused before narrow-angle ones along the optical axis. The corrector in this case acts in the opposite way; it provides a greater focusing action on the narrow angle electrons, focusing them before the wide-angle trajectories. Now since wider-angle electrons travel further off-axis, the strategy of placing a corrector at two successive line foci should not only correct for chromatic aberrations, but can in principle also correct for spherical aberrations. The problem is however, the rate at which the quadrupole field strength rises is proportional to the off-axis distance, r, while spherical aberration in a round lens is proportional to r^3. It turns out that octupole field strengths are mainly proportional to r^3, and by superimposing suitable octupole field distributions at the two line foci (in y-z and x-z), spherical aberration can also be corrected for.

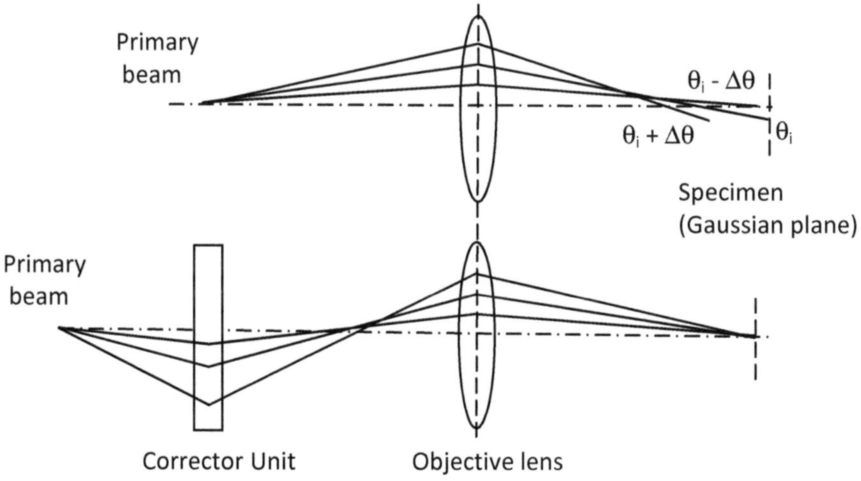

Fig. 4-13. Schematic illustration of spherical aberration correction.

Figure 4-14 show equipotential levels for idealized quadrupoles and octupoles derived from the solution of Laplace's equation for an infinitely long wedge structure, black indicates one extreme potential value (say +1 volts), while white denotes the other extreme value (say -1 volts). The grey area in the middle of the multi-pole represents a region of low field strength around the optical axis. Details of this field solution, and an introduction to the nature of field solutions in multi-pole lenses in general are given in Appendix 2.

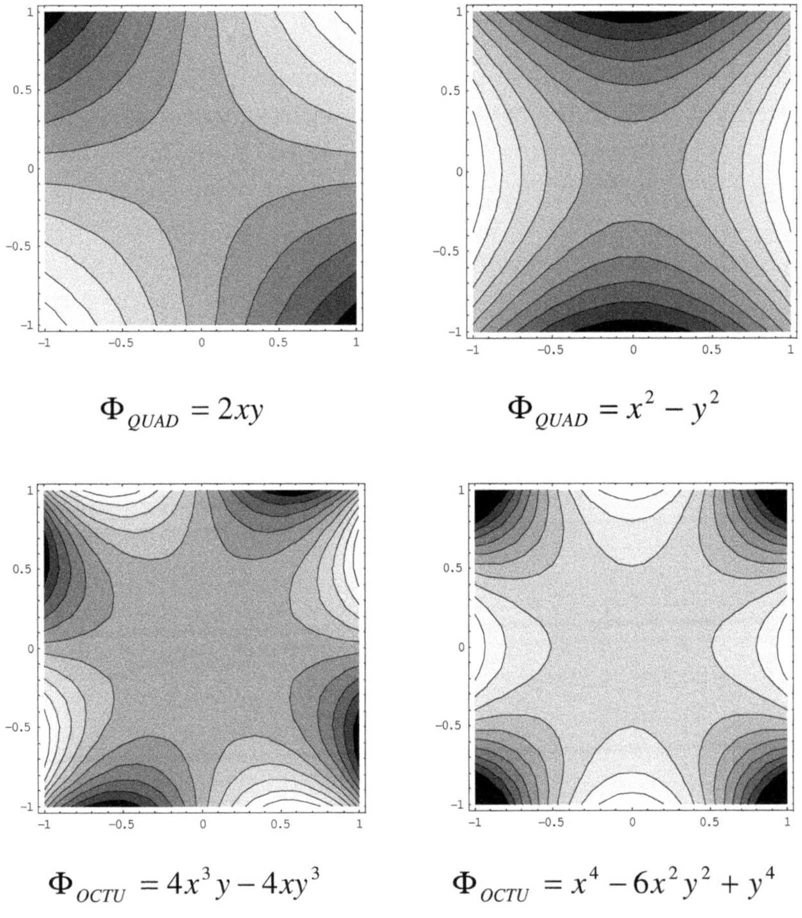

Fig. 4-14. Graphs showing equipotential *x-y* plots for equations describing quadrupoles and octupoles.

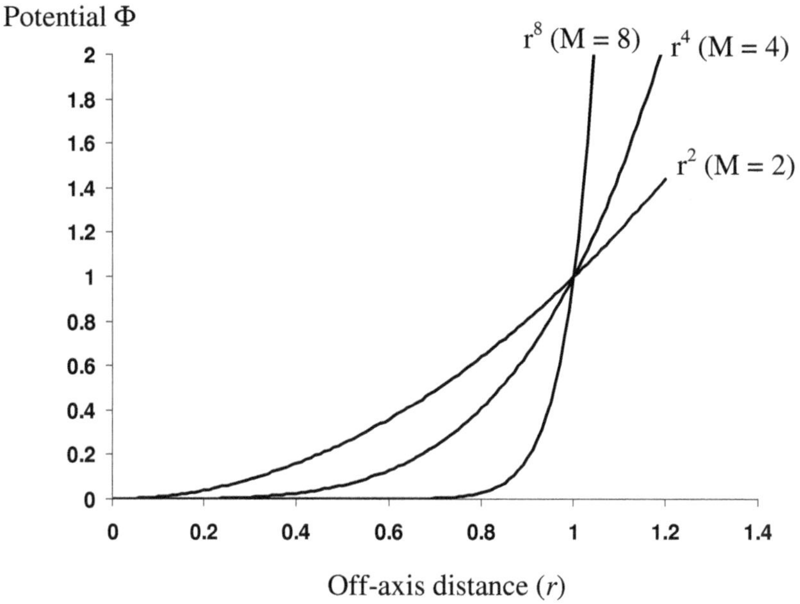

Fig. 4-15. A two-dimensional analytical representation of multi-pole field distributions. M refers to the number of symmetry lines in the multi-pole. $M = 2$ for quadrupoles and $M = 4$ for octupole (see Appendix 2 for more details)

Figure 4-15 illustrates the potential dependence on the exponent of r, showing how r^2, r^4, and r^8 grow as r increases for different multi-poles. M refers to the number of symmetry lines in the multi-pole, which is the number of poles divided by 2. $M = 2$ for quadrupoles and $M = 4$ for octupole (see Appendix 2 for more details).

Relating this to Fig. 4-14, it demonstrates that r^8 stays low (grey) longer than r^4 or r^2 but also that it increases much faster once it begins to rise, which among other things means that quadrupoles affect paraxial (close to axis) electrons while octupoles (and higher symmetries) do not. The physics behind this behavior comes from the fact that when the M-fold symmetry increases, the distance between neighboring electrodes becomes smaller, leading to the situation where most field changes are concentrated in the area between electrodes, far away from the axis, and leaving only a weak field strength around the optical axis. In the region close to the axis, we therefore have the following radial field dependence for the electric field strength:

$$\left.\begin{aligned} E_{r,M=2} &\propto r \\ E_{r,M=3} &\propto r^2 \\ E_{r,M=4} &\propto r^3 \end{aligned}\right\} \text{which is easily generalized to } E_{r,M} \propto r^{M-1}. \tag{4.3}$$

These expressions illustrate how the octupole (M = 4), giving rise to a force proportional to r^3, can act on an off-axis electron in such a way as to correct for the spherical aberration of round lenses.

It can be shown that the above conclusion also holds for a more complete (fully three-dimensional) solution of Laplace's equation for multi-poles, which is formulated in terms of the potential distribution $\Phi(r,\theta,z) = P_M(r,z)\cos M\theta$ around the axis, where M = 0 for a round lens, and M = 4 for an octupole. If $P_0(r,z)$ denotes a round lens scalar potential solution with an on-axis function variation of $f_0(z)$, and $P_4(r,z)$, denotes the octupole solution with an on-axis function variation of $f_4(z)$, then it can be shown that (Liu *et al.*, 2002)

$$P_0(r,z) = f_0(z) - \frac{f_0''(z)}{4}r^2 + \frac{f_0^{(4)}(z)}{64}r^4 - \frac{f_0^{(6)}(z)}{2304}r^6 + O(r^8) \tag{4.4}$$

$$P_4(r,z) = f_4(z)r^4 - \frac{f_4''(z)}{20}r^6 + \frac{f_4^{(4)}(z)}{960}r^8 - \frac{f_4^{(6)}(z)}{80640}r^{10} + O(r^{12}). \tag{4.5}$$

The round-lens expression has been also derived in this book (see Appendix 1.0). The functions $f_0(z)$ and $f_4(z)$ are related to on-axis scalar potentials. The focusing forces for both round lens and octupole (F_{lens} and F_{octu}) are therefore proportional to the following expressions:

Round lens: $$F_{lens} \propto -\frac{f_0''(z)}{2}r + \frac{f_0^{(4)}(z)}{16}r^3 - \ldots \tag{4.6}$$

Octupole: $$F_{octu} \propto 4f_4(z)r^3\cos 4\theta - \ldots \tag{4.7}$$

The round lens sets up a focusing field with *r*-dependence, but has an aberration term with r^3-dependence. The octupole can only setup an aberration (correcting) term with r^3-dependence, within its symmetry planes. It does not have a proper focusing field by itself. The round lens cannot change its r^3-term without also changing the *r*-term (both depend upon $f_0(z)$. The sign and magnitude of the octupole term, however, can be varied, according to how it is excited.

Figure 4-16 shows the schematic diagram of a quadrupole-octupole corrector unit which compensates for both spherical and chromatic aberrations. Φ_{o1}, and Φ_{o4} are superimposed octupole fields on the outer quadrupole units, while Φ_{o2}, and Φ_{o3} denote superimposed octupole fields on the inner mixed field quadrupole units. This corrector unit is similar from a first-order optics point of view to the 4 quadrupole unit for chromatic aberration shown in Fig. 4-11, but in addition, it superimposes octupole field distributions to correct for spherical aberration. The strategy for using the corrector is to first set the quadrupole units to correct for chromatic aberration, and then turn on the inner octupole units Φ_{o2}, and Φ_{o3} to correct for spherical aberration. It turns that this spherical aberration correction introduces four-fold astigmatism (off-axis aberration), which can then be eliminated by suitably exciting the outer octupoles, Φ_{o1}, and Φ_{o4}. The whole procedure is applied iteratively until the final spot size is minimized.

A simple example simulating the performance of the quadruopole-octupole corrector for a test SEM objective lens is depicted in Fig. 4-17. Quadrupole and octupole potential distributions were generated from the idealized model derived in Appendix 2, as depicted in Fig. 4-14. These fields are assumed to be step-functions in the *z*-direction. The objective lens is assumed to be magnetic, where the on-axis field distribution varies according to the square of the well known Glaser function, as shown in Fig. 4-18. Distances between each focusing element, and lens excitations necessary to achieve on-axis aberration correction are given in Fig. 4-17. The parameter *w* in the Glaser function and its peak on-axis field strength were adjusted to provide a working distance of 12 mm and a probe demagnification of 10, this occurred for $w = 0.006$ m and an on-axis peak field strength of 28.3 mT. Direct electron trajectory paths were simulated for initial semi-angles 0.01, 0.05, 0.1, 0.2 and 0.5 mrad, and energies 999.5, 1000 and 1000.5 eV. The semi-angles of primary beam electrons at the specimen were then 0.1, 0.5, 1, 2, and 5 mrad.

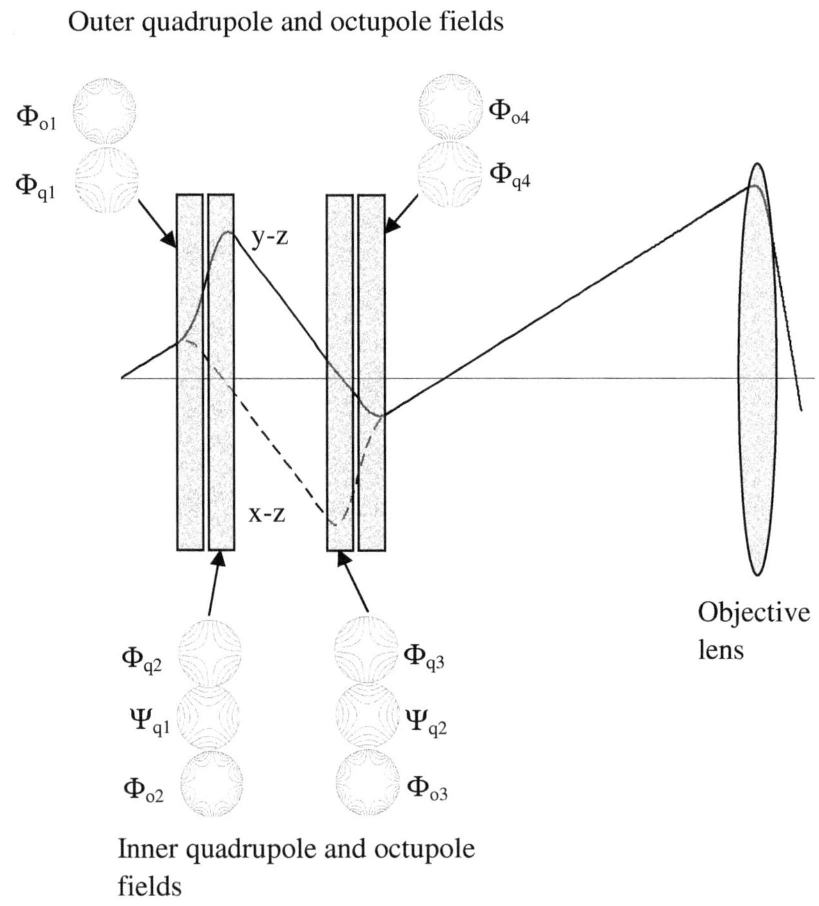

Fig. 4-16. Schematic diagram of a qudrupole-octupole corrector unit which compensates for both spherical and chromatic aberrations. Φ_{o1}, and Φ_{o4} are superimposed octupole fields on the outer quadrupole units, while Φ_{o2}, and Φ_{o3} denote superimposed octupole fields on the inner mixed field quadrupole units.

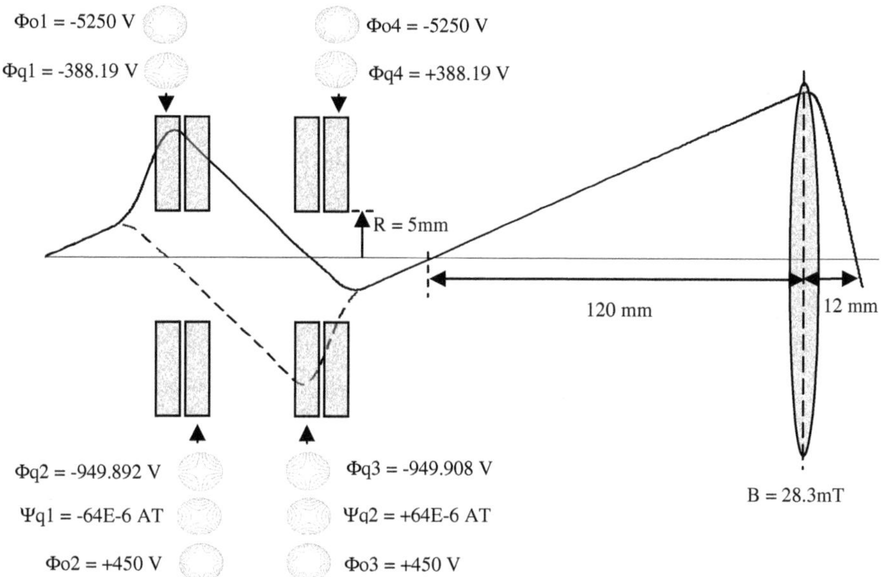

Fig. 4-17. Simulation model of a quadrupole-octupole corrector for a test SEM objective lens operating at a working distance of 12 mm.

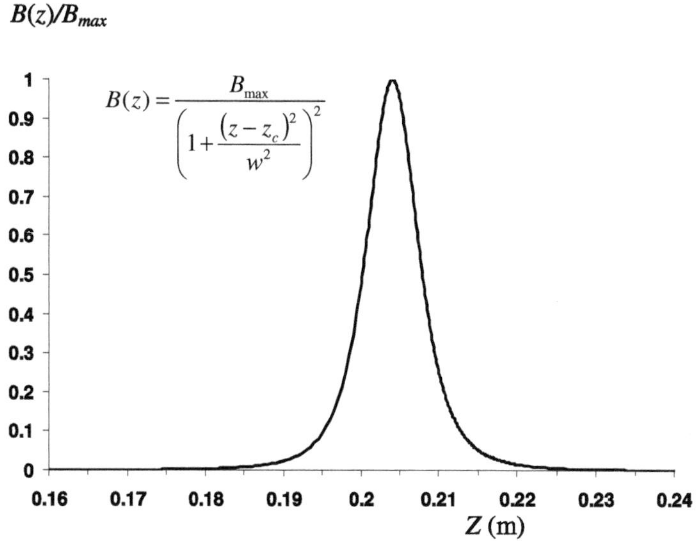

Fig. 4-18. Squared Glaser function used to generate the test SEM objective lens on-axis field distribution (w = 0.006 m used in the simulation).

The simulated aberration limited spot diagram with and without correction for the layout shown in Fig. 4-17 is shown in Fig. 4-19. It shows that the relative size of on-axis aberrations is expected to be reduced approximately by an order of magnitude through the use of a quadrupole-octupole corrector. The full-width of the simulated aberration limited spot (diameter), is 35 nm without correction, and 3 nm with correction. Just exactly how the image resolution is expected to improve is a separate question, which depends on the source spot size and brightness, diffraction aberration, as well as the precise current distribution in the final spot. If the image resolution is taken to be approximately half the full-width value, we can conclude that these simulation results predict that the corrector can reduce the contributions of on-axis aberrations to be between 1–2 nm for this simple test objective lens example. For smaller variations in initial energy/angular spread, this predicted value can be brought down to the sub-nano scale level. Alternatively, the aberration limited probe size can be kept small enough to provide a moderately good resolution, say below 3 nm, while at the same time, increasing the probe current, by using larger final apertures, leading to brighter images for the same degree of aberrations in the final spot.

Figure 4-19 indicates that aberration coefficients are predicted to fall by over an order of magnitude through correction. Zach and Haider (1995) report experimental results that indicate on-axis aberration coefficients (C_s and C_c) can be reduced to 1 to 2.5% of their original values. How does the final predicted probe size of corrected aberration lenses therefore compare with other objective lenses? Let us assume for simplicity, that the on-axis lens aberration coefficients can be reduced to 1% of their original values through correction, and that the uncorrected lens is the Cambridge S100 objective lens examined in Chapter 1, where simulations predict that $f = 15.69$ mm, $C_s = 30.4$ mm and $C_c = 13.29$ mm at a working distance of 5 mm. Let us further assume the use of a field emission Schottky source for which the energy spread is 0.4 eV and the beam energy is 1 keV.

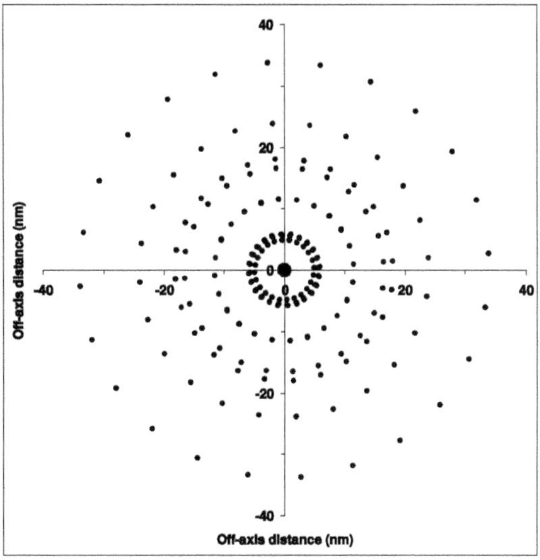

(a)

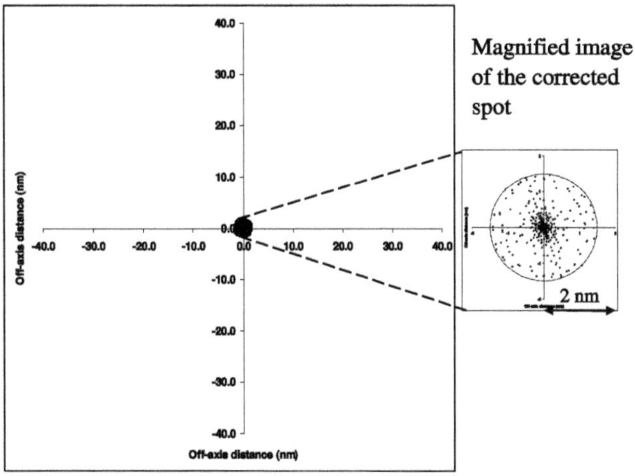

(b)

Fig. 4-19. Simulated aberration limited spot diagrams of the primary beam at the specimen with and without correction for the test SEM objective lens operating at a working distance of 12 mm: (a) Without correction, the spot diameter is 35 nm, (b) With correction, the spot diameter is 3 nm.

A comparison of the predicted final probe diameter with and without correction as a function of final semi-angle for the S100 lens is shown in Fig. 4-20(a). Here the projected probe from the source is assumed to be zero, and the final probe consists only of diffraction and on-axis aberration contributions. Figure 4-20(a) indicates that the optimum operating final semi-angle now is significantly increased, from 3.4 mrad (uncorrected probe diameter of 8.67 nm), to 22.9 mrad (corrected probe diameter of 1.052 nm). The increase in optimum semi-angle position comes directly from the fact that diffraction aberration is smaller for larger semi-angles, and that lower aberration coefficient values allow for larger semi-angles to be used. The optimum aperture radius is 53.4 μm for uncorrected lens, compared to 359.4 μm for the corrected lens.

Figure 4-20(a) indicates that not only does aberration correction reduce the size of the final probe, but it also produces much brighter images. In this simulation example, the final probe current is increased approximately by a factor of 45, since the probe current is proportional to the area of the final aperture. This highlights perhaps the most significant aspect of aberration correction in the SEM, aberration correction provides a way of obtaining brighter high resolution images, well over an order of magnitude brighter than conventional lenses. In the present context of spectrometers, this also has considerable significance. There will be well over an order of magnitude better signal-to-noise characteristics for spectrometers in SEMs that incorporate on-axis aberrations correctors.

Figure 4-20(b) shows the predicted aberration limited probe diameter (taking into account diffraction) for the Cambridge S100 SEM corrected objective lens (reduction to 1% of aberration coefficients) compared with the 5 kV/mm mixed field immersion lens simulated earlier as a function of final aperture radius. The energy spread is 0.4 eV and the beam energy is 1 keV. Figure 4-20(b) shows that although the immersion lens has a minimum predicted aberration probe diameter of 0.809 nm, lower than the corresponding value of 1.052 nm for the corrected S100 lens, it occurs at a much smaller aperture. The useful range of apertures for immersion lenses in order to achieve high resolution is typically over one

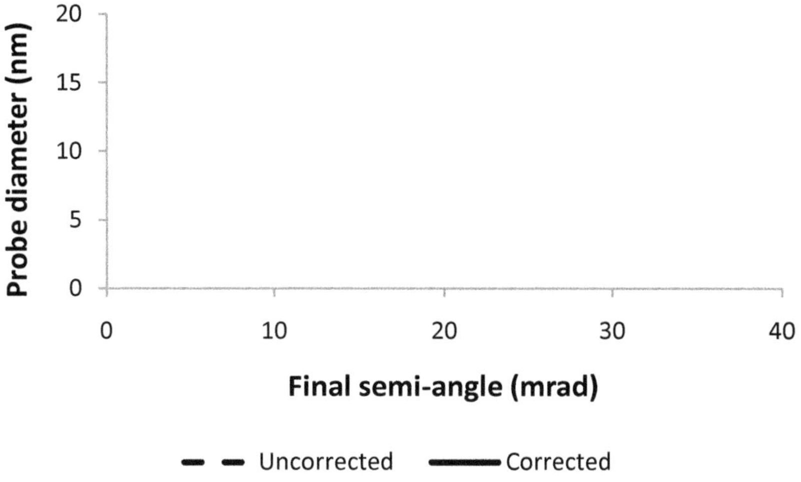

(a)

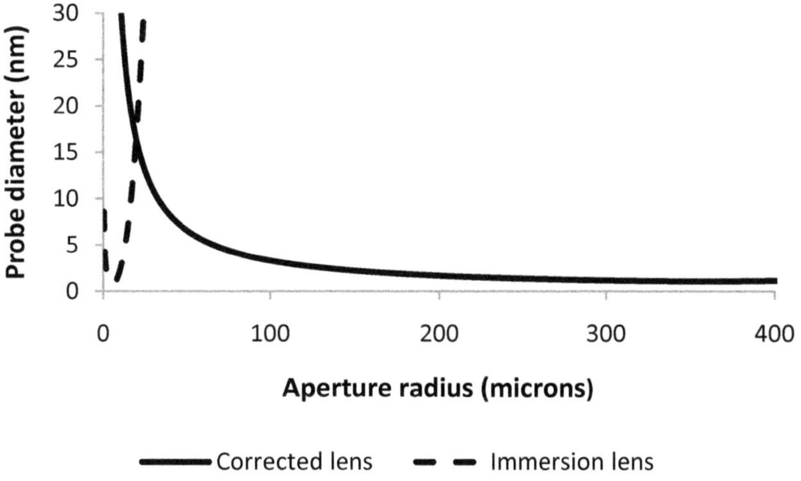

(b)

Fig. 4-20. Simulated probe diameter comparison between aberration corrected and uncorrected lenses for a Schottky field emission source spread of 0.4 eV: (a) Comparison for the S100 lens (f = 15.69 mm, C_s = 30.4 mm and C_c = 13.29 mm at WD = 5 mm), (b) Comparison with a mixed field immersion lens (5 kV/mm, f = 0.205 mm, C_s = 0.1 mm and C_c = 0.084 mm) and S100 lens with aberration coefficients reduced to 1% of their original values.

order of magnitude smaller than the optimum apertures required for aberration corrected lenses, and since the semi-angles at the specimen are of the same order (20 to 30 mrad), the probe currents of aberration corrected lenses are expected to be around two orders of magnitude higher than immersion lenses (for nm image resolution). Based upon these kinds of observations, it is safe to say that aberration corrected lenses are not only predicted to provide nano-meter probe diameters, but they are also expected to do it with over an order of magnitude higher probe currents, compared to other types of objective lenses.

Just how the depth of focus (DOF) of an aberration corrected image compares with an uncorrected one is an important point to establish. This is because the relatively long depth of focus of the conventional SEM (several μm) gives it a significant advantage over other imaging instruments, such as conventional light microscopes. Also, earlier in this chapter, it was noted that although immersion lenses were able to improve the image resolution to a few nano-meters or less, this came at the price of correspondingly reducing the depth of focus (to less than 50 nm). Although aberration corrected lenses can maintain the same semi-angle at the specimen as an uncorrected lens, when diffraction is taken into account, the depth of focus reduces, because a lower probe size can only be obtained for larger semi-angles. This has the desirable property of increasing image brightness, but it also means that the depth of focus will inevitably decrease. For a probe resolution of 1 nm, the depth of focus for the S100 lens is predicted to be 43.6 nm (optimum semi-angle of 22.9 mrad), this value is similar to the depth of focus of high resolution immersion lenses. It is interesting to note however, that the predicted probe diameter rises relatively slowly as the semi-angle is reduced below its optimum value (Fig. 4-20(b)), so with a little compromise in probe diameter and reduction in probe current, the depth of focus can be substantially improved. For instance, for the reduction of the semi-angle to 10.5 mrad, reducing the probe current by approximately 4 times, the simulated aberration probe diameter increases to 2 nm, and the depth of focus is predicted to be 95.2 nm. These kinds of characteristics of aberration corrected lenses have also been noted by Joy (2008).

Figure 4-21 show sub-nanometer aberration corrected images taken from the JEOL 7700F/JWS-7555S SEM for two gold-on-carbon test specimens. They show uncorrected and corrected images side by side, illustrating a significant improvement in the sharpness of the corrected images compared to the uncorrected ones.

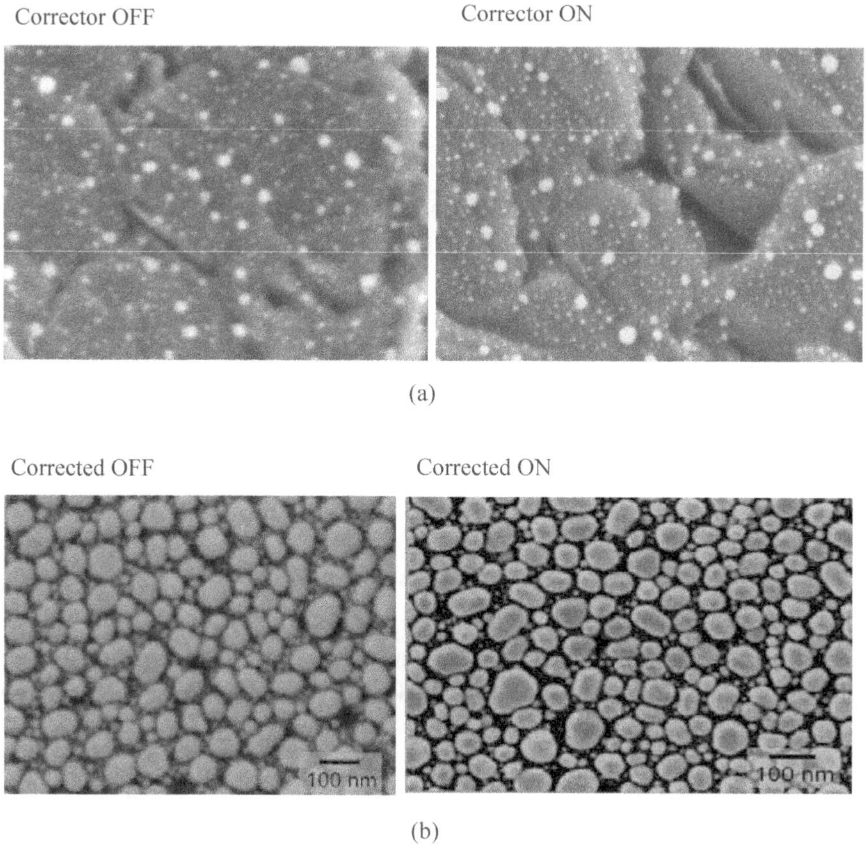

Fig. 4-21. Aberration corrected SEM images 7700F SEM for a primary beam energy of 1 keV (courtesy of JEOL Inc.): (a) Gold particles on a graphite matrix with image demagnification of 300,000, (b) Gold on carbon test specimen (from the JWS-7555S SEM).

4.3 The helium ion microscope

The Helium Ion Microscope is not strictly speaking, a Scanning Electron Microscope, and yet, it forms a natural part of the topic under discussion here. Its recent success largely derives from developments of a stable high performance Helium ion source. This gun is reported to have a brightness of 4×10^9 A/cm^2 sr), around 2 orders of magnitude better than a Schottky field emitter, an energy spread of less than 1 eV, and a very small virtual source diameter of 0.3 nm (Ward *et al.*, 2006). In terms of its column optics, it resembles both a Focused Ion Beam (FIB) and a SEM. It has most of the same features as the SEM, except that it uses a Helium ion beam. Like the FIB, it uses an Einzel electrostatic lens, necessitated by the fact that Helium ions are nearly 8,000 times heavier than electrons. Unlike magnetic lenses, the focusing properties of electrostatic lenses are independent of charge-to-mass ratio, as already noted in Chapter 1. The Helium Ion Microscope creates much less sputtering at the specimen surface than the FIB, and is used for inspection imaging, like the SEM, rather than milling or deposition as is the case for the FIB. It does however cause some permanent damage to the specimen surface, which can be minimized by limiting the primary beam current by beam blanking plates, switching the beam on/off.

Figure 4-22(a) shows simulated equipotentials of the type of electrostatic objective lens used for the Helium Ion Microscope. The potential on the central electrode was adjusted to give a working distance of 10 mm for a 30 keV primary ion beam. The lens layout here is not taken from the actual Helium Ion Microscope design, but is only used by the author for the purpose of a simulation model. The focal length and on-axis aberration coefficients for these conditions were simulated to be f = 27.06 mm, C_s = 670.4 mm, and C_c = 83.5 mm. Since Helium ions are nearly 8,000 times heavier than electrons, they give rise to a much lower diffraction aberration spot, around 85 times smaller than for electrons (at the same voltage and semi-angle). This reduced diffraction aberration in the primary beam is one of key reasons why the Helium Ion Microscope can obtain sub-nanometer image resolution. Figure 4-22(b) shows how the aberration limited probe diameter is expected to vary as a function of

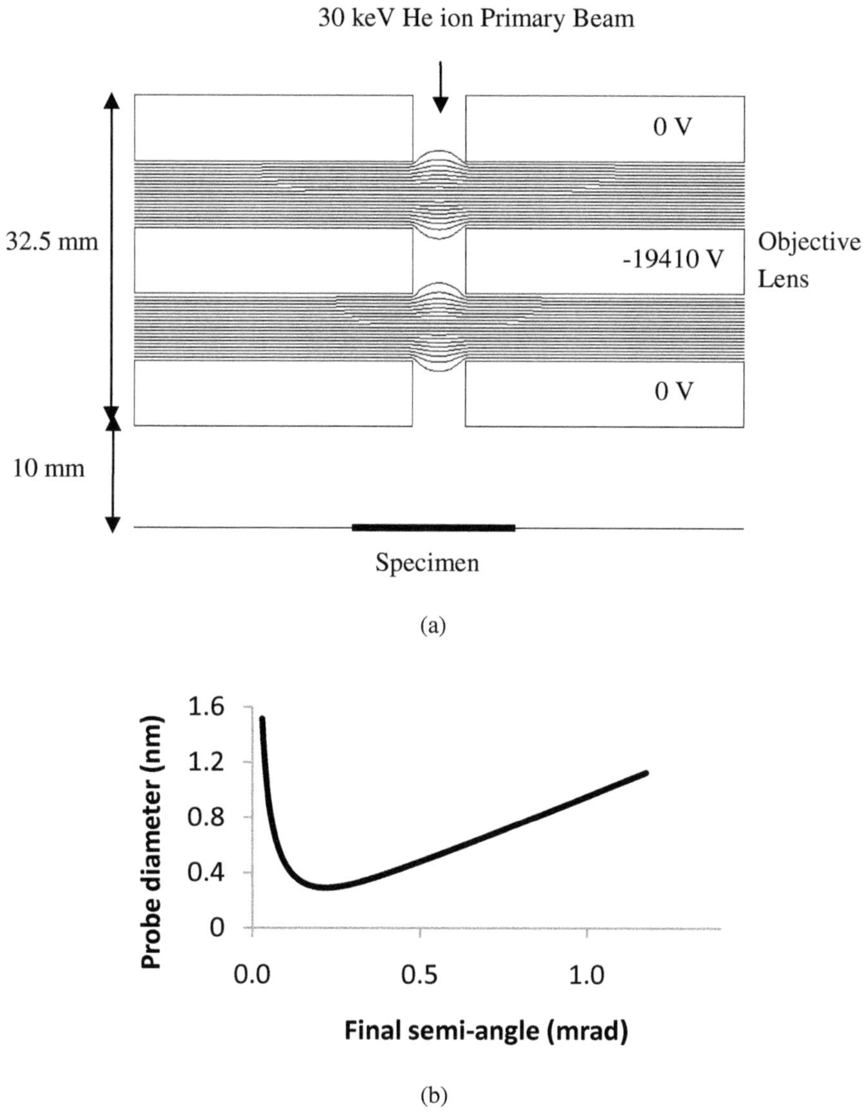

Fig. 4-22. Helium ion microscope optics: (a) Simulation model of an Einzel electrostatic objective lens to focus a 30 keV primary beam at a working distance of 10 mm, (b) Simulated aberration limited probe diameter as a function of final semi-angle.

semi-angle. A minimum probe diameter of 0.29 nm occurs at a semi-angle of 0.218 mrad, corresponding to an aperture radius of 5 μm. In this case, the optimum semi-angle operating point compared to the conventional SEM objective lens becomes smaller, necessitating the use of smaller apertures. This is opposite to the aberration corrected SEM objective lens case, where the optimum semi-angle becomes larger (than for the conventional SEM). The subsequent loss of current in the case of the Helium Ion Microscope is compensated by the fact that its source is much brighter, its projected source size at the specimen is negligible, and that Helium primary ions have a much larger secondary electron yield than primary electrons (Bell, 2009). Other aspects of the Helium ion beam that act in its favor is that it has a much smaller transverse interaction volume than for electrons. Helium atoms penetrate into the specimen to much deeper levels than electrons for the same primary beam energy, therefore, the Helium Ion Microscope can be operated at relatively high primary beam energies, typically between 20 to 30 keV, and still generate a relatively narrow interaction volume. At 20 keV, its interaction volume at the specimen surface is comparable to the one created by a 0.5 keV primary electron beam (Bell, 2009).

The relatively low semi-angles required to obtain high resolution with the Helium Ion Microscope give it an important advantage over the previous two sub-nanometer imaging techniques described in this chapter, namely, a relatively large depth-of-focus, comparable to that of conventional SEMs. With an estimated minimum probe size of 0.29 nm at a final semi-angle of 0.218 mrad (for the lens model depicted in Fig. 4-22(a)), the depth-of-focus is approximately 1.33 μm.

Figure 4-23 shows 200 nm field-of-view images obtained by the Carl Zeiss SMT Orion Helium Ion Microscope on 2 test specimens. Figure 4-23(a) shows an image of a gold on carbon substrate, while Fig. 4-23(b) depicts an image of an asbestos fiber on a holey carbon substrate. Both these images display the Helium Ion Microscope's ability to provide sub-nanometer resolution images, a resolution of 0.25 nm has been estimated for the image shown in Fig. 4-23(b).

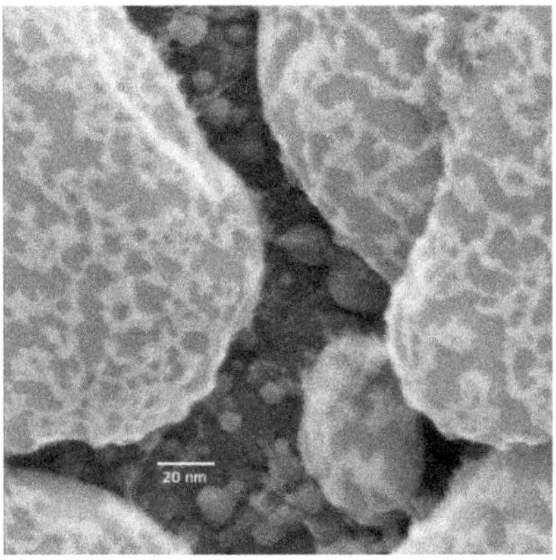

(a)

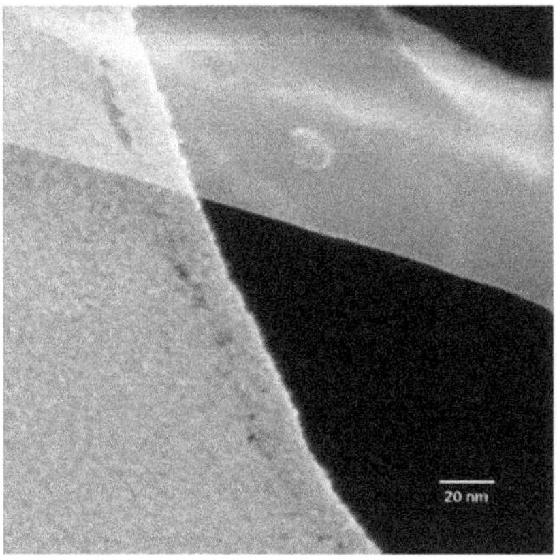

(b)

Fig. 4-23. Sub-nanometer resolution images from the Orion Helium ion Microscope (courtesy of Carl Zeiss STM Inc.): (a) A 200 nm field-of-view image showing evaporated gold on a carbon substrate, (b) A 200 nm image of asbestos (crocidolite) fiber on a holey carbon substrate.

Chapter 5

Secondary Electron Spectrometers

Spectrometers are best understood in terms of the parameters they set out to measure. Most energy spectrometers proposed for the scanning electron microscope have been made for the purpose of quantifying voltage contrast, and this chapter gives a brief summary of how such spectrometers were developed.

5.1 Early deflection analyzers

The first published work on voltage contrast was by Oatley and Everhart (1957), who noticed that changes in specimen voltage caused brightness variations in its corresponding SE image. They demonstrated for example, how the SE image of a back-biased p-n junction specimen varied in brightness as the p-n junction voltage changed. Generally speaking, if a region on the specimen is biased positively with respect to its surroundings, the region in the SE image darkens, while if it is biased negatively with respect to its surroundings, the region in the SE image brightens.

The first energy spectrometer designed to quantify the voltage contrast effect in the SEM was made by Wells and Bremer (1968), their analyzer design is shown in Fig. 5-1(a). This spectrometer consisted of a modified Cylindrical Mirror Analyzer (CMA), with its axis of symmetry orientated horizontally. The specimen was positioned next to the inner plate, biased to 40 V, and the outer plate was biased to 27 V. The analyzer voltage resolution was estimated to be around 5 V. Since the specimen was placed directly at the entrance of the analyzer, with no shielding, large errors were incurred through the presence of electric fields between the specimen, inner and outer plates, which acted to displace the primary beam. Additionally, there were errors due to specimen misplacement.

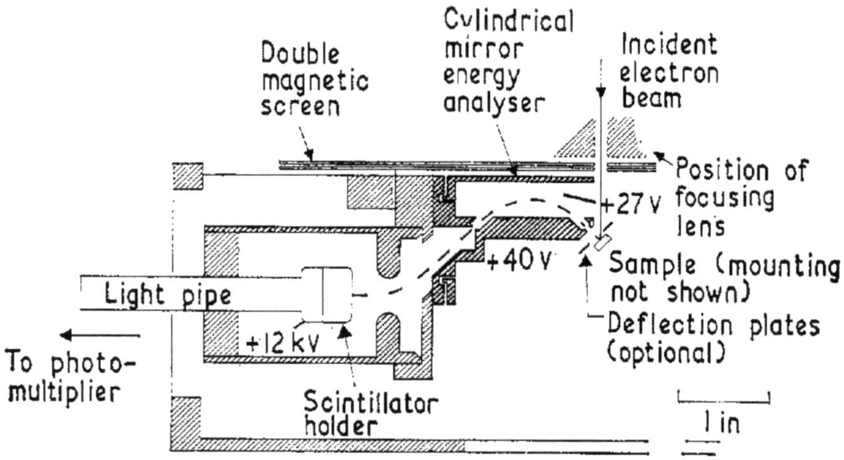

(a)

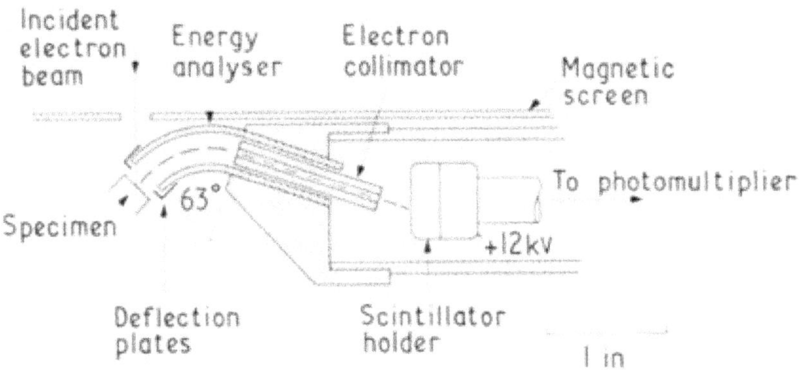

(b)

Fig. 5-1. The Wells and Bremer voltage contrast spectrometers: (a) Initial CMA design (1968), (b) Improved 63° CDA (1969).

Wells and Bremer (1969) improved on their initial design by proposing the use of a 63° Cylindrical Deflector Analyzer (CDA), as shown in Fig. 5-1(b). Although the primary beam still travels through analyzer deflection fields just before reaching the specimen, these fields are much smaller than in the original CMA design. The CDA is designed so that electrons at the pass energy travel parallel to the exit plates, and energy filtering is achieved through extending the length of these plates. In this way, the use of a slit is avoided, electrons below and above the pass energy exit the analyzer at an angle relative to the central ray, and therefore do not reach the secondary electron detector. The voltage resolution for this spectrometer was measured to be 1 V and the objective lens working distance for both the initial and improved Wells and Bremer spectrometer was 20 mm.

The work of Wells and Bremer was further developed by Hannah (1974). He used the improved 63° CDA, but re-orientated it for horizontal specimens, as shown in Fig. 5-2(a). In addition he used an accelerator lens assembly to help extract secondary electrons emitted from the specimen surface, counteracting the adverse effect of transverse surface fields. By simulating the effect of these lenses on the trajectory paths of secondary electrons through the analyzer, he was able to optimize the analyzer's performance, achieving a voltage resolution of 0.1 V. Hannah was able to obtain shifted SE spectra for different specimen voltages, as shown in Fig. 5-2(b), and proposed a peak detection scheme to speed up data-acquisition times. One disadvantage of Hannah's spectrometer design is that it involved the use of a relatively large working distance, 65 mm. Apart from this early work on deflection analyzers, further development in quantitative voltage contrast was largely confined to retarding field spectrometers. In the years from 1970 to 1990, a wide variety of different retarding field voltage contrast analyzers were reported. However, towards the end of this period, there was a renewed interest in deflection analyzers, mainly in the context of multi-channel analyzer designs. Reviews on voltage contrast analyzer design can be found in a book chapter written by Dubbledam (1993), and articles written by Plies (1990), Garth (1987) and Menzel and Kubalek (1983).

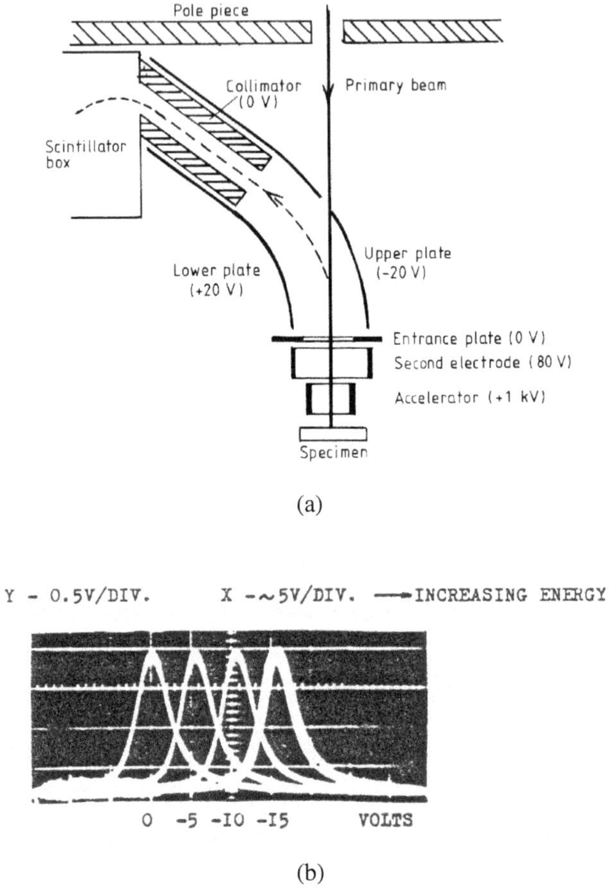

Fig. 5-2. The 63° CDA Hannah voltage contrast spectrometer (1974): (a) Spectrometer layout, (b) Experimentally acquired SE spectra for different specimen voltages.

5.2 Retarding field analyzers

The principle of voltage contrast retarding field analyzers aimed at making signal measurements on integrated circuits is shown in Figs. 5-3(a) and 5-3(b). These spectrometers function by detecting only those secondary electrons that can surmount a potential barrier created within the analyzer, and are, in effect, high pass energy filters. The potential barrier

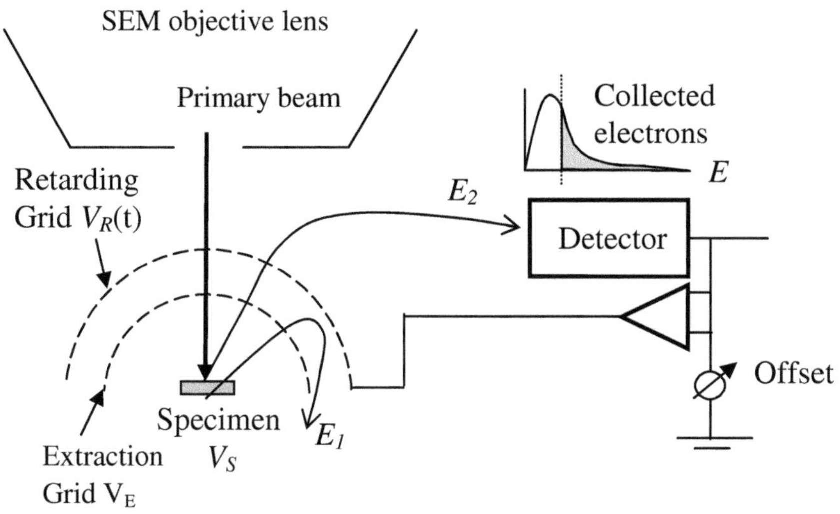

(a)

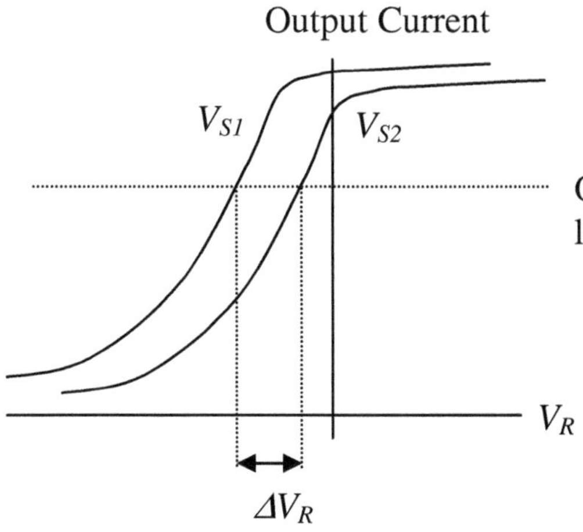

(b)

Fig. 5-3. Principle of closed loop retarding field spectrometers for voltage contrast: (a) Spectrometer layout, (b) Output S-curve signals.

is created by biasing a retarding electrode or grid to a voltage V_R. The strength of the potential barrier is given by V_R-V_S, where V_S is the specimen voltage. Only electrons having initial energies greater than $e(V_R\text{-}V_S)$ will surmount the potential barrier and contribute towards the output signal (represented by the shaded area in Fig. 5-3(a). When V_R is varied, an integrated form of the secondary electron energy spectrum is obtained, as illustrated in Fig. 5-3(b). This output signal is sometimes referred to as an "S-curve". Shifts in the output signal are therefore obtained when the specimen voltage changes (indicated by V_{S2} and V_{S1} in Fig. 5-3(b)).

For quantitative voltage measurements, retarding field spectrometers are normally operated in the closed feedback loop mode, a method first suggested by Flemming and Ward (1970), where the strength of the potential barrier is adjusted to keep the output current constant, as illustrated in Fig. 5-3(a). The output current is continually compared to an off-set current level. Adjustments in the retarding electrode voltage to keep the output current constant are therefore equal to specimen voltage changes, $\Delta V_R = V_{S2}\text{-}V_{S1}$. This feedback loop method of surface specimen voltage measurement has the merit of simplicity and is easily automated. It avoids the need to generate the SE spectrum and interpret it, but it has the disadvantage of discarding useful information within the SE spectrum (lower energy part) and suffers from the presence of unwanted background noise (generated from the higher energy part).

An early voltage contrast retarding field analyzer using a hemispherical retarding grid was proposed by Fentem and Gopinath (1974), while ones using planar grids, were reported by Plows (1969), Flemming and Ward (1970), and Gopinath and Sanger (1971). For over a decade, from 1970, retarding field analyzers were placed below the objective lens, and used as add-on attachments for the SEM. Some selected analyzer designs of this type are shown in Figs. 5-4(a)-5-4(c).

Figure 5-4(a) shows a spectrometer by Goto *et al.* (1981) that uses a hemispherical grid retarding electrode. The detector in this case is placed to one side, other earlier spectrometers of this type used a hemispherical grid/plate as the detector (Fentem and Gopinath, 1974; Tee and Gopanath, 1976), which although ideal in shape, had poor signal to noise characteristics.

Figure 5-4(b) shows a 127° CDA design, followed by planar retarding grids, as proposed by Balk *et al.* (1976). Instead of using the CDA to function as a band-pass spectrometer, it is used as a reliable way of transporting secondary electrons to one side, so despite its appearance, this design functions as a retarding field analyzer. By locating the retarding grid to one side, well away from the primary beam, this spectrometer ensures that the primary beam is not disturbed while voltage measurements are being made.

Figure 5-4(c) shows an example of a planar retarding grid analyzer reported by Feuerbaum (1979). Like most other retarding field spectrometers that use planar extraction/retarding grids directly above the specimen, it deflects secondary electrons to one side on to an E-T detector, once they have surmounted the internal potential barrier. The deflector in this case consists of a positively biased cylindrically shaped grid located on the left hand side of the analyzer. Planar grids are easier to place closer to the specimen surface than hemispherical shaped ones, and can therefore apply a stronger extraction field. Extraction field strengths of over 1 kV/mm were found to be necessary in order to minimize the influence of adverse effects of surface fields on Integrated Circuit specimens. The Feuerbaum spectrometer was simulated by Khursheed (1984) to have a collection efficiency of around 97% for all electrons below 10 eV (neglecting scattering on grids). This high collection efficiency, along with its relatively short working distance of 10 mm, makes it one of the better performing spectrometers amongst the early retarding field voltage contrast analyzers.

Also of interest is the analyzer design shown in Fig. 5-4(d), which was made by Plows (1981). In this case, the analyzer is rotationally symmetric, and there is detection in the full 2π azimuthal angular direction via the use of a ring scintillator and optical fiber link. The retarding electrode is a cylindrical plate, and SEs are deflected off-axis by the electric field created between this plate, a suppressor grid (-50 V) and the 10 kV scintillator voltage. Despite it having rotational detection, this analyzer was simulated to have a collection efficiency of only around 57% for SEs having energies below 10 eV (Khursheed, 1984).

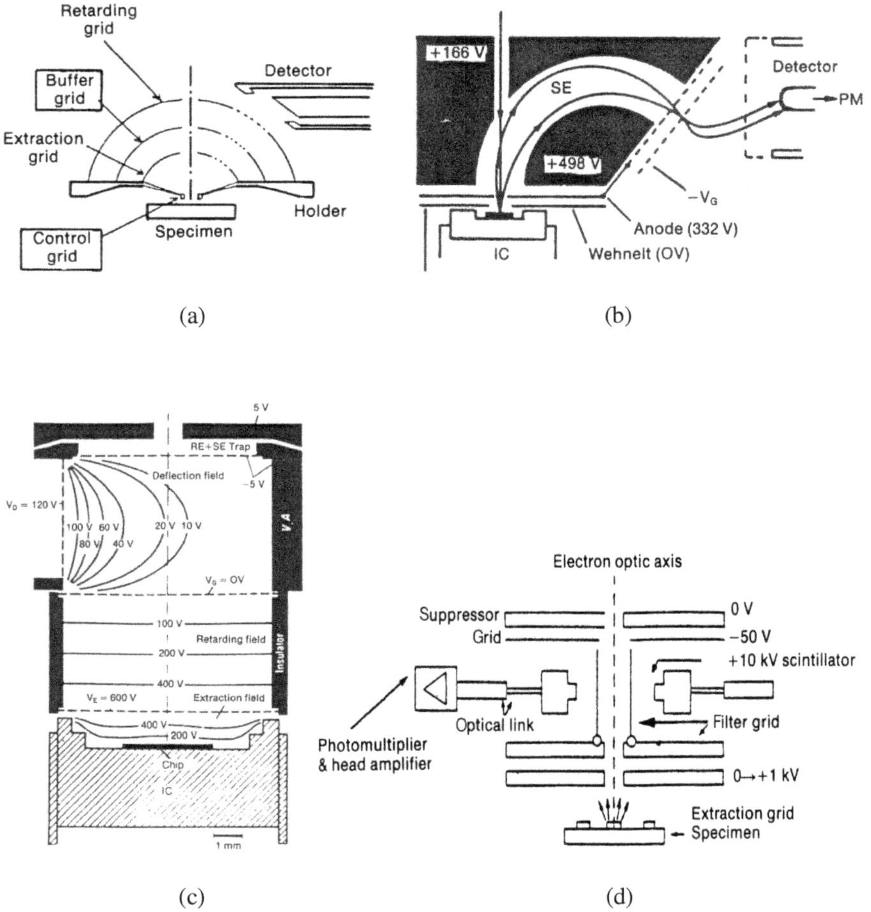

Fig. 5-4. Selected examples of some voltage contrast retarding field analyzers: (a) The hemispherical analyzer of Goto *et al.* (1981), (b) The 127° CDA planar grid design of Balk *et al.* (1976), (c) The planar grid analyzer of Feuerbaum (1979), (d) Cylindrical filter electrode analyzer of Plows (1981).

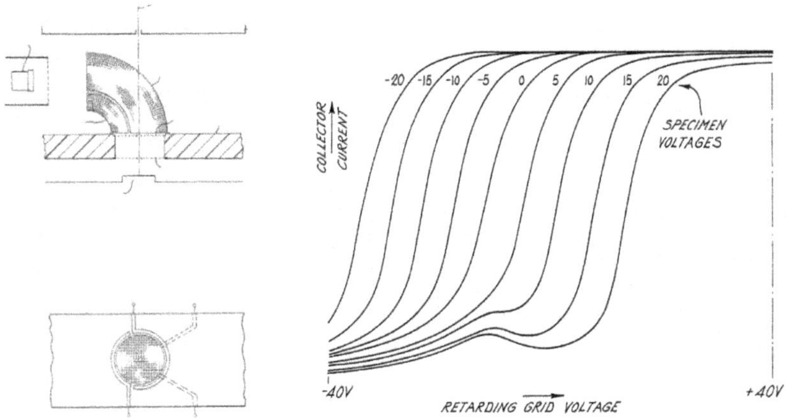

(a) (b)

Fig. 5-5. Experimental results from a retarding field voltage contrast spectrometer (Khursheed, 1983): (a) Spectrometer layout, (b) Experimentally measured S-curves.

Typical experimental S-curves for the kind of retarding field analyzers described so far are shown in Fig. 5-5(b). These results were generated from a retarding field analyzer designed by Khursheed (1983), shown in Fig. 5-5(a), which uses planar extraction/retarding grids above the specimen and deflects SEs to one side by upper and lower grids that lie on a curved tube geometry. This spectrometer was simulated to have a collection efficiency close to 99% (neglecting scattering on grids) (Khursheed, 1983).

One major drawback of the voltage contrast spectrometers described so far is that they needed to be placed below the SEM objective lens, and as a result, inevitably increased the working distance, limiting the image resolution that could be attained. As better objective lenses were proposed for the SEM, such as the magnetic immersion designs described in Chapter 3, researchers started to design in-lens voltage contrast spectrometers. The first spectrometer of this type was proposed by Menzel and Buchanan (1984), a schematic diagram of how this analyzer functioned is illustrated in Fig. 5-6(a). They placed a planar extraction grid at the entrance of a wide-bore SEM final lens, and a planar filter grid at the top of the lens. Secondary electrons are accelerated up into the lower part of the lens and are progressively slowed down as they travel upwards through it. As the secondary electrons slow down, they typically describe tight spirals around the lines of magnetic flux created by the objective lens. Figure 5-6(b) shows simulated 10 eV SE trajectory paths illustrating this effect where the extraction electrode is 100 V. A detector located off-axis on top of the lens attracts the SEs away from the optical axis to form the output signal. The filter grid voltage is varied in the normal way, keeping the final output current constant.

Many subsequent through-the-lens analyzer designs were based upon the initial work of Menzel and Buchunan (1984). Garth (1987) reported an analyzer design, shown in Fig. 5-7(a), where the extraction grid accelerates the SEs through the peak magnetic field of the objective lens,

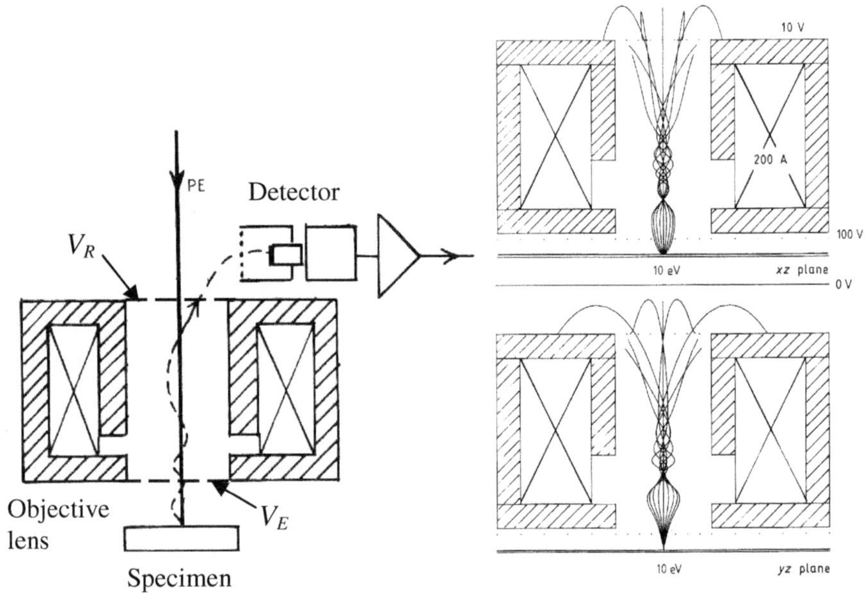

Fig. 5-6. Through-the-lens retarding field analyzer design (Menzel and Buchanan, 1984): (a) Schematic layout, (b) 10 eV simulated SE direct ray paths for a wide variety of emission angles.

but then slows them down with an additional electrode, so that they spread out and trace parallel trajectory paths as they exit the lens. Air-gaps are introduced into the bore of the magnetic lens, so that the rate at which the magnetic field drops can be controlled, providing a way to control the degree to which electrons are allowed to spread out. Garth biased the extraction electrode so that electrons move through the region of increasing magnetic field with energies exceeding 500 eV.

Figures 5-7(b) and 5-7(c) shows analyzer designs that accelerate the SEs away from the specimen, and uses the objective lens magnetic field to bring them to a focus shortly afterwards, so that they spread out in a radial direction. The kinetic energies of secondary electrons as they pass through the objective lens air gap are typically around 2 keV. A hemispherical retarding field grid is then used to filter their energies. Two detectors are placed off-axis opposite to one another, increasing the overall collection efficiency and minimizing the effect of asymmetric fields acting on the primary beam. These analyzers are designed to focus the SEs so that they appear to come from a point that lies at the center of the hemispherical retarding grid (Plies, 1990). The analyzers shown in Figs. 5-7(b) and 5-7(c) were developed independently, although they have similar optics for the secondary electrons. The pancake objective lens analyzer by Todokoro *et al.* (1986), in Fig. 5-7(b), builds upon the previous work of Goto *et al.* (1981) who reported an earlier hemispherical grid analyzer design, and the compound (both electric and magnetic fields) analyzer of Frosien and Plies (1987), in Fig. 5-7(c), follows on from previous suggestions for locating a secondary electron focal point at the centre of retarding hemispherical grids made by Plies (1984).

Dinnis (1987) reported an analyzer design that made use of an objective lens doublet that is normally used to cancel magnetic rotation, shown in Fig. 5-7(d). Here the objective lens is constructed in two successive halves with opposite directions of magnetic field. Low energy electrons spiral up into the central region where the magnetic field is zero, and with the help of suitable radial electrostatic fields (created by grids, indicated by the dotted lines in Fig. 5-7(d)) they are channeled efficiently to the periphery of the lens system.

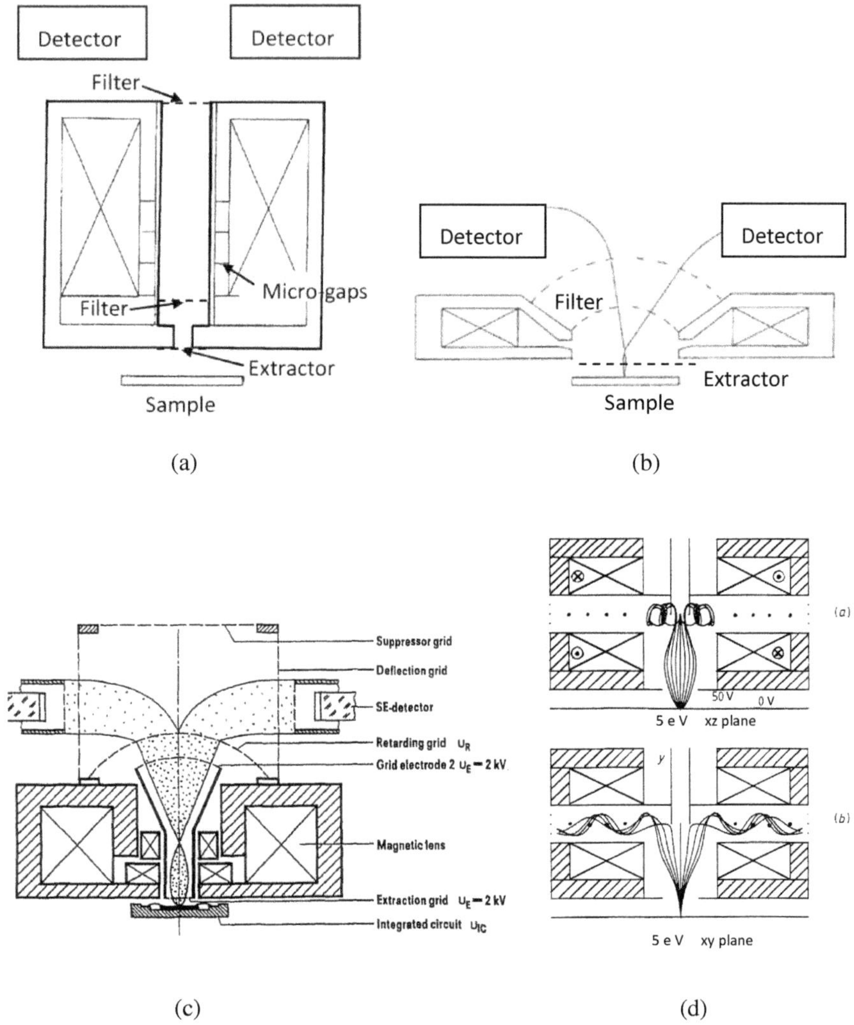

Fig. 5-7. Through-the-lens retarding field voltage contrast spectrometers: (a) Garth (1987), (b) Todokoro *et al.* (1986), (c) Frosien and Plies (1987), (d) Dinnis (1987).

The analyzer designs shown in Figs. 5-7(a)–5-7(d) have much smaller working distances (typically 2–5 mm) than the previous electrostatic analyzers which are placed below the objective lens. Figure 5-8 shows examples of analyzer designs that go one step further, they place the specimen in the peak magnetic field (immersion lens), and use very small (or no) extraction fields. In such cases, secondary electrons travel through a decreasing magnetic field as they leave the specimen, and experience an immediate collimating action as they leave the specimen.

As described at the end of Chapter 2, low energy electrons that traverse a sharply decreasing magnetic field spiral around magnetic field lines and follow trajectory paths that become more parallel as they reach the region of low magnetic field strength (Kruit and Read, 1983). Figure 5-8(a) shows this collimating effect in the lower part of a voltage contrast analyzer designed by Dubbledam and Kruit (1987a) which places the specimen below a single-pole lens. Figure 5-8(b) shows an analyzer where the single pole lens is placed below the specimen (Garth *et al.*, 1986), and Fig. 5-8(c) depicts an analyzer design for a wide bore immersion/single-pole lens above the specimen reported by Richardson and Muray (1988). For the analyzers shown in Figs. 5-8(a) and 5-8(c), Variable Axis Immersion Lens (VAIL) deflections coils are mounted on the lens pole-pieces (see Chapter 1), this method, first proposed within Electron Beam Lithography, is a way of increasing the field of view of immersion lenses, by setting the deflection field of the scan coils to be proportional to the gradient of the objective lens magnetic field (Pfeifer and Langner, 1981). These analyzers also incorporated a solenoid within the upper part of the lens bore in order to control the rate at which the objective lens magnetic field decreases, preventing them from spreading out too far and counteracting the adverse effect of stray magnetic fields.

Before the mid 1980s, voltage contrast analyzers were used as SEM add-on attachments, functioned by creating electric retarding fields, and were invariably placed below the objective lens. Towards the late 1980s, several companies emerged which manufactured dedicated scanning electron microscope columns, known as Electron Beam Testers (EBTs) which were specifically designed and optimized to make quantitative voltage measurements on Integrated Circuits. The company, Integrated Circuit Testing, utilized the Frosien and Plies (1987) through-the-lens analyzer design, while the company, Sentry Schlumberger, employed the wide-bore immersion/single-pole lens analyzer designed by Richardson

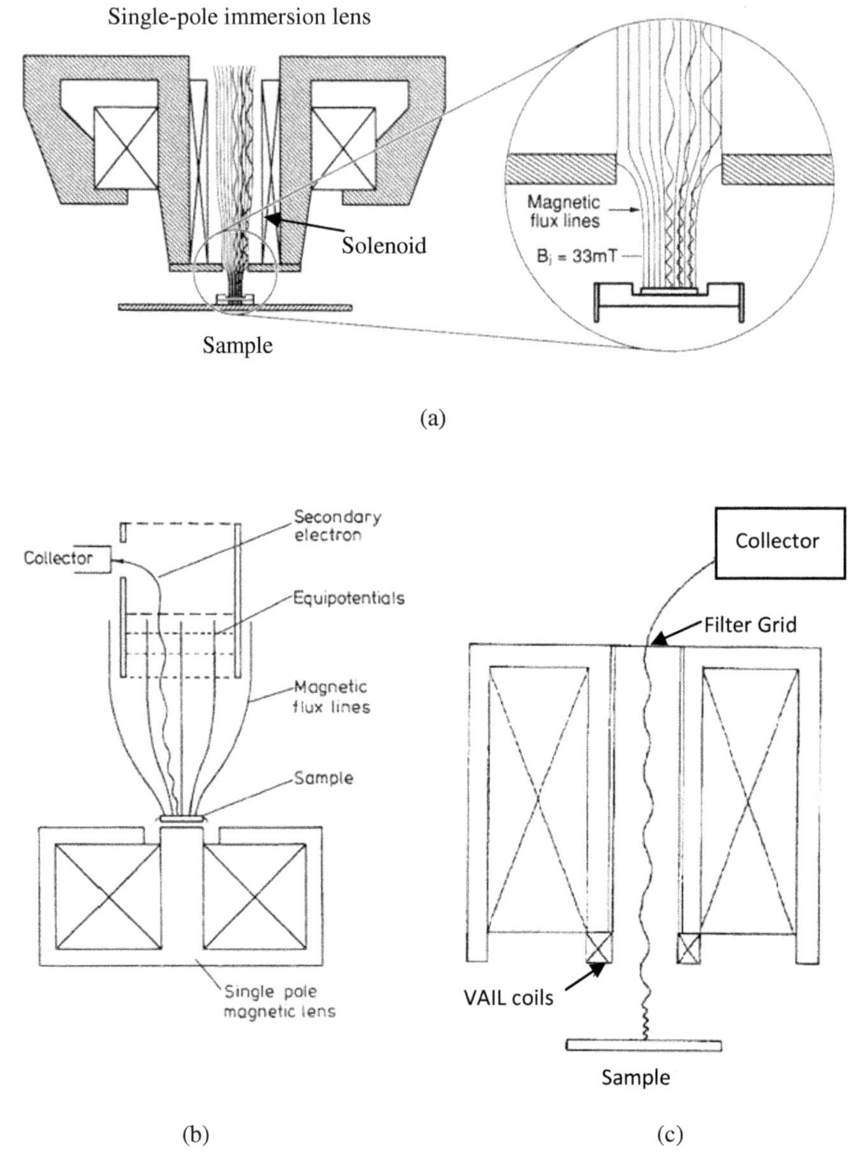

Fig. 5-8. Immersion lens retarding field analyzer designs: (a) Single pole lens above specimen, Dubbledam and Kruit (1987a), (b) Single-pole lens below specimen, Garth *et al.* (1986), (c) Wide-bore immersion/single-pole lens, Richardson and Muray (1988).

and Muray (1988). A more detailed description of EBTs can be found in a book edited by Thong (1993).

5.3 Surface fields and signal-to-noise characteristics

The absence of a strong extraction field for the immersion magnetic retarding field analyzers shown in Fig. 5-8 is partially compensated by the collimating effect of the decreasing magnetic field, since the final trajectory paths of the SEs will be relatively insensitive to the emission angle at the specimen. The adverse effect of transverse fields (cross-fields) at the specimen will therefore be greatly reduced. On the other hand, the lack of a strong extraction field will make the signal output much more vulnerable to the kind of surface fields that form potential barriers on the specimen surface, as illustrated in Fig. 5-9. Figure 5-9(a) shows simulated equipotential lines above a three track conductor layout in the presence of a 1 kV/mm extraction field strength, where the outer two are grounded and the middle is biased to a positive voltage. The track widths and separation distances are all set to 1 μm. For this example, a potential barrier of 1.4 V is created, in general, the size of the potential barrier depends on the extraction field strength and the distance between conductor tracks. Figure 5-9(b) shows simulated 1 eV SE ray paths emitted over a wide range of initial angles from the middle track surface, indicating that they are pulled back to the specimen surface. Figure 5-9(c) shows how the strength of the potential barrier varies with extraction field strength in order to suppress the effect of this kind of surface field on lower energy electrons, the extraction field strength needs to be greater than 2 kV. For very low strength extraction fields, such as those used in the immersion lens retarding field analyzer designs shown in Fig. 5-8, the potential barrier will be approximately equal to the specimen voltage, ie, for a middle track potential of +5 V, SEs having energies less than 5 eV will not reach the detector. In this case, the retarding (filter) electrode needs to be biased to a more negative voltage, since it cannot make use of the lower energy electrons that are returned to the specimen surface. If the specimen voltage is +5 V, the magnitude of the filter voltage needs to be set 1 to 2 V above it, to say -7 to -6 V.

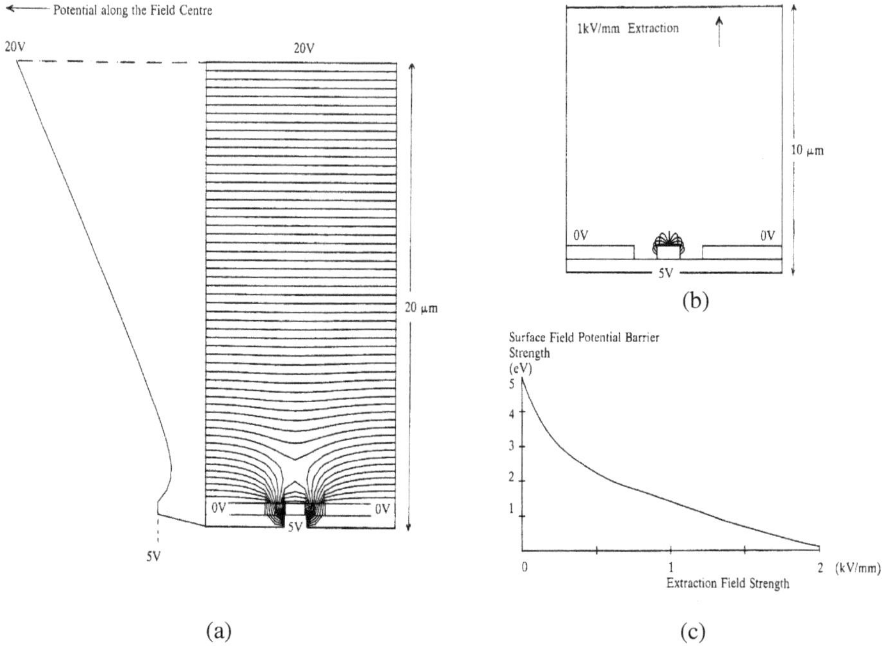

Fig. 5-9. Simulated influence of surface fields: (a) Equi-potential distribution, (b) Ray tracing of 1 eV SEs emitted over a wide range of angles, (c) Dependence of potential barrier strength with extraction field strength.

In general, the voltage resolution of retarding field voltage contrast spectrometers, like the energy resolution of surface science retarding field spectrometers, is limited not by spectrometer optics, but by shot noise. Scattered electrons that have energies above the potential barrier set by the retarding electrode and which reach the detector, will contribute to background noise on the output signal. From this perspective, it is very important to prevent backscattered electrons or the tertiary electrons that they generate, from reaching the SE detector.

Gopinath (1977) proposed a useful guide to characterize the voltage resolution of retarding field analyzers, he defined voltage resolution ΔV_S in terms of the detector current, I_D, bandwidth Δf, and a spectrometer constant K, as

$$\Delta V_S^2 = K^2\left(\frac{\Delta f}{I_D}\right). \tag{5.1}$$

This formula helped establish a figure of merit for retarding field analyzers and is useful for estimating the data-acquisition time required to attain a given voltage resolution. Dubbledam and Kruit (1987b) derived a theoretical value of the spectrometer constant K, based upon secondary electrons having a Chung–Everhart energy distribution (Chung and Everhart, 1974), and assuming ideal filtering within the retarding field spectrometer. Khursheed and Dinnis (1990) reformulated this analysis in terms of a voltage resolution constant χ_r which links together voltage resolution ΔV_S and the total number of electrons N_0 that leave the specimen by

$$\Delta V_S^2 = \frac{\chi_r}{N_0}. \tag{5.2}$$

If it is assumed that the noise at the detector is approximately governed by Gaussian statistics, that is the rms value of the noise is $\sqrt{N}$, then a change in specimen voltage ΔV_S gives rise to a change ΔN in the number of electrons that reach the detector, and that the minimum measureable voltage is determined by the magnitude of the noise (signal to noise ratio of unity), then the voltage resolution is approximately given by

$$\Delta V_S = \frac{\sqrt{N}}{\left(\frac{\partial N}{\partial V_S}\right)}. \tag{5.3}$$

In an ideal retarding field analyzer which has 100% collection efficiency for all electrons that pass through the filter grid, changes in specimen voltage directly create changes in electron kinetic energy E, that is

$$\frac{\partial N}{\partial V_S} = -e\frac{\partial N}{\partial E}. \tag{5.4}$$

The number of electrons N arriving at the detector is given by

$$N = \int_{E_C}^{\infty} \left(\frac{dN}{dE} \right) dE \tag{5.5}$$

where E_C is the cut-off energy created by the voltage difference between filter grid and specimen. The number of electrons arriving at the detector can therefore be estimated by taking into account the Chung–Everhart energy distribution,

$$\begin{aligned} \frac{dN}{dE} &= \frac{6W^2 N_0 E}{(E+W)^4} \\ \Rightarrow N &= 6W^2 N_0 \left(\frac{1}{2(E_C+W)^2} - \frac{W}{3(E_C+W)^3} \right) \end{aligned} \tag{5.6}$$

where W is the specimen work function. The above formula assumes that both W and E_C are measured in eV. Substituting both these equations into the expression for voltage resolution (Eq. 5.3),

$$\Delta V_S^2 = \frac{(E_C+W)^5 (3E_C+W)}{36W^2 N_0 E_C^2}. \tag{5.7}$$

The voltage resolution constant χ_r for an ideal retarding field spectrometer is therefore given by

$$\chi_r = \frac{(E_C+W)^5 (3E_C+W)}{36W^2 E_C^2}. \tag{5.8}$$

It is instructive to compare this expression with the one predicted for deflection or multi-channel analyzers, that is, to the situation where voltage measurements are made using the entire SE spectrum (output signal formed over many energy channels). In this case, changes in voltage are approximately given by a quadrature sum of the minimum measureable voltage in each channel

$$\Delta V_S^2 = \sum_{j=1}^{n} \frac{N_j}{\left(\dfrac{\partial N}{\partial V_S}\right)_j^2} \tag{5.9}$$

where j is the channel index. The corresponding voltage resolution constant χ_m is found from

$$\chi_m = \sum_{j=1}^{n} \frac{P_j}{\left(\dfrac{\partial P}{\partial V_S}\right)_j^2} \tag{5.10}$$

where $P(E)$ is the probability density function for energy distribution, and P_j represents the probability of an electron being in channel j. By substituting the Chung–Everhart expression into this expression, a theoretical prediction for the multi-channel/deflection analyzer voltage resolution constant χ_m can be obtained. The above expression was integrated numerically as a function of E_C (the cut-off energy) for a channel width of 0.1 eV and a work function value of 4 eV (Aluminium) and is plot together with the corresponding prediction for ideal retarding field analyzers in Fig. 5-10. Note that in the case of multi-channel analyzers, the variations in the cut-off energy corresponds to the specimen being biased positively, while for retarding field analyzers, the cut-off energy is related to the difference in voltage between the specimen and retarding grid. Figure 5-10 shows that retarding field analyzers have an optimum cut-off voltage of 1.6 eV, where χ_r has the minimum value of 32 V^2. If electrons having energies below 1 eV can be collected, multi-channel analyzers are predicted to have significantly better signal to noise characteristics. The voltage resolution constant for multi-channel analyzers is predicted to be more than an order of magnitude better than that for retarding field analyzers where electron energies below 0.15 eV can be detected. The theoretical prediction of the voltage resolution constant falling to zero as the cut-off energy goes to zero, is of course, unphysical, and comes from the assumption that the noise has a Gaussian distribution in every channel. This assumption breaks down as the number of electrons per channel decreases (which it must do if the channel width tends to zero). Khursheed and Dinnis (1990) used a Monte-Carlo non-linear least square method to simulate the number of electrons collected per channel in a multi-channel spectrometer, for around 10,000 secondary electrons collected,

$$\frac{\chi_r}{\chi_m} = 32. \tag{5.11}$$

This factor of improvement goes up as the number of collected electrons increase, Khursheed and Dinnis (1990) estimates that it rises up to 86 for a million electrons collected.

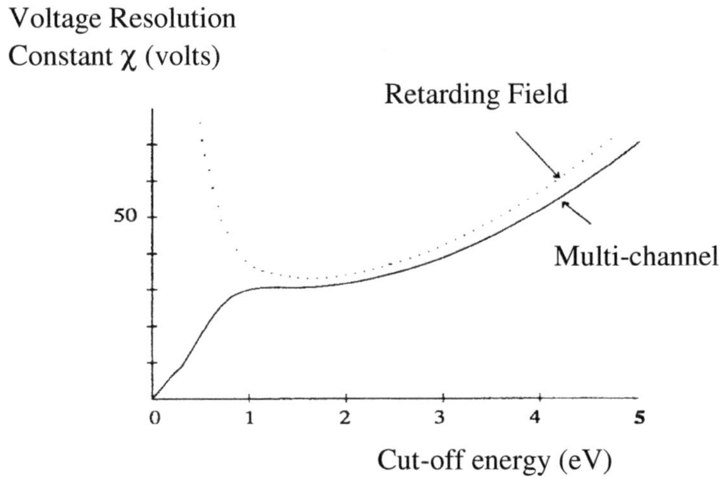

Fig. 5-10. Theoretical predictions of the voltage resolution constant χ for ideal retarding field and multi-channel spectrometers.

The presence of surface fields, where a point on the specimen is positive with respect to its surroundings, will certainly bring down the factor of improvement that multi-channel spectrometers have over retarding field spectrometers, from a signal to noise point of view. If for instance, the local voltage rise of a point on the specimen is above 2 V, and the extraction field strength is relatively small (or entirely absent), the cut-off energy will in turn be above 2 eV, and from Fig. 5-10, ideal retarding field analyzers and multi-channel analyzers are predicted to have similar voltage resolution characteristics (for the same number of collected electrons) for these conditions. However, collecting the entire secondary energy spectrum yields much more information that can be obtained through the integrated signal of a retarding field analyzer. Consider the simple feedback loop method of keeping the total output current signal constant. If the retarding field grid is zero and the specimen is biased to positive voltages, the overall current cannot be kept constant (electrons are returned to the specimen), and a large error in the output voltage measurement will be obtained. This is why many voltage contrast spectrometers functioned with a high extraction field. On the other hand, multi-channel analyzers can analyze the shape of the spectrum, not only detect changes in the overall number of collected electrons in its channels. Furthermore, band-pass/multi-channel analyzers have a distinct advantage over retarding field secondary electron analyzers in situations where the point to be measured takes high negative potential values. For insulators, the specimen surface can charge to large negative potentials, comparable in magnitude to the primary beam voltage. Monitoring changes in the total collected current does not reveal much about the way the specimen is charging in such situations. A deflection analyzer can track the SE peak as it shifts higher up in energy, and suitably adjust its pass energy, making estimates of the specimen potential as the charge builds up. Within the context of voltage contrast analyzers, multi-channel analyzers have the advantage of being inherently open-loop, that is, they do not need a feedback loop to make a voltage measurement. The feedback measurement system of retarding field voltage contrast analyzers is relatively slow, limited to the kHz range. Multi-channel spectrometers can in principle acquire spectra much faster, as long as the energy spectrum can be acquired and stored rapidly (in the MHz range).

5.4 Deflection/multi-channel analyzers

The work of Dubbledam and Kruit in the late 1980s, highlighted important signal to noise advantages of deflection/multi-channel analyzers over retarding field secondary electron analyzers (Dubbledam and Kruit, 1987b). They proposed a dual channel trochoidal analyzer, based upon the use of a magnetic immersion lens, the lower half of which has already been presented in Fig. 5-8(a). The full dual channel analyzer is shown in Fig. 5-11. This type of spectrometer uses the trochoidal field principle, consisting of crossed electric and magnetic fields where electrons travel in the direction of the magnetic field (Granneman and Van Der Wiel, 1983). This analyzer is in effect, a hybrid multi-channel-retarding field analyzer, since a retarding grid is placed in front of each detector and the output signal of each channel is an integrated signal. One disadvantage of this voltage contrast spectrometer is that the trochoidal analyzer can only be optimized for a given secondary electron energy, limiting its energy resolution.

The first fully multi-channel voltage contrast spectrometer was designed by Khursheed and Dinnis (1990). Their proposal was based upon using the time-of- flight of secondary electrons. The primary beam is pulsed, as is normally required for time resolved voltage contrast measurements on integrated circuits. The specimen is placed in a magnetic immersion lens, and the secondary electrons are collimated as they travel up the objective lens bore, as shown in Fig. 5-12. A high bandwidth detector system collects the secondary electrons at the top of the objective lens. The lens bore acts as a drift-tube region for the secondary electrons. The sharply decreasing axial field strength through which the secondary electrons initially travel will not only collimate the secondary trajectory paths, but make their time-of-flight relatively independent of their initial emission angle (as discussed at the end of Chapter 2). Under these conditions, the output signal will be directly related to the form of the initial secondary electron energy spectrum (Kruit and Read, 1983).

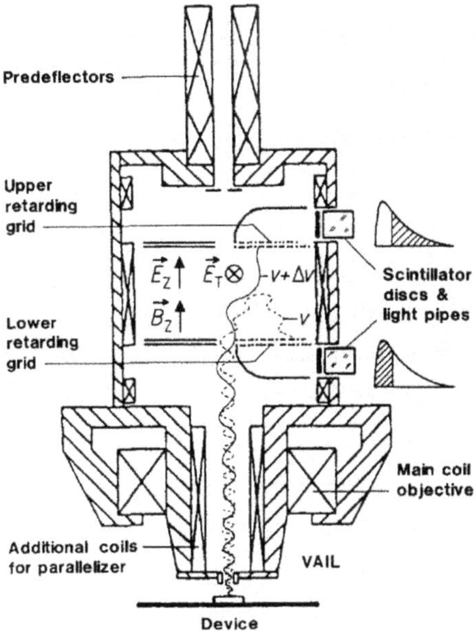

Fig. 5-11. Schematic diagram of a dual channel voltage contrast analyzer (Dubbledam and Kruit, 1987a).

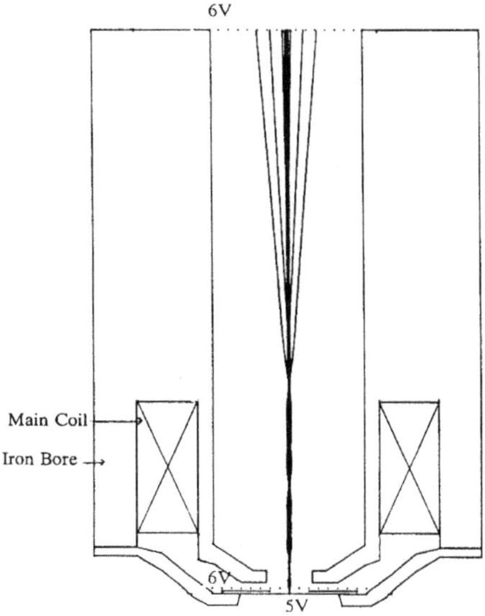

Fig. 5-12. Simulated 3 eV SE trajectory paths through a time-of-flight voltage contrast analyzer (Khursheed and Dinnis, 1990) for a wide-variety of different emission angles.

Neglecting the effect of initial emission angle, a secondary electron having an initial energy E (in eV), has a time-of-flight T given approximately by

$$T = \frac{d}{v} = \frac{c}{\sqrt{E + e(V_d - V_s)}} \tag{5.12}$$

where c is a constant directly related to the drift tube distance d, V_d is the drift tube voltage, and V_s is the specimen voltage. If the energy spectrum of secondary electrons at the specimen is $Q(E)$, then the time-probability density function for electrons arriving at the electron detector is given by

$$P(T) = \frac{dN}{dT} = \frac{dN}{dE}\frac{dE}{dT} = 2\frac{c^2}{T^3}Q(E). \tag{5.13}$$

The limited bandwidth of the detection system will tend to distort the output signal, and its influence may be approximated by a low pass filter response with a single time constant τ, in which case, the output signal $S(T)$ is given by

$$S(T) = (1/\tau)\exp(-T/\tau)\int_0^T P(x)\exp(x/\tau)dx. \tag{5.14}$$

The Chung–Everhart energy distribution (Eq. 5.6) can be used for the secondary electron energy distribution $Q(E)$. By using the above equations, the output signal of the time of flight spectrometer can be predicted.

Dinnis *et al.* (1992) used a modified Cambridge S100 SEM to test the time-of-flight spectrometer concept. A schematic drawing of the experimental layout they used is shown in Fig. 5-13. They pushed a small copper stub specimen into the middle of the S100 objective lens magnetic gap, and used a multi-channel plate detector with a high bandwidth trans-impedance head amplifier to obtain the output signal.

The final signal was displayed on a 400 MSa/s digitising oscilloscope that was also linked to a personal computer for data capture. The system bandwidth was limited to be around 10.6 MHz (τ=15 ns). Experimental results, for the specimen voltage changing in 1 V steps from –1 to –5 V and a drift tube voltage of 0 V are shown in Fig. 5-14(a) (Khursheed, 1992), 500 time channels were used and the effective drift distance was 11.9 cm. Figure 5-14(b) shows that the theoretical predictions based upon using the above equations match quite well with experiment. Using the same kind of results depicted in Fig. 5-14, Khursheed (1992) demonstrated that the time-of-flight voltage contrast analyzer gives around 30 times better signal-to-noise characteristics than retarding field contrast spectrometers for the same voltage resolution (30 times lower voltage resolution constant).

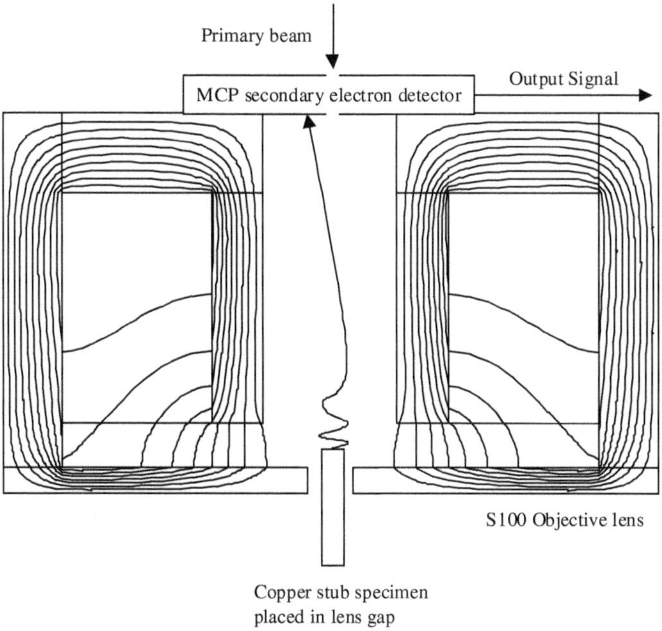

Fig. 5-13. Experimental setup of a time-of-flight voltage contrast analyzer using a modified Cambridge S100 SEM (Dinnis *et al.*, 1992).

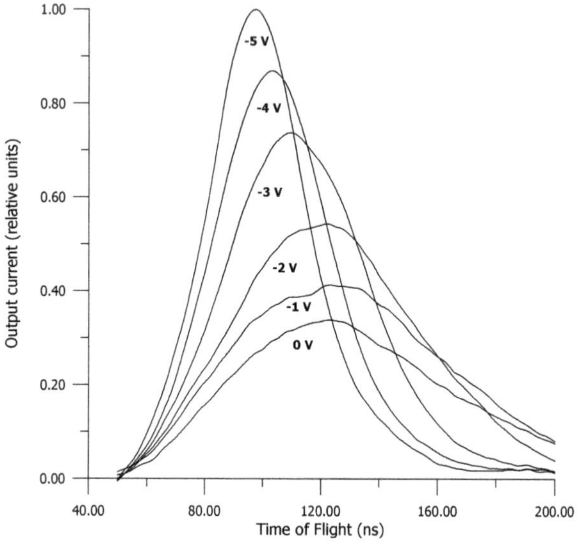

(a)

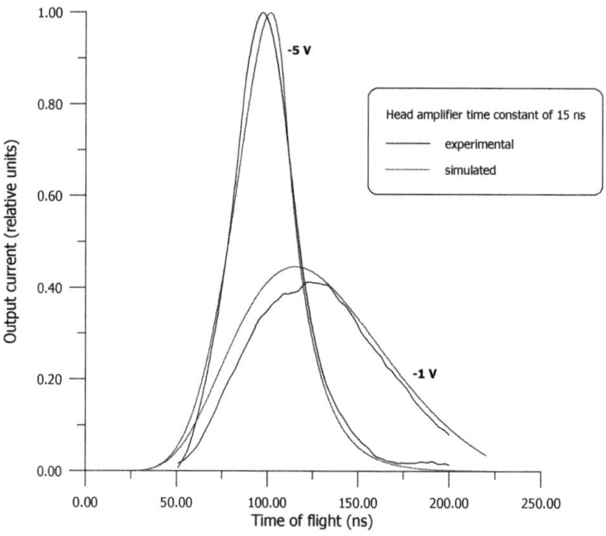

(b)

Fig. 5-14. Experimental results from a time of flight voltage contrast analyzer (Khursheed 1992): (a) Specimen voltage change from 0 to -5 V in steps of 1V, (b) Comparison of experimental signals and theoretical predictions for specimen voltages of -1 and -5 V.

Convenient ways to quantify the specimen voltage for multi-channel spectrometers can make use of the signal mean μ and variance σ

$$\mu = \int_0^{T_c} T\, f(T)dT \tag{5.15}$$

$$\sigma^2 = \int_0^{T_c} (T - \mu)^2 f(T)dT \tag{5.16}$$

where *f(T)* is the output signal probability density function. This function is obtained by normalizing each output signal *S(T)* by its total area. Obviously, with suitable post-processing, variations of signal mean μ and variance σ can be linearized, and for a given output signal, an estimate of the specimen potential can be obtained.

The time-of-flight voltage contrast analyzer appeared towards the late 1980s, at a time when interest in Electron Beam Testers began to diminish. It was from this time that integrated circuit manufacturers started to make circuits covered by a top ground plane, making it difficult for Electron Beam Testers to directly probe conductors. Since the need for e-beam quantitative voltage measurements had been the main driving force spurring developments in secondary electron SEM spectrometer designs up to that point, from 1990 onwards, very few new voltage contrast analyzer designs were reported. Since the 1990s, only two new multi-channel/deflection secondary electron analyzers have been reported.

Khursheed and Karuppiah (2001) proposed a multi-channel/deflection secondary electron analyzer in the context of developing permanent magnet high resolution add-on lens attachments. Figure 5-15 shows a schematic diagram of the experimental setup used in a conventional SEM chamber. This diagram also indicates simulated flux lines of the permanent magnet lens, designed to focus a 1 keV primary beam. The distance from the specimen to the top plate is only 0.8 mm, and the top plate hole has a diameter of 4 mm. Although the add-on lens attachment is essentially an immersion magnetic lens, it differs in several ways from previous voltage contrast analyzers that made use of magnetic immersion lenses, such as those shown in Figs. 5-8, 5-11 and 5-12. In this case, the lens field is confined to a much smaller region than in any of these previous immersion lens spectrometers, and the lens field falls by over an order of magnitude within the space of only a few millimeters, so that there is a much higher degree of collimation for low energy secondary electrons. Secondly, since the specimen is placed very close to the lens top plate (0.8 mm), this helps in filtering out wide-angle higher energy electrons (those that travel more than 2 mm off-axis through the top plate hole), reducing the presence of background noise. There is no extraction field, so when the specimen is biased with positive voltages, a potential barrier between the specimen and top plate (0 V) causes lower energy secondary electrons to return to the specimen.

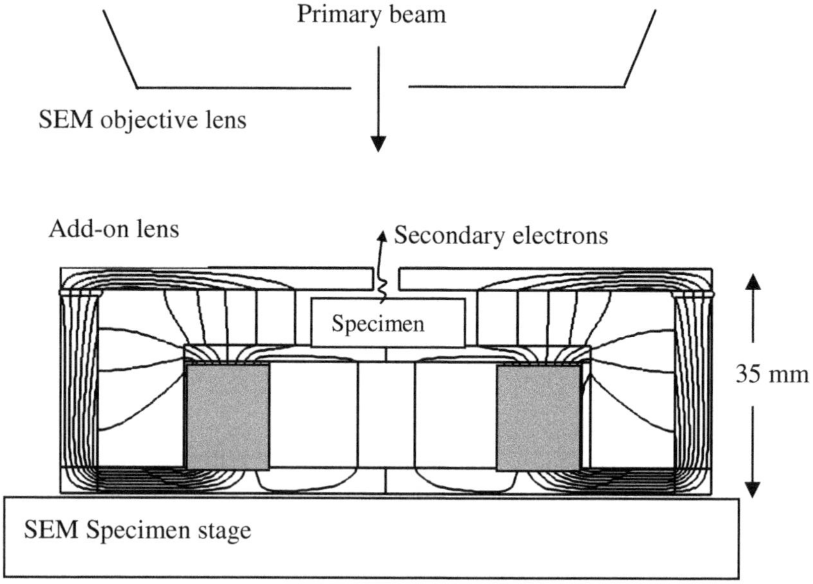

Fig. 5-15. Schematic diagram of an add-on immersion lens experimental setup within the specimen chamber of a conventional SEM. Simulated flux lines for the add-on lens are indicated for the focusing of a 1 keV primary beam. The ray path of the SE trajectory is deliberately exaggerated for reasons of scale.

Figure 5-16 shows simulated trajectory paths for secondary electrons having initial energies of 1, 3 and 5 eV that are emitted over a wide-range of emission angles at the specimen and that are subsequently deflected by a Wien filter located above the lens top-plate. The Wien filter is designed to deflect secondary electrons while leaving the primary beam unaffected. The trajectory paths shown in Fig. 5-16 indicate that since low energy electrons experience a high degree of collimation, dispersion due to initial emission angle is expected to be small, and this gives rise to the possibility of detecting the energy spectrum in parallel via the use of a multi-channel detector. On the other hand, high energy secondary electrons or backscattered electrons will not be collimated, and not able to exit from the lens. Nor will they be deflected on to the detector by the Wien filter. The background noise level on the output signal is therefore expected to be relatively small.

Khursheed and Karuppiah (2001) generated secondary electron spectra from the add-on immersion lens spectrometer by using it in the band-pass energy mode, where an effective slit is cut into a foil that surrounds

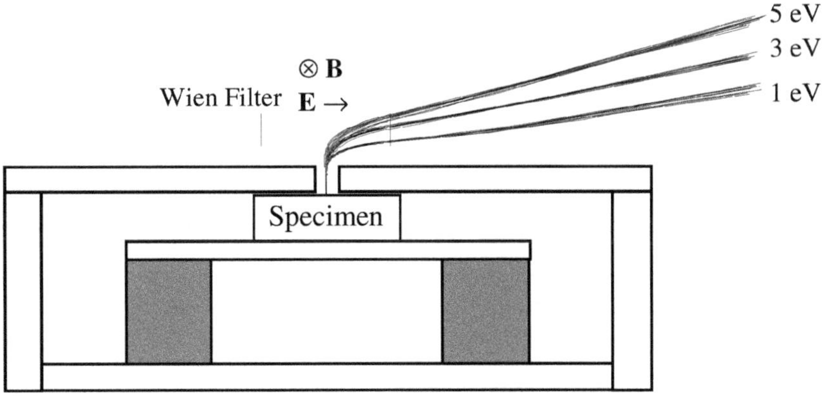

Fig. 5-16. Direct ray tracing of low energy secondary electrons through the immersion add-on lens with a Wien filter placed above it. The secondary electrons are emitted over a wide-range of emission angles at the specimen. The Wien filter has a radius of 1cm, with 1 AT magnetic excitation.

the scintillator cage, as shown in Fig. 5-17. Instead of a Wien filter, an electrostatic deflector was found to be sufficient. Since only low energy secondary electrons are collected, relatively small deflection voltages on electrostatic deflectors are required (less than 12 V), which typically shift the primary beam, but do not degrade the image resolution. These shifts can be minimized by designing the deflector to have a compensating electric field in its upper part, as shown in Fig. 5-17. The deflector consisted of a single cylindrical wall plate having a diameter of 30 mm, which had a rectangular opening in its lower portion to allow secondary electrons to exit and be attracted to the scinitllator cage. An opening on the top portion compensates for deflection on the primary beam, see Khursheed and Karuppiah (2001) for more details on the deflector.

Experimental secondary electron spectra obtained by varying the deflector voltage from 0 to -13 V for specimen voltages of -5, 0 and 5 V are shown in Fig. 5-18(a). They show that a truncated secondary electron spectrum is obtained, where high energy electrons are filtered. Not only is there considerable voltage contrast for the left-hand edge of the spectrum when the specimen changes from 0 to -5 V, but there is also significant changes to the right-hand side, indicating that the total number of secondary electrons reaching the scintillator as the specimen is biased more negatively is higher. This most likely comes from that fact that the pass energy is higher for negative specimen potentials, effectively allowing a wider band of energies to be transmitted through the slit.

To obtain an estimate of the minimum detectable specimen voltage difference associated with the output signals shown in Fig. 5-18(a), 16 signals were recorded for a specimen voltage of –5 V. The mean and variance of each signal were calculated, and an rms noise value for the mean was found (Khursheed and Karuppiah, 2001). The rms noise value for the signal mean, 0.01485 V, was then compared with the change in mean value for the specimen voltage switching from –5 to –4 V. The output signals corresponding to the specimen voltages of –5 and –4 V are shown in Fig. 5-18(b).

The change in the mean value for the specimen changing from -5 to -4 V was found to be 0.2751 V. This provides a signal to noise ratio of 18.52, giving a minimum detectable specimen voltage difference of 54 mV. The noise level is of course, dependent on data acquisition time, which in this case was 0.16 seconds for each signal. Since the spectrometer's signal to noise ratio depends on many operating factors such as the primary beam current and detector efficiency, it is difficult to compare it to other voltage contrast systems. However, these results obtained from the add-on magnet immersion lens spectrometer show that it can provide significant voltage contrast and that it should be able to operate as a multi-channel secondary electron analyzer.

Figure 5-19 shows that the secondary electron immersion add-on lens spectrometer is able to provide quantitative material contrast. Experimental secondary electron spectra were taken from a specimen having copper and brass regions separated by glue. The specimen is coated with a 25 nm layer of carbon. In this case, the deflection voltage is synchronized with the vertical slow scan, and varies from 2 to -18 V.

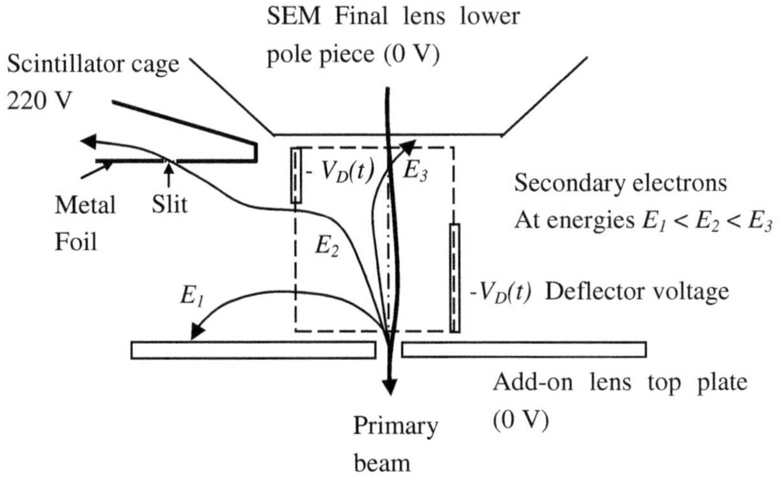

Fig. 5-17. Schematic layout of an experiment operating the immersion add-on lens spectrometer in band-pass mode (Khursheed and Karuppiah, 2001). The central cross-section of the deflector is shown.

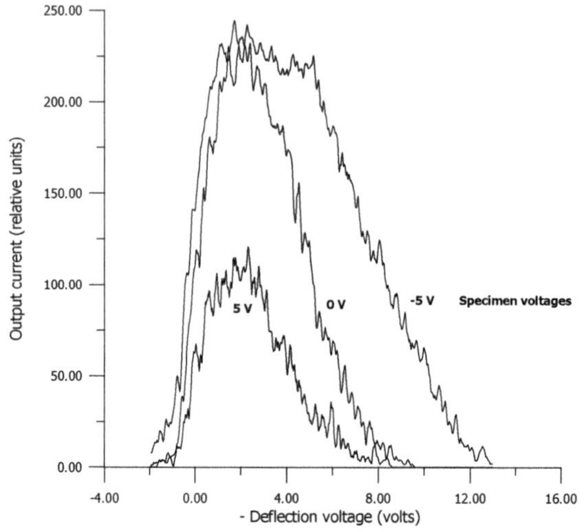

(a)

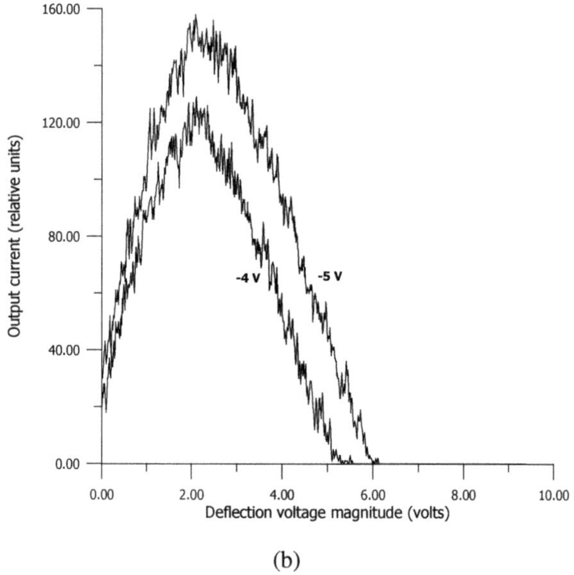

(b)

Fig. 5-18. Experimentally measured secondary electron spectra for the immersion add-on lens analyzer (Khursheed and Karuppiah, 2001): (a) Specimen voltages of -5, 0 and +5 V, (b) Specimen voltages of -4 and -5 V.

The secondary electron image, shown in Fig. 5-19(a), shows that the maximum intensity is obtained for a deflector voltage of around -2.5 V. By examining vertical line scans taken on either side of the glue junction, shown in Fig. 5-19(b), considerable contrast is obtained between brass and copper regions, particularly with respect to the change of the output signal width. This contrast between spectra was found to be invariant to the sample's orientation, confirming that it is a genuine material contrast effect, and is not an artifact of the spectrometer. The contrast was quantified by recording 16 such lines scans from the secondary electron image, and calculating the sample mean and standard deviation. The standard deviation was found to be 0.0256 V for copper line scans, while the change in standard deviation between the copper and brass signals was calculated to be 0.3, giving an approximate signal to noise ratio of 11.72.

Apart from the secondary eletron multi-channel/deflection magnetic immersion lens spectrometer attachment reported by Khursheed and Karuppiah (2001), the other main contribution towards multi-channel SE analyzer design since the 1990s has come from Kienle and Plies (2004). Kienle and Plies report a secondary electron analyzer design for a mixed field objective lens, where the primary beam is retarded just before it reaches the specimen, and additional focusing is provided by a single-pole/wide bore immersion lens. As discussed in Chapter 4, this type of mixed field lens gives the highest image resolution of all immersion lens designs. In the analyzer reported by Kienle and Plies, the primary beam travels through most the column at around 9 kV, and is decelerated down to a landing voltage of 1 kV at the specimen. Secondary electrons are typically accelerated to 8 kV as they travel back up the column, and are deflected off-axis by a Wien filter, as shown in the schematic diagram shown Fig. 5-20.

Another Wien filter is placed further up the column, in order to cancel out adverse energy dispersion and second-order geometrical aberration effects on the primary beam, the remaining astigmatism is corrected using a conventional stigmator. The two Wien filters are excited in

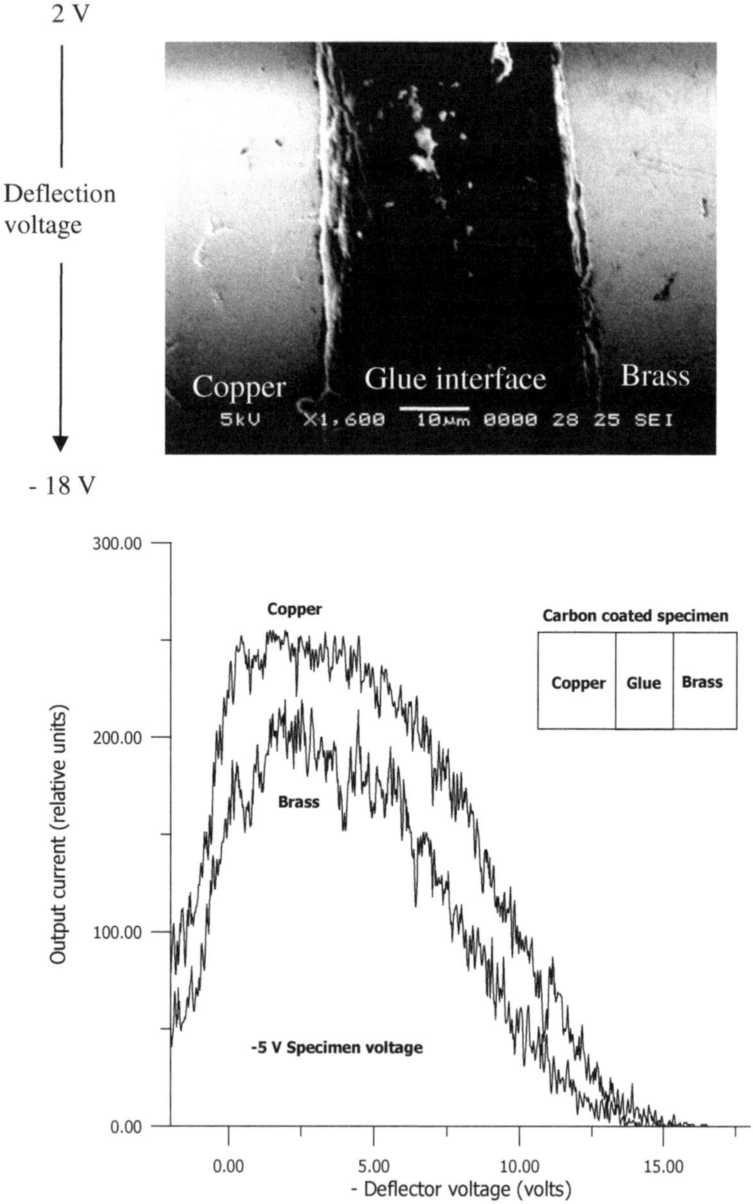

Fig. 5-19. Experimental material contrast results with the immersion add-on lens secondary electron spectrometer on a copper-brass specimen (Khursheed and Karuppiah, 2001): (a) Secondary electron image, (b) Vertical line scans.

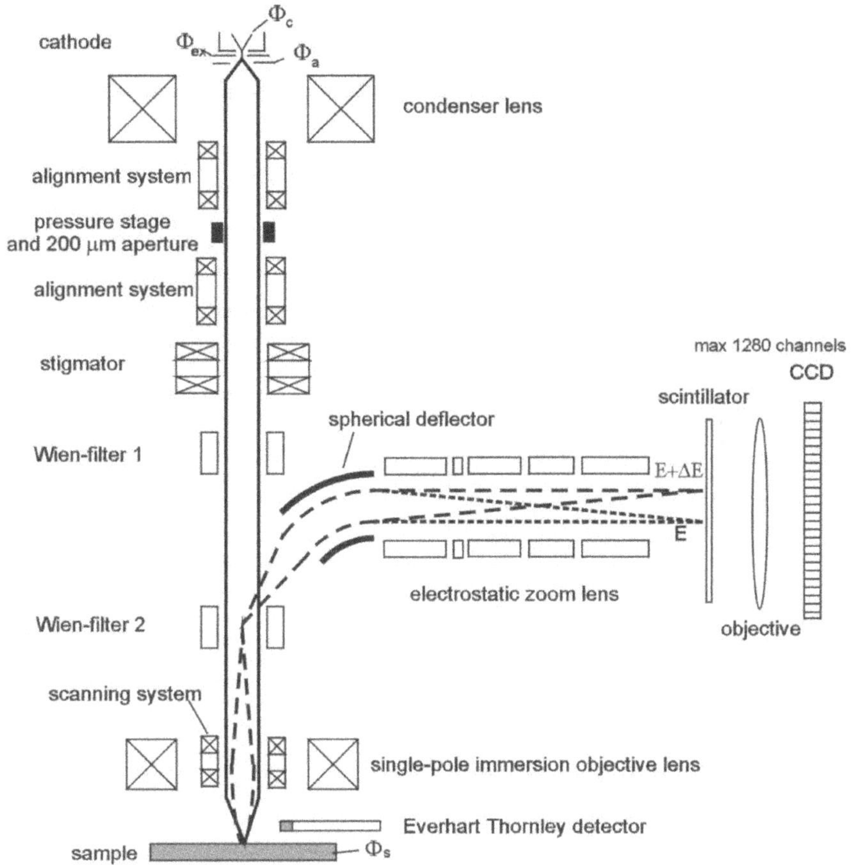

Fig. 5-20. Schematic latout for the multi-channel secondary electron off-axis analyzer reported by Kienle and Plies (2004).

opposing directions with respect to one another. The secondary electrons after having been deflected by the lower Wien filter are further deflected and dispersed by an electric spherical deflector analyzer (SDA), after which their image is magnified and focused on to a YAG scintillator. The photon image from the YAG scinitillator is further magnified by a light optical objective lens before it is detected on a CCD camera. The CCD camera acts as a position sensitive detector which simultaneously captures the secondary electron spectrum from 0 to 20 eV over 1,280 channels.

Several stages of magnifying the energy dispersion are required in this analyzer, since the lower Wien filter only deflects the secondary electron beam by 15°, and the SDA is limited to a deflection angle of 75°.

Simulated trajectory paths for secondary electrons through the analyzer, reported by Kienle and Plies 2004), are shown in Fig. 5-21, while experimental secondary electron spectra for the specimen voltage changing between 0 to -5 V in -1 V steps are shown in Fig. 5-22. The experimental voltage resolution is around 50 mV with an average data acquisition time of 0.5 s per spectrum. The analyzer is able to completely separate secondary electron and backscattered electron signals from one another. This analyzer, like the secondary electron time-of-flight analyzer, is fully multi-channel and is able to capture spectral energy information in parallel.

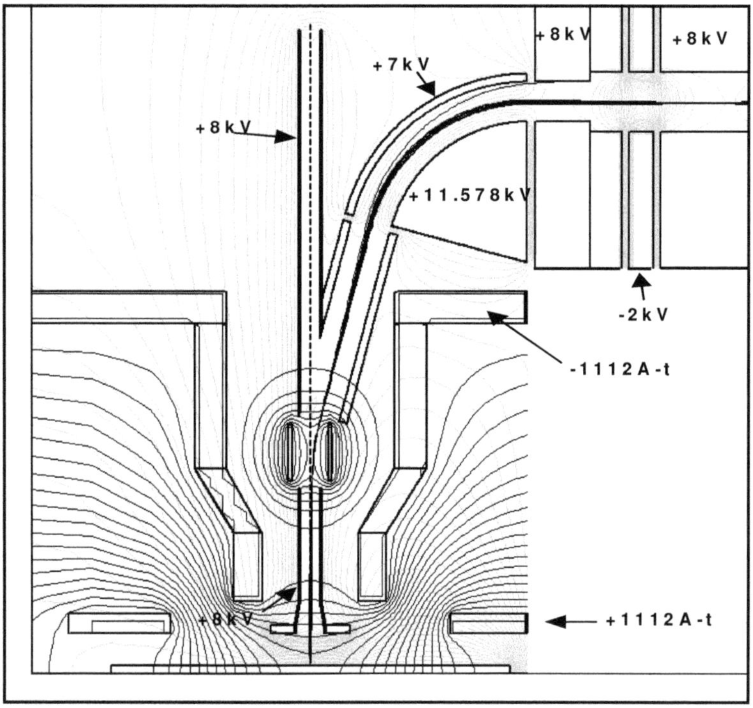

Fig. 5:21. Simulated trajectories through the multi-channel off-axis analyzer, as reported by Kienle and Plies (2004), courtesy of E. Plies.

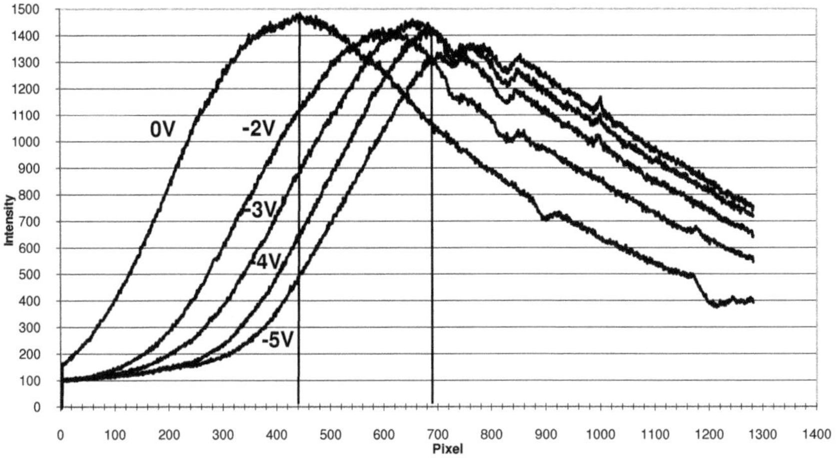

Fig. 5-22. Experimental secondary electron spectra for the multi-channel off-axis analyzer as reported by Kienle and Plies (2004), courtesy of E. Plies.

Some recent work by Kazemian *et al.* (2006, 2007) on measuring shifts in the secondary electron spectra has been carried out in the context of quantitative two-dimensional dopant mapping of *pn* junction specimens. Kazemian *et al.* use a field emission magnetic immersion SEM, where an electrostatic deflector and off-axis detector are inserted into the upper-pole piece, as shown in Fig. 5-23. The specimen is placed below the upper pole-piece either at the peak position of the axial magnetic field or just below it. In this SEM, the magnetic gap of the lens is the specimen chamber, with chamber walls defining the lower pole-piece. An extraction tube is placed into the conical tip of the upper pole-piece, and the deflector consists of three parts, a lower, middle and upper section, where the middle section is biased oppositely to the lower and upper parts. This deflector design provides an electric deflection field for secondary electrons, but has little or negligible effect on the primary beam. The analyzer functions as a retarding field one, where the deflector voltage is swept over a certain range, and shifts in S-curves (integrated secondary electron spectra signals) are detected. The S-curves are quite irregular in shape, since the action of the electric deflection field filters the higher energy secondary electrons.

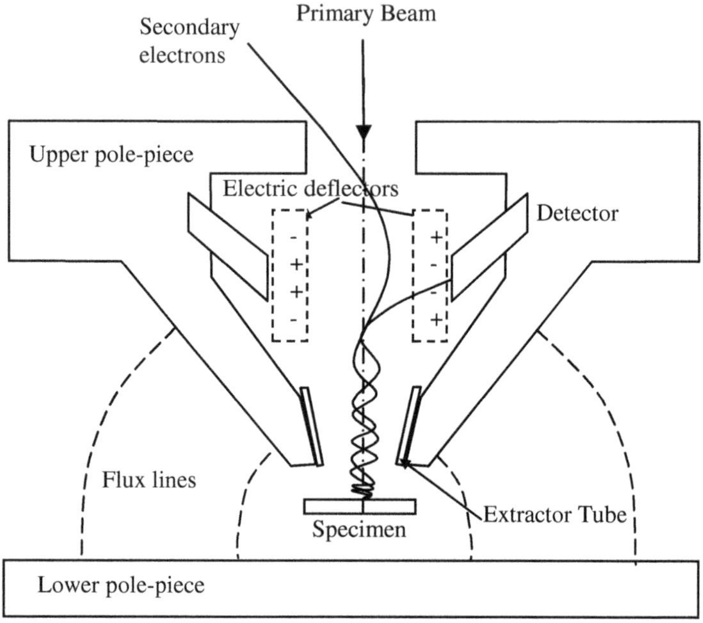

Fig. 5-23. Schematic magnetic immersion lens SE analyzer layout of Kazemian *et al.* (2006, 2007) used for quantitative dopant mapping.

Although simulations have been carried out (Kazemian *et al.*, 2007), very few details are given about the lens field distribution, exactly where the specimen is placed in relation to the peak magnetic field, and the form of the extraction field distribution created between the specimen and extractor tube located in the lens upper bore. S-curves were generated for extractor tube voltages of 20 V and 250 V and the deflector voltage typically ranged from 0 to 60 V. Simulated S-curves confirmed the experimentally measured ones, and showed that they were irregular in shape (deviating from a regular S-shape), therefore monitoring of shifts in the energy spectrum was limited to a linear region. Experimental results were able to map dopant concentrations ranging from 2.8×10^{17} to 7.5×10^{19} cm^{-3} with an accuracy of ±8.5%, and the surface potential across the *pn* junction was measured to be 0.72 ± 0.15 V (Kazemian *et al.*, 2006). A later attempt measured the surface potential to be 0.81 V ± 0.1 V (Kazemian *et al.*, 2007).

Although the work of Kazemian *et al.* do not make any significant advances on secondary electron energy spectrometer design, they demonstrate that by monitoring changes in the secondary electron spectrum, dopant concentrations in semi-conductor specimens can be quantitatively mapped. These results highlight the possible emergence of a new application for quantitative secondary electron voltage contrast, beyond the measurement of signals on Integrated Circuits. With better analyzer design, based upon the use of say band-pass or multi-channel analyzers, improvements in signal-to-noise can be expected, yielding more accurate dopant concentration measurements.

Chapter 6

Full Range Deflector Spectrometer Designs

There have been very few full range energy spectrometers proposed for the detection and capture of scattered electrons inside the SEM. The full energy range typically lies between 0 to 30 keV in most conventional SEMs, starting with sub eV secondary electrons, extending up to elastic backscattered electrons (primary beam energy). A spherical retarding field analyzer was used by Matsukawa *et al.* (1974) to capture the BSE spectrum between 10–30 keV in a custom built specimen chamber. Retarding field analyzers are however, generally not ideally suited for BSE spectral acquisition due to the need to employ high negative voltages on the filter electrode, which for elastic BSEs, are comparable in magnitude to the primary beam voltage.

Although it is possible to use either a 90° spherical electric/magnetic sector deflector, to record the full energy spectrum on flat plane detectors, with tilted/untilted specimens, as shown in Figs. 6-1(a) and 6-1(b), the overall collection efficiency is small (fractions of a percent). This comes from limited collection in the azimuthal angular direction, which must typically be less than ±3° for relative energy resolutions below 1%. Making an array of such analyzers in the azimuthal angular direction is in principle possible, but due to space constraints within the SEM chamber, practical ways of integrating and miniaturizing them are also required.

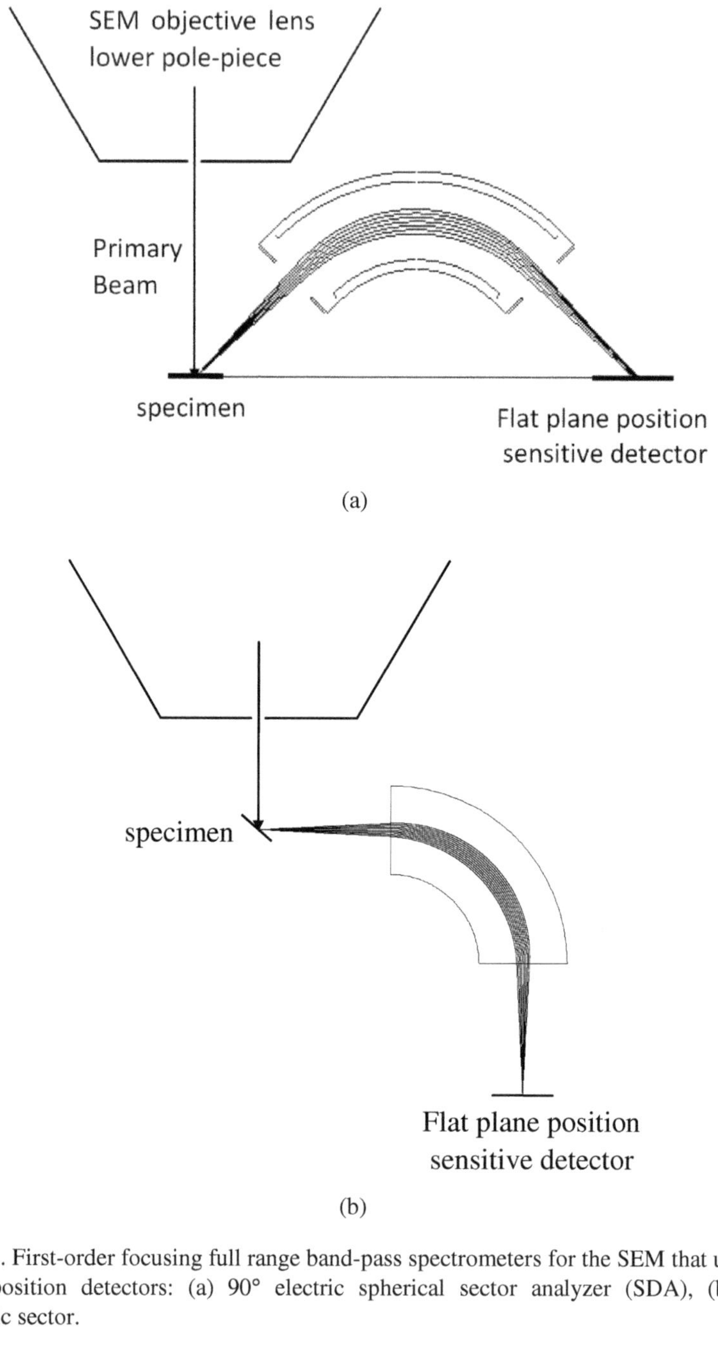

Fig. 6-1. First-order focusing full range band-pass spectrometers for the SEM that use flat plane position detectors: (a) 90° electric spherical sector analyzer (SDA), (b) 90° magnetic sector.

6.1 First-order focusing toroidal analyzers

Rau and Robinson (1996) proposed an electrostatic toroidal deflection analyzer attachment for SEMs designed to capture BSE energy spectra up to 30 keV. The geometry of this analyzer is a 65° deflection toroidal one, as shown in Fig. 6-2(a). Unlike most previous Toroidal Deflector Analyzers (TDAs), this analyzer has 2π transmittance in the azimuthal direction, and electrons are deflected in the r-z plane. In the past, most TDAs were designed to accept a beam of electrons over small polar and azimuthal angles (a few degrees), and deflection occurred in the azimuthal plane.

Direct ray tracing of 10 keV BSEs emitted over an angular range from -3° to 3° are shown in Fig. 6-2(b) (Rau *et al.*, 2002) for an inner deflection plate radius of 16 mm, and an outer deflection plate radius of 18 mm (central radius of 17 mm). Simulated deflection plates voltages of ±1177.8 V were required, and unlike the SDA or CDA, the central ray does not travel along the central radius of the deflector. The analyzer design was optimized through simulations of the kind shown in Fig. 6-2(b), which predict that the best analyzer relative energy resolution is around 0.25% (half width), corresponding to a base resolution of around $2\alpha^2$ (0.5% for ±3°), similar to the theoretical base energy resolution of an ideal 90° SDA.

Figure 6-3(a) shows simulated spectral predictions of the BSE spectrum for two different design layouts of the analyzer. Here, BSE trajectory paths are simulated for a large number of emission angles ranging between -3° to 3° and emission energies from 9.9 to 10.1 keV (pass energy at 10 keV). Curve 1 corresponds to the situation of the analyzer deflection angle Φ being 65°, while for curve 2, it is 59°. The output focal point distance (beyond the analyzer exit) was found to be 4 mm in the first design, while for the second design, it is moved to a distance of 3.8 mm. The energy width at half height is 70 eV for curve 1, and 25 eV for curve 2, corresponding to respective relative energy resolutions of 0.7 and 0.25%. The best base (full-width at base) resolution is approximately 0.5% (for an angular spread of ±3°).

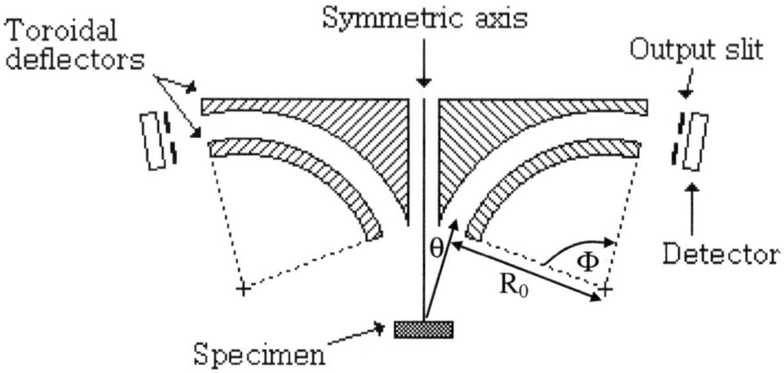

(a)

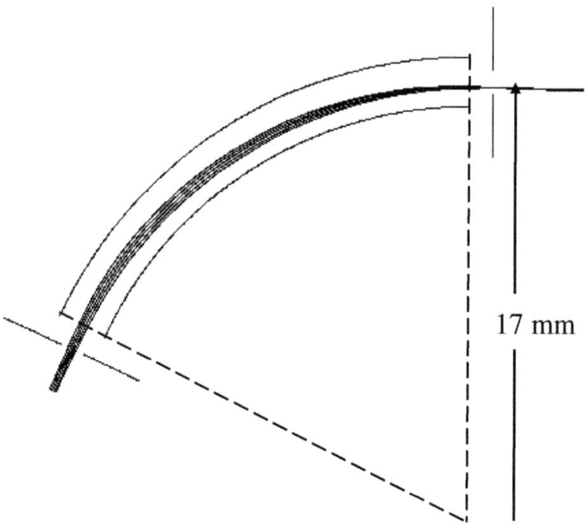

(b)

Fig. 6-2. First-order toroidal focusing toroidal analyzer: (a) Schematic layout diagram (Rau and Robinson, 1996), (b) Direct ray tracing of trajectory paths over a -3 to 3° input angular range (Rau *et al.*, 2002).

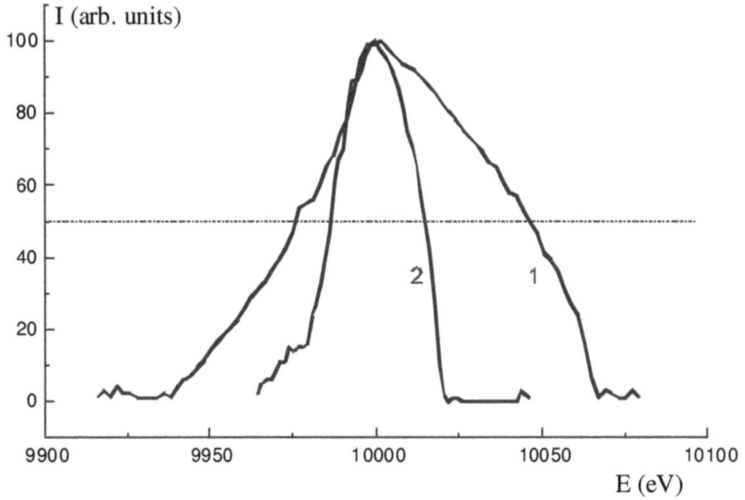

(a)

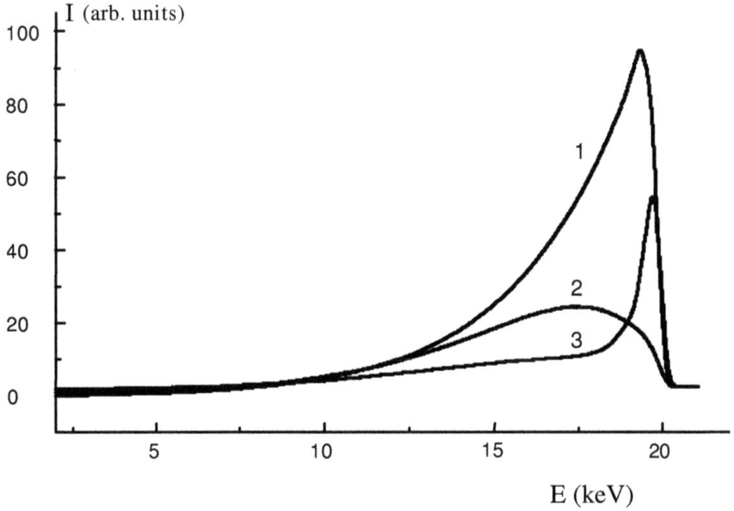

(b)

Fig. 6-3. Energy spectra from the first-order focusing toroidal analyzer (Rau *et al.*, 2002): (a) Simulated spectra. Curve 1 has a deflection angle of 65°, curve 2 has a deflection angle of 59°, (b) Experimentally measured spectra for different specimens using the toroidal analyzer. Curve 1 is for bulk gold, curve 2 for copper, and curve 3 is for a 25 nm gold layer on silicon substrate.

Figure 6-3(b) depicts experimental BSE spectra from the first-order focusing toroidal spectrometer designed by Rau *et al.* (2002) for different samples at a primary beam energy of 20 keV. The values of the detected signal are displayed in relative units. The more elastic BSEs are represented by the right-hand side of Fig. 6-3(b) and correspond to electrons emitted close to the sample surface. The more inelastic BSEs are represented by the left-hand side of Fig. 6-3(b), and correspond to electrons that are emitted from deeper within the sample. Curve 1 is the BSE spectrum for bulk gold, curve 2 for copper, and curve 3 is obtained from a sample consisting of a silicon substrate on which a thin gold layer 25 nm thick was deposited by sputtering. Rau *et al.* (2002) used these spectral results to estimate the thickness of the gold layer on the silicon specimen. The gold-silicon interface approximately corresponds to the position of maximum gradient in curve 3, which occurs at 18.76 keV, and from this, the thickness of the gold layer was estimated to be 29 nm (Rau *et al.*, 2002), accurate to within 16%. These results demonstrate that acquisition of the BSE spectrum can help provide depth information of multi-layer specimens. The same spectrometer design was also recently used to acquire the energy spectrum of highly negatively charged specimens, where surface potentials can charge to around -5 kV (Belhaj *et al.*, 2008).

The output slit in the first-order focusing toroidal analyzer design reported by Rau and Robinson (1996) cannot easily be replaced by a detector in order to capture multi-channel energy spectral information, since it needs to lie on a cylindrical wall shaped surface. Figure 6-4(a) shows simulated ray tracing of trajectory paths through an alternative 2π collection toroidal analyzer design (proposed by the author), one in which the focal plane is flat. In this case, a flat ring shaped detector is placed at the analyzer output focal plane and an energy pass range of typically -5 to 5% of the pass energy can be captured in parallel (multi-channel), similar to hemispherical deflector analyzers. Trajectory paths of electrons are plot through this analyzer over an angular range of ±3° (about the polar angle of 45°) at the pass energy E_P. The pass energy was found to be 5.01 times the voltage on the deflection plates, $\pm V_D$. Detailed examination of the ray path around the exit focal plane shows the analyzer is predicted to have a relative base resolution (full width) of approximately 0.56%, as shown in Fig. 6-4(b), similar to the toroidal analyzer designed by Rau and Robinson (1996), and close to the theoretical value predicted for the 90° SDA ($2\Delta\alpha^2$). The transmittance of this analyzer is around 10% (for an angular spread of ±3°).

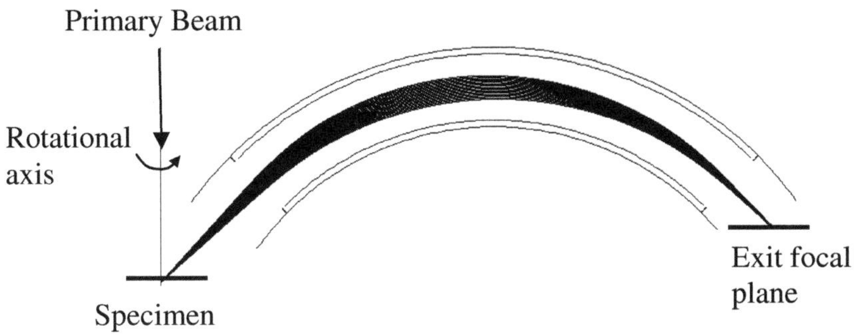

(a)

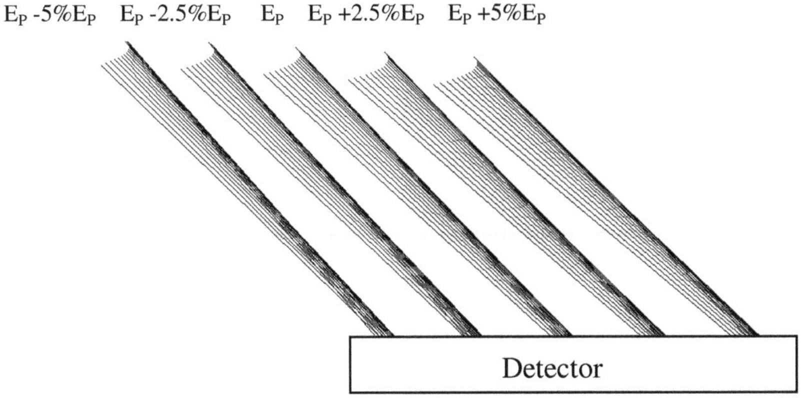

(b)

Fig. 6-4. Simulated trajectory paths for a flat detector plane toroidal deflector analyzer design. 17 trajectories are plot in uniform steps for an angular spread between -3 to +3°: (a) Through analyzer at the pass energy E_P, (b) Around the detection plane for energies ±5% around the pass energy E_P.

6.2 A Second-order focusing toroidal analyzer design

A second-order focusing 2π collection toroidal analyzer design was reported by Khursheed and Hoang (2008a) for the SEM. The principle of the analyzer is to form an intermediate focus point in the r-z plane, and to have an exit slit at a second focal point. This is illustrated by simulated trajectory paths in Fig. 6-5(a). The spectrometer's plates are centered at point 0. This design allows for second-order spherical aberration contributions accumulated before and after the intermediate focus to cancel, leaving third-order spherical order aberrations at the exit focal point, as shown in Fig. 6-5(b).

In the first version of the spectrometer (Khursheed and Hoang, 2008a), the analyzer was simulated to have a relative energy resolution of 0.146% (around 0.3% base resolution) for an input angular spread of $\pm 6^{o}$ (corresponding to 20% transmission), which is about 7 times better than that of first-order focusing electrostatic toroidal spectrometers (for the same input angular spread). This energy resolution is comparable to the energy resolution of the second-order focusing cylindrical mirror analyzer (CMA) for the same entrance acceptance angles (between $\pm 6^{o}$).

The analyzer also has the capability of operating in a parallel energy acquisition mode with a maximum band energy of ±15% of the pass energy. Figure 6-6(a) shows simulated ray tracing of central rays with initial energies ranging from 90 to 110% of the pass energy, predicting the existence of an achromatic point beyond the analyzer exit, while Fig. 6-6(b) shows the corresponding simulated dispersion at the exit plane due to energy spread. Note that the spectrometer trace-width and energy dispersion are normalized to the radius of the inner plate, R_1.

Figure 6-7 shows simulated trajectory paths at the analyzer detector plane for 11 emission energies ranging from 95% to 105% of the pass energy and 11 input angles from -52 mrad to 52 mrad around the central ray in uniform steps, the detector plane has an angle of 26.4° with respect to the horizontal axis, a relatively shallow angle, which allows for the

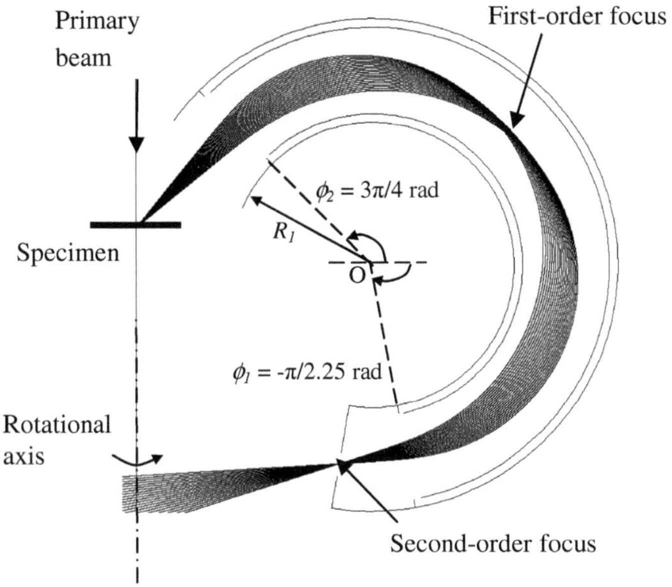

(a)

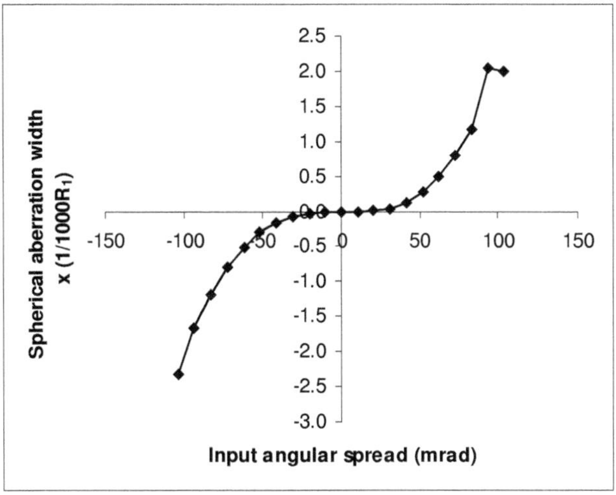

(b)

Fig. 6-5. Spherical aberration simulations for a second-order toroidal spectrometer design (Khursheed and Hoang, 2008a) normalized to R_1, the inner plate radius: (a) Direct tracing at pass energy for an angular spread between -6 to 6°, (b) Trace-width as function of input angular spread.

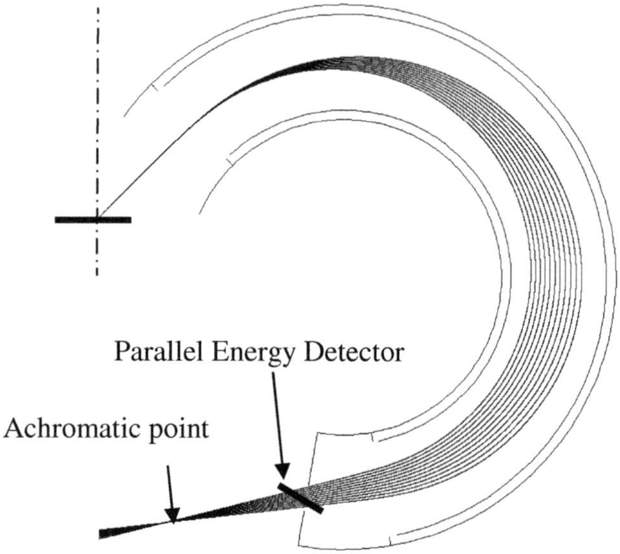

(a)

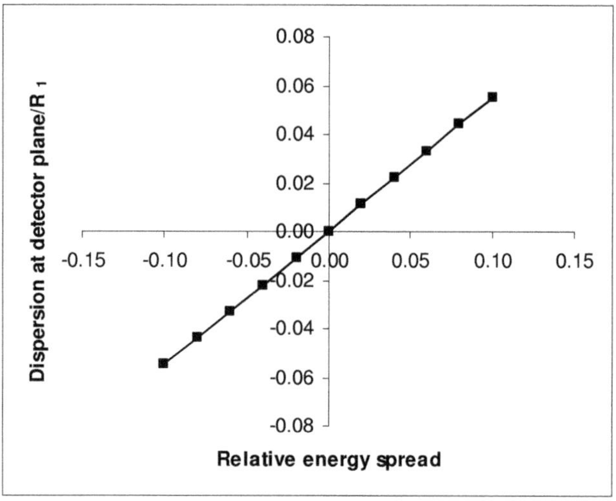

(b)

Fig. 6-6. Energy dispersion simulations for the Khursheed and Hoang (2008a) second-order focusing toroidal spectrometer: (a) Direct tracing at for energies ranging from 90 to 110% of the pass energy, (b) Trace-width as function of energy dispersion.

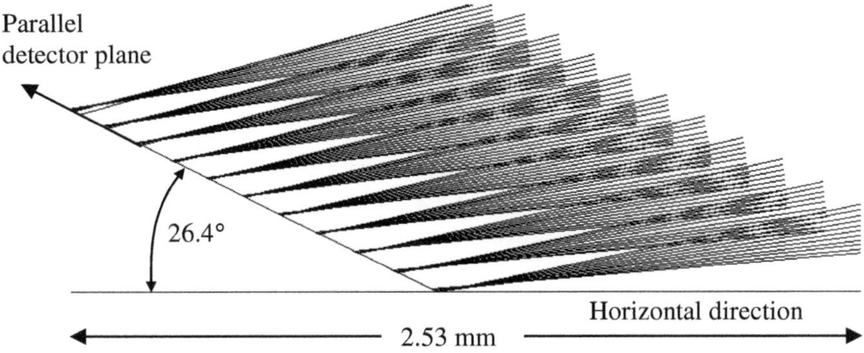

Fig. 6-7. Simulated trajectories around the output focal plane of the Khursheed and Hoang (2008a) toroidal spectrometer. There are 11 emission energies ranging from 95% to 105% of the pass energy and 11 input angles from -52 mrad to 52 mrad around the central ray in uniform steps.

placement of an array of straight detectors in the azimuthal direction, without incurring significant errors (Khursheed and Hoang, 2008a).

In a second version of the spectrometer design (Hoang and Khursheed, 2009), an accelerating pre-focusing lens is used to improve the predicted energy resolution for a given entrance angular spread by an order of magnitude (0.02% for $\pm 6^{\circ}$ entrance angular spread).

Khursheed and Hoang (2008a) propose a scheme for deflecting low energy electrons (less than 50 eV) on to a flat plane detector for parallel energy detection located below the specimen, where electrons are mirrored up towards the specimen after exiting the analyzer.

Figure 6-8 shows the layout of a second-order focusing toroidal analyzer prototype attachment in the SEM. In this case, collection in the azimuthal direction is restricted to around 90°, due to restrictions in space imposed by the specimen stage. A minimum working distance of 15 mm can be achieved with this spectrometer layout. Scattered electrons are detected by the normal SEM SE detector, while electrons that pass through the spectrometer are simultaneously collected by a Photo-multiplier (PMT) detector placed on the specimen stage.

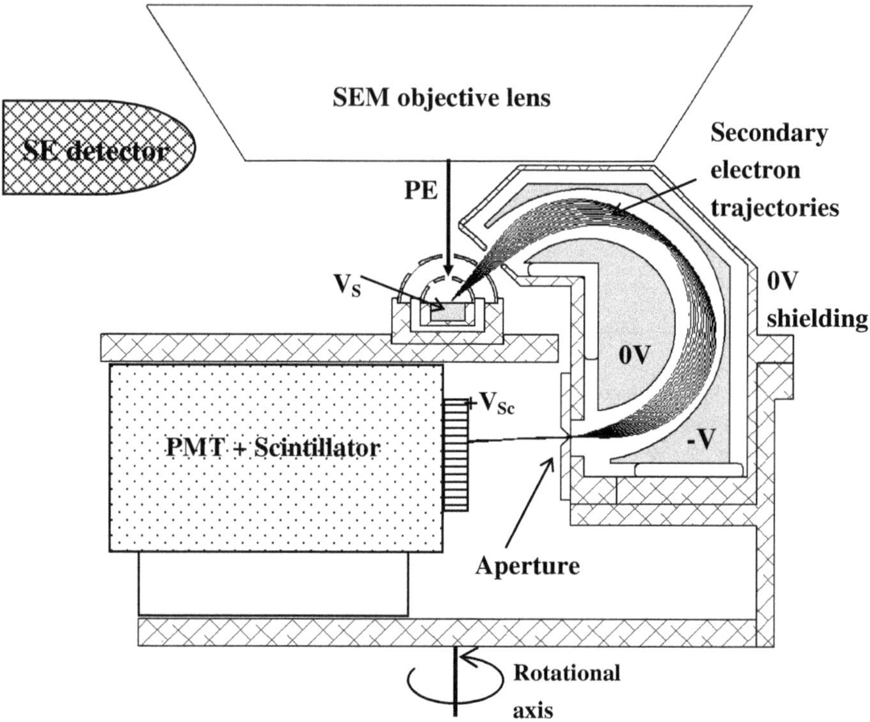

Fig. 6-8. Layout of a second-order focusing toroidal analyzer prototype attachment in the SEM.

Also shown in Fig. 6-8 are small hemispherical metal caps with slit apertures placed above the specimen. They help to define the spectrometer entrance angular spread, and allow for specimen biasing to improve secondary electron extraction. The specimen is electrically connected to the inner hemispherical electrode, so that as the specimen is biased negatively, secondary electrons are pre-accelerated into the spectrometer. The hemispherical cap electrodes therefore act as an electrostatic extractor for secondary electrons.

Figure 6-9 shows some initial experimental secondary electron spectra obtained for the second-order focusing analyzer attachment used in a JEOL 5600 SEM. The SE spectra is shown in Fig. 6-9(a). Figure 6-9(b) illustrates that the noise level on the steep part of the SE spectrum (for

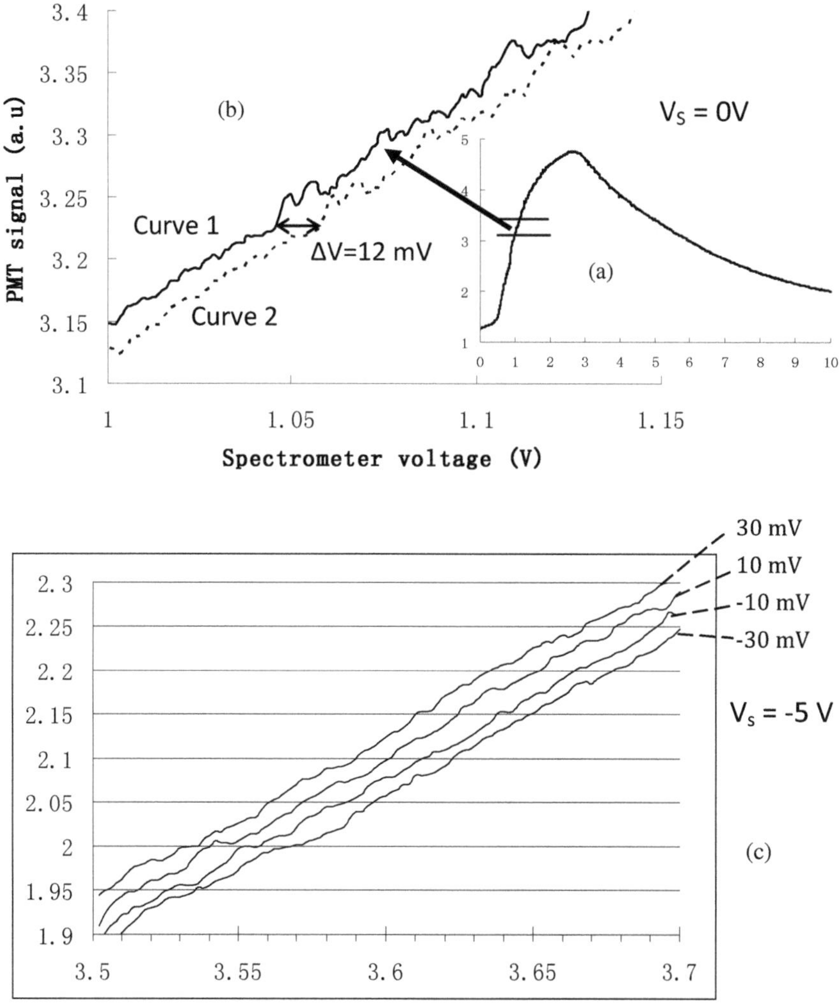

Fig. 6-9. Experimental SE spectra obtained from a second-order focusing toroidal analyzer attachment: (a) SE spectra up to 10 eV at 0V specimen bias, (b) Magnified SE spectra. Dotted line represents a shift of 12 mV, (c) Magnified SE spectra for the specimen bias changing in 20 mV steps around -5 V.

zero volt specimen bias) allows for shifts below 12 mV to be detected, while Fig. 6-9(c) shows how a magnified part of the SE spectrum shifts when the specimen bias voltage changes in 20 mV steps around a specimen bias of -5 V. These results indicate that minimum detectable specimen voltage changes below 5 mV can be monitored.

Figure 6-10(a) shows how the output signal changes as the specimen and inner electrode are biased negatively, ranging from 0 to -24 V in steps of -4 V. Compared to the 0 V bias case, the output signal not only shifts to the right (as expected), but it grows significantly in height, developing an increasingly sharp symmetric peak value. The increase in signal height comes from the fact that negative specimen voltages accelerate secondary electrons, giving them a higher pass energy, effectively broadening the energy range of electrons that pass through the output slit. There is also an increase in the number of secondary electrons that enter the spectrometer, due to the focusing action of the first entrance aperture. The electric field generated between the first and second hemispherical electrodes tends to penetrate into this aperture, pulling through secondary electrons that would otherwise strike the inner surface of the first hemisphere electrode. The output signal now comprises of a wider range of electron energies, no longer directly related to the secondary electron spectrum.

The signal-to-noise ratio when using negatively biased specimen voltages is significantly better than for the unbiased case. This is illustrated by experimental signals shown in Fig. 6-10(b), where the bias voltage on the specimen/inner hemispherical electrode changes from -10 to -10.1 V. In this case, the deflector voltage is restricted to a small range (1 V) around the peak signal value. The presence of shot noise on these signals is clearly relatively small.

An estimate of the signal-to-noise ratio for the experimental signals shown in Fig. 6-10(b) can be made by calculating the expectation function value (mean) for each signal, 7.49405 V for V_s = -10 V, and 7.511865 V for V_s = -10.1 V, giving a change of 17.8154 mV. In comparison, the change in expectation value due to shot noise from a smoothed fit for the signal at V_s = -10.1 V is calculated to be 6.272 μV,

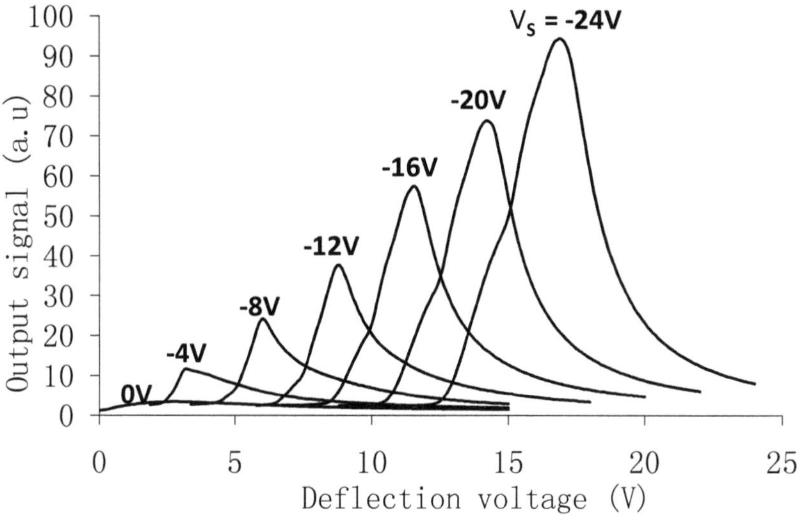

(a)

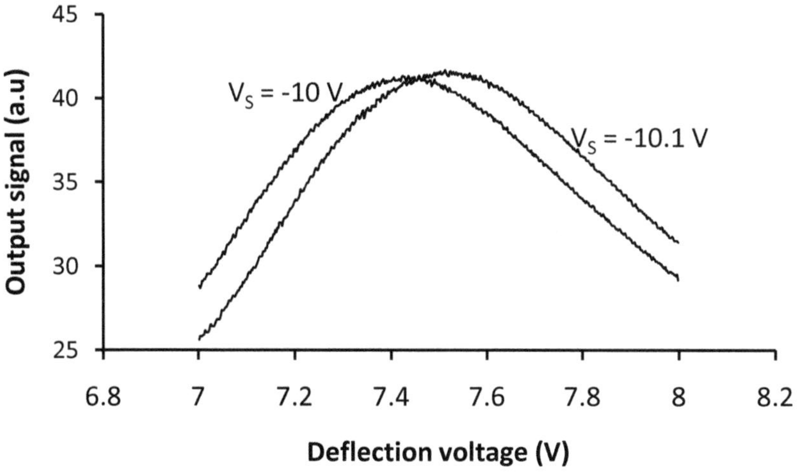

(b)

Fig. 6-10. Experimental SE toroidal spectrometer signals on a Cr specimen for different extraction voltage conditions: (a) Specimen voltage changing from 0 to -24 V in -4 V steps, (b) Specimen voltage changing from -10 to -10.1 V.

giving a signal-to-noise ratio of 2,840. Note that for a perfectly symmetric signal, the expection function value is equal to the peak position, and is largely unaffected by shot noise, whose contributions to it will, on average, cancel out. Assuming that the minimum measurable voltage occurs at a signal to noise ratio of unity, this sets the minimum detectable shift due to shot noise to be around 32.2 μV. These kind of low noise spectra are expected to be useful for mapping dopant carrier concentrations in semiconductor devices, where shifts in SE spectra of several mV need to be monitored (Kazemian *et al.*, 2006).

Although the SE spectrometer output signals no longer represent the SE energy spectrum as the extractor field increases (more negative specimen voltages), they are in a more useful form for the purposes of quantifying specimen contrast effects. Not only does the output signal develop a sharp symmetric peak, which can be precisely tracked as a function of specimen potential, but the signal-to-noise ratio is more than an order of magnitude higher than the case when there is no specimen bias. These results illustrate an important point about the acquisition of the SE spectra in general: high energy resolution is not required, and accurate recording of the SE energy spectrum is unnecessary. By biasing the specimen/inner hemispherical electrode negatively, the SE spectrum is transformed into a much more convenient form for open-loop specimen voltage measurement than the original SE spectrum. If the surface potential does not change by large amounts over the area scanned, it is possible to store several images, each at a different analyzer deflector voltage, thereby obtaining an estimate of the peak position (or signal mean) in the output signal for each pixel in the image. In this way, a quantitative voltage map can be superimposed upon the normal SE image, rapidly acquired in open-loop mode.

Figure 6-11(a) displays experimental BSE energy spectra obtained from the use of the second-order focusing toroidal analyzer attachment for different metal specimens, illustrating that there is considerable material contrast in the output signal. Figure 6-11(b) shows that the experimental energy spectrum for gold closely matches the one predicted

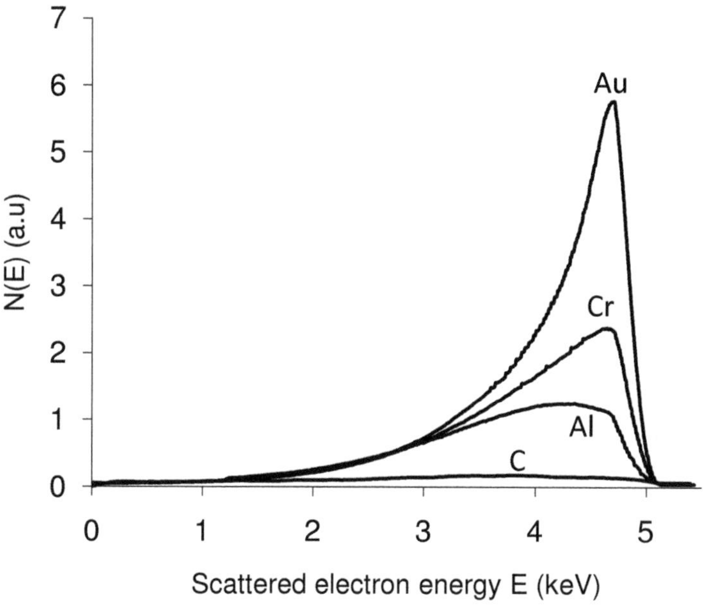

(a)

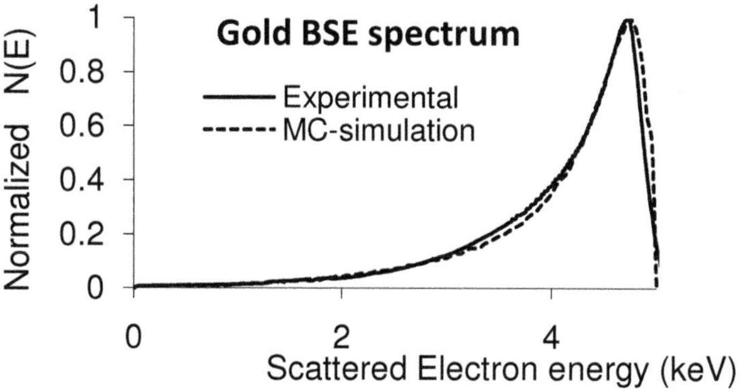

(b)

Fig. 6-11. Experimental BSE spectra from the second-order focusing toroidal analyzer attachment for a primary beam voltage of 5 kV for different types of metal specimens: (a) Different metal specimens, (b) Comparison with Monte-Carlo simulation for gold specimen.

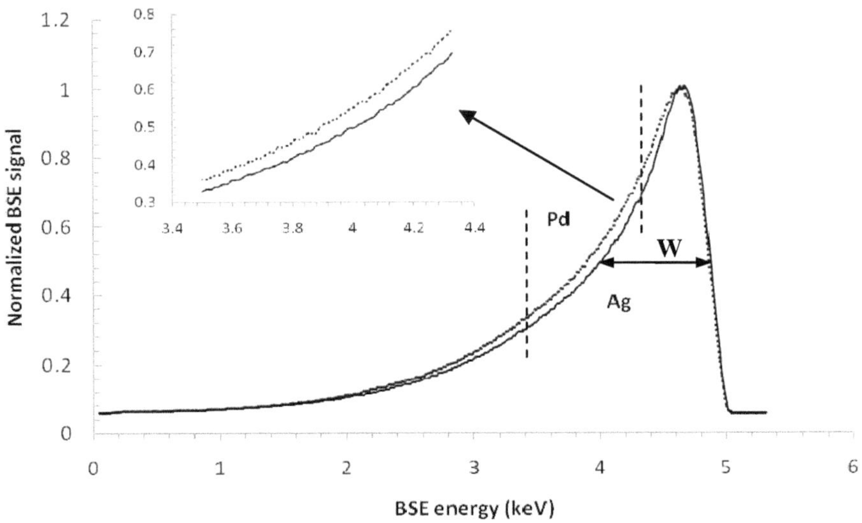

(a)

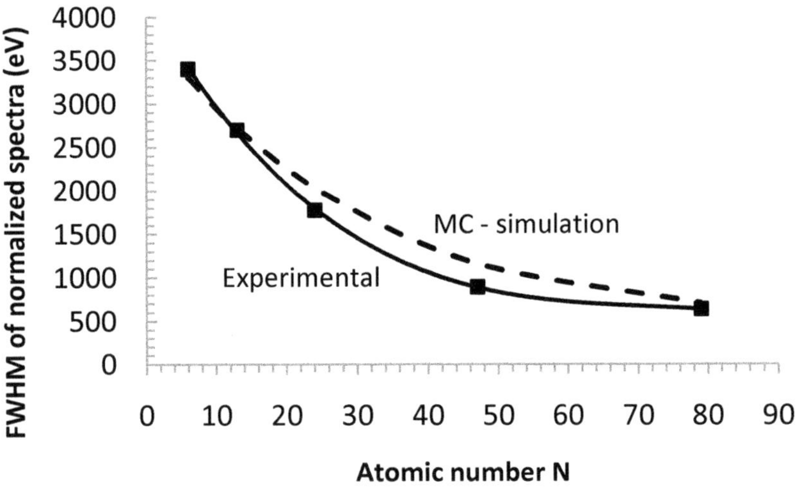

(b)

Fig. 6-12. Experimental BSE spectra from the second-order focusing toroidal analyzer attachment for a primary beam voltage of 5 kV for different types of metal specimens: (a) Silver and palladium specimens, (b) Comparison with Monte-Carlo simulation for a range of different atomic numbers.

by Monte-Carlo simulation. Figure 6-12(a) shows that close atomic number specimens can be resolved, a silver specimen (atomic number 47) is distinguished from a palladium specimen (atomic number 46) with a signal-to-noise ratio of 23.4. The signal in this case is the full-width at half-maximum height (*W*) for each spectrum. Figure 6.12(b) shows a graph of how *W* varies with specimen atomic number, both by simulation and experiment. The graph shows increased sensitivity for the low atomic number range, and demonstrates how quantitative elemental mapping might be achieved through backscattered spectrum detection.

6.3 A modified Fountain Analyzer design

The Fountain Analyzer (discussed in Chapter 2.4.1) can in principle combine relatively high transmission, with high energy resolution. This comes from the fact that it is rotationally symmetric, allowing for 2π collection in the azimuthal angular direction, and that for a given input angle (30° with respect to the horizontal axis), it has second-order focusing properties. The main problem preventing the use of the Fountain analyzer inside the SEM as an add-on spectrometer attachment is that there is no central hole through which the primary beam can pass in order to strike the specimen. The introduction of such a central hole will inevitably change its internal field distribution and alter the analyzer's energy dispersion and focusing properties. The following field-free axis Fountain spectrometer design allows for an even smaller working distance than the previously described second-order focusing toroidal spectrometer. Its minimum working distance is typically of the same size as the objective lens lower pole-piece radius, which can in principle be as small as 5 mm. It has the advantage of being located above the specimen, much like the conventional backscattered detector.

Figure 6-13(a) shows a modified Fountain analyzer design made by Hoang (2010), suitable for use as a SEM spectrometer attachment. A large central conical hole is introduced into the analyzer design, enabling it to fit around the lower pole-piece of a SEM objective lens. Inside the analyzer, around the conical hole, segmented plates, V_1, V_2, and V_3,

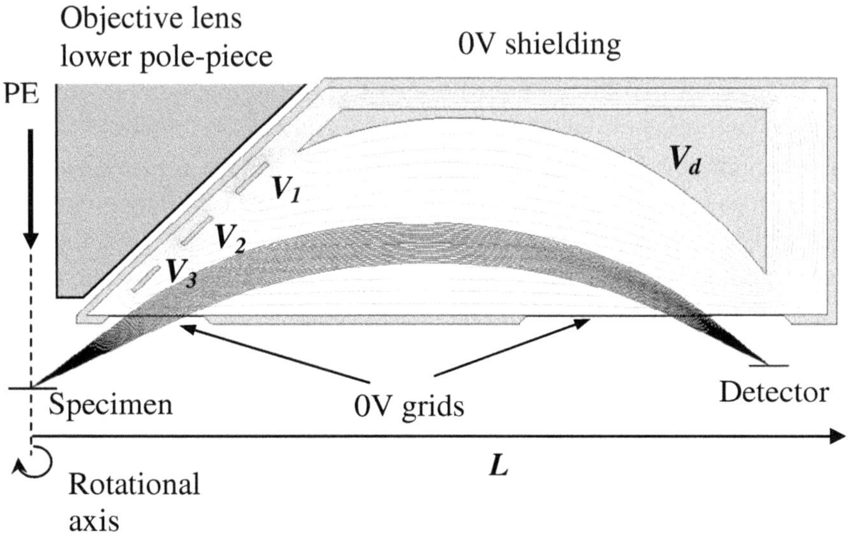

(a)

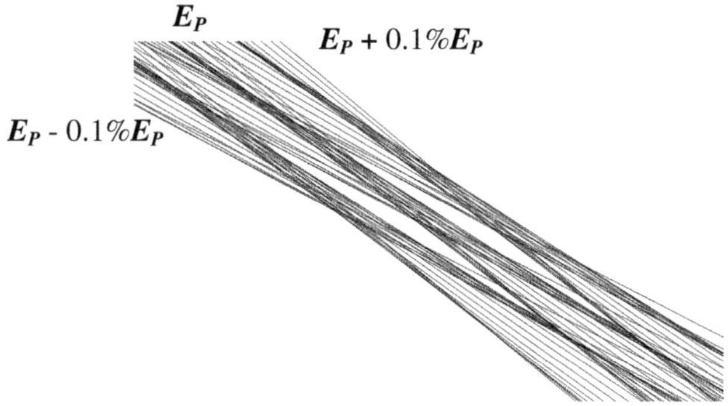

(b)

Fig. 6-13. Simulated trajectory paths through a modified Fountain Analyzer suitable for use as a SEM attachment. In terms of the deflection voltage V_d, the pass energy is $E_P = 1.75V_d$, $V_1 = V_d$, $V_2 = 0.711V_d$, and $V_3 = 0.302V_d$. The central ray enters in at 32.6° and 21 trajectories are plot over uniform steps for an input angular spread varying from 6° to 6°: (a) Through spectrometer, (b) Around detector plane (0.052% full-width relative energy resolution).

control how the electric potential falls, suitably adjusted so that the analyzer maintains its mirroring/focusing action on incoming electrons. Conditions predicted for second-order focusing are found to be at a central ray entrance angle of 32.6°, $V_1 = V_d$, $V_2 = 0.711V_d$, and $V_3 = 0.302V_d$, where V_d is the upper deflector voltage. The pass energy is set to $1.75V_d$. The simulated best base (full-width) relative energy resolution at the detector plane is found to be 0.0514%, as shown by Fig. 6-13(b). Assuming that the energy resolution is given by half the base energy resolution, the best relative energy resolution expected from this analyzer design is therefore 0.025%. This is around a factor of 6 times better than the corresponding idealized conventional Fountain Analyzer design (see Chapter 2.4.1).

Simulated parallel energy detection characteristics of the analyzer are shown in Fig. 6-14. The height of a flat detection plane is adjusted in order to maximize the pass band energy range, while minimizing aberrations on the final spot. In Fig. 6-14, there are 16 emission energies ranging from 92.5% to 107.5% of the central pass energy and 11 input angles from -104 mrad to 104 mrad around the central ray in uniform steps.

Use of a flat plate detector inevitably involves some defocusing of the final spot, as illustrated by Fig. 6-14b, lower energies tend to focus a little before the detection plane, while high energies focus a little beyond it. Figure 6-15 shows that the relative energy resolution is expected to lie between 0.06 to 0.07% for an energy pass range of 95% to 105% of the pass energy, around two times worse than if a sequential slit/detector analyzer arrangement is used (Fig. 6-13).

The modified Fountain spectrometer proposal by Hoang (2010) is a promising design not just as an SEM attachment, but also as a new energy spectrometer for Auger spectroscopy. Its best base energy resolution of around 0.11% is around 4 times smaller than the corresponding one for the CMA. Moreover, unlike the CMA, it can operate in parallel detection mode, allowing for a band-width of greater than ± 5% of the energy pass energy.

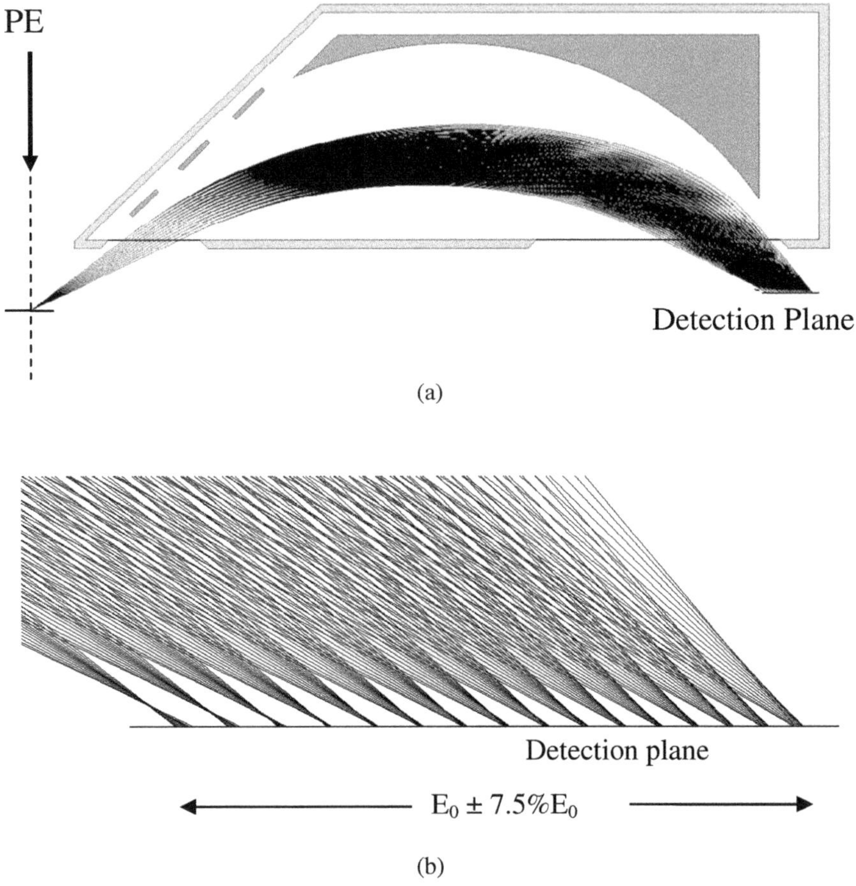

Fig. 6-14. Simulated trajectory paths illustrating parallel energy detection for the redesigned Fountain Analyzer suitable for use as a SEM attachment. There are 16 emission energies ranging from 102.5% to 107.5% of the central pass energy and 11 input angles from -104 mrad to 104 mrad around the central ray in uniform steps: (a) From specimen to detector, (b) Around the detector plane.

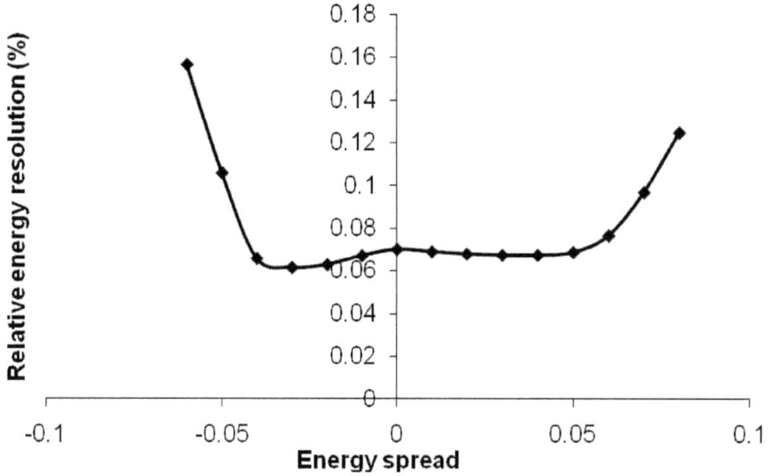

Fig. 6-15. Simulated relative base energy resolution as a function of energy spread around the pass energy for the modified Fountain Analyzer.

Chapter 7

Full Range Parallel Energy Spectrometer Designs

The full range energy spectrometer designs presented in the last chapter are sequential in nature, that is, at any moment in time, they function by acquiring a single point on the energy spectrum. In order to generate the whole spectrum, deflection voltages within the analyzer must be suitably ramped. Parallel energy acquisition has the obvious advantage of speeding up data-acquisition times, particularly where the energy changes by many orders of magnitude. The search for a parallel electron energy spectrometer, one that can simultaneously analyze energies varying by several orders of magnitude, is of interest for a wide variety of different applications. Within the category of electrostatic Auger spectrometer designs, Jacka *et al.* (1999) reported a spectrometer design based upon using a two-dimensional hyperbolic electric field (HFA), while Read (2002) proposed a Parallel Cylindrical Mirror Analyzer (PCMA), an extension of the Cylindrical Mirror Analyzer (CMA). In the category of parallel magnetic spectrometer designs, Kelly (1999) reported a magnetic rotation analyzer using magnetic meshes for Auger Electron Spectroscopy (AES), Photoemission Spectroscopy (PES) and Electron Energy Electron Loss Spectrometry (EELS) applications.

Figure 7-1(a) shows direct ray tracing through the hyperbolic electric field analyzer, as reported by Jacka (2001). The electric field lines inside the analyzer are approximately hyperbolic in shape, controlled in part, by the voltage distribution on an array of top plate electrodes. Practical design parameters for the analyzer as given by Jacka *et al.* (1999) are: an

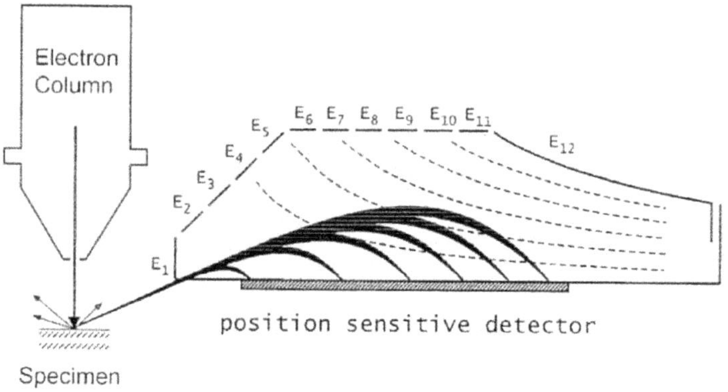

(a)

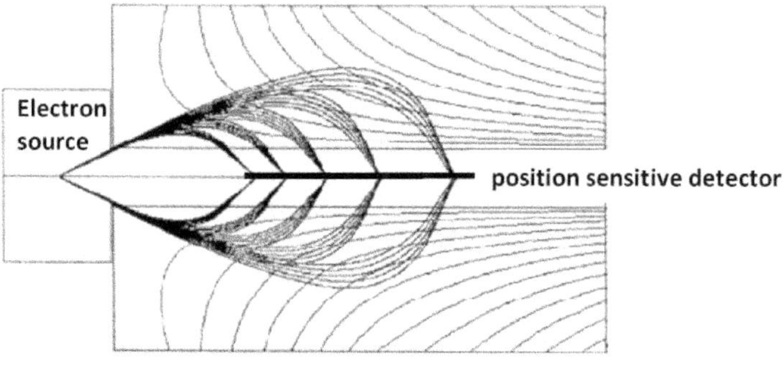

(b)

Fig. 7-1. Parallel electrostatic energy analyzer designs: (a) The hyperbolic field analyzer (Jacka *et al.*, 1999), (b) The parallel cylindrical mirror analyzer (Read *et al.*, 2004).

energy range from 57 to 2050 eV, an entry polar angle of 20.76°, and an input polar angular spread of ±1.1°. If it is assumed that the azimuthul angular spread is also ±1.1° (reasonable assumption for first-order focusing, similar to characteristics of the parallel-plate analyzer), the overall transmittance (as given by equation 1.45), is approximately 0.00776%. Jacka (2001) later reported the transmittance to be 0.05%. Although this level of transmittance is relatively small, the analyzer's main attraction lies in its ability to capture the entire Auger energy spectrum simultaneously, greatly speeding data-acquisition times. Jacka

(2001) reports the HFA base energy resolution to be 0.8 eV at 100 eV, and 4 eV at 2500 eV. Figure 7-1(b) shows direct ray tracing of electrons through the parallel cylindrical mirror analyzer in the axis to axis configuration (Read *et al.*, 2004). The electron energy ranges from 300 to 1500 eV. This analyzer design is not so suitable for the SEM, since the detector plane lies on the analyzer axis of symmetry, making it difficult for the primary beam to reach the specimen (which must be placed at the source position in Fig. 7-1(b)).

For most SEMs, parallel energy acquisition necessarily involves the simultaneous detection of scattered electrons in the 0 to 30 keV range, consisting of secondary electrons and backscattered electrons. Although the parallel energy spectrometer designs to be discussed in this chapter are equally applicable to AES, it is primarily parallel acquisition of SE and BSE spectra that will be discussed here, since they form the two most commonly used output signals in SEMs. Parallel spectral acquisition in the SEM has the attraction of avoiding the need to synchronize spectrum capture with the primary beam scan signal. As long as multi-channel detection/storage can respond fast enough, that is, faster than the scan rates of the SEM, each pixel in the final image will have its associated energy spectrum.

In this chapter, three different full-range parallel energy spectrometer designs will be presented, each adapted for a different objective lens design. The time-of-flight spectrometer, to be described in the next section, is suitable for use with magnetic immersion objective lenses. The second spectrometer design, the Gaussian field magnetic sector spectrometer, is suitable for use with most conventional SEMs, where the specimen is placed in a field-free region below the lower pole-piece of the final lens. The third spectrometer to be described, the magnetic beam separator spectrometer, is designed to operate together with a mixed field objective lens, where the specimen is immersed in a magnetic field and at the same time, utilizes an electric field to retard the primary beam to low landing energies. As shown in Chapter 4, this latter type of objective lens design is capable of providing the highest image resolution within the category of non-aberration corrected SEMs.

It should perhaps be noted here that a conventional objective lens SEM, one in which the specimen is placed below the lower lens pole-piece in a field free region, can incorporate an aberration correction unit in order to improve its image resolution and increase its probe current. Therefore, all different spectrometer designs to be outlined in this chapter are of equal importance, the main difference between them is that they are adapted to function with a different objective lens design.

7.1 The time-of-flight spectrometer

Up to now, the time-of-flight spectrometer is, to the author's knowledge, the only parallel spectrometer that has been used to capture the complete energy spectrum of scattered electrons inside a SEM. The spectrometer, designed by the author (Khursheed and Dinnis, 1990) in the context of developing voltage contrast analyzers, has already been described in Chapter 5. Although strictly speaking, the time-of-flight spectrometer output signals are sequential in nature, that is, they involve the recording of a signal that varies in time (with the arrival time of electrons), they do not involve any analyzer electrode voltage being ramped. The arrival time spread is generally many orders of magnitude smaller than the rate to which analyzer deflection electrodes can be ramped. For this reason, time-of-flight spectrometers are usually referred to as multi-channel or parallel spectrometers.

In general, when compared to other spectrometer types, time-of-flight spectrometers have the additional constraint of the primary beam needing to be pulsed, where pulse durations need to be small compared to the total electron arrival time spread, restricting them to be typically a few nano-seconds or less. In the context of Electron Beam Testers, where the voltages on the sample are periodic and rapidly changing, blanking of the primary beam to sub-nanosecond pulse widths is routine (Thong, 1993). For more general applications, where the goal is to acquire SE or BSE energy spectra (or both), blanking of the beam will inevitably reduce the output average current by one to two orders of magnitude. This need not necessarily lead to lower quality SE or BSE images, since blanking of the primary beam can be carried out only when the energy spectrum is required, however, the blanker must be carefully designed so that it does not limit the beam current when it is switched off. One way of avoiding this problem is to use a high brightness laser pulsed cathode, the kind proposed by Batiné *et al.* (1995) for the SEM, capable of providing 100 ps pulses, with nano-ampere average beam currents.

An example of a full range experimental time-of-flight energy spectrum obtained inside the SEM is shown in Fig. 7-2. This signal was obtained inside a Cambridge Instruments S100 SEM (Dinnis *et al.*, 1992, and Khursheed, 1992). Chopping plates were inserted below the electron gun unit, in order to blank the primary beam to pulses of a few nano-seconds duration, with a repetition rate of 1 MHz. A small specimen (less than 8 mm diameter) was pushed up into the bore of the objective lens to the point of maximum magnetic field, and a micro-channel plate (MCP) detector was placed above the objective lens. Scattered electrons emitted from the specimen surface therefore experience a sharply decreasing magnetic field as they travel up through the bore of the lens, and low energy secondary electrons are collimated, an effect that makes their transit times, to the first order, independent of initial emission angle and dependent only on initial energy (as described in Chapter 2.6).

For the backscattered electrons, they will travel back up through the bore in many different directions and collide into the inner walls of the objective lens bore. This is illustrated in Fig. 7-3, in which 1keV BSE trajectory paths are simulated through an example magnetic immersion objective lens. The lens field strength is adjusted so that it focuses a 1 keV primary beam on to the specimen, located at the peak magnetic field strength position. Although BSEs are lost via collisions into the lens bore, the BSEs that reach the MCP detector have trajectories that are approximately vertical, so that their transit times are determined

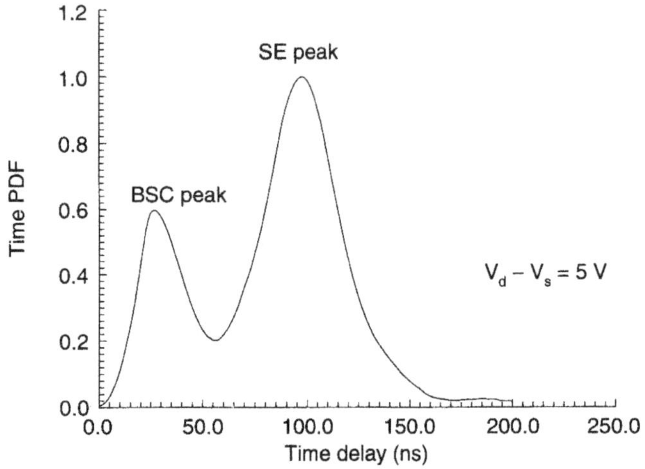

Fig. 7-2. Example of an experimental time-of-flight signal captured inside a Cambridge Instruments S100 SEM (Dinnis *et al.*, 1992 and Khursheed, 1992).

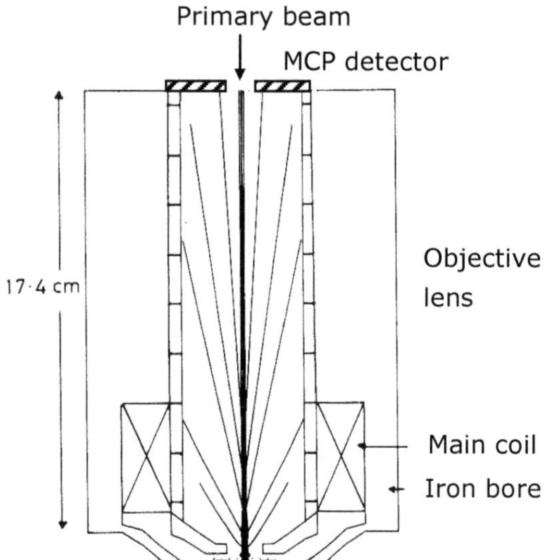

Fig. 7-3. Simulated 1 keV BSE trajectory paths emitted over a wide range of angles inside an immersion lens designed to focus a 1 keV primary beam on to a specimen located at the peak magnetic field strength position.

primarily by initial energy. This means that the BSE signal is predicted to have information related to the BSE spectrum. The precise shape of the spectrum for high energy electrons (above say 10 eV to elastic BSEs) is not expected to be accurate, since the MCP detector captures electrons over a relatively small emission angular range. The spectrum is perhaps more useful for monitoring specimen elemental changes, from point to point in the image. The spectrometer is also likely to be useful for monitoring the spectrum of secondary electrons that are accelerated by the specimen charging negatively.

7.2 A Gaussian field magnetic sector

Inside a magnetic sector spectrometer, the magnetic deflection field strength is usually constant. However, by allowing the deflection field strength to vary in such a way that it steadily increases in the path of incoming electrons, electrons having a wide range of different energies can be deflected and focused on to a single flat plane detector. A deflection field distribution of this type, proposed by the author, is one that takes the form of a two-dimensional asymmetric Gaussian deflection function, $B_{z0}(x,y)$ on the odd-symmetry plane at $z = 0$

$$B_{z0}(x, y) \approx B_0 \exp\left[-\left(\frac{x - x_0}{\sigma_x(x)}\right)^2 - \left(\frac{y - y_0}{\sigma_y(y)}\right)^2 \right] \tag{7.1}$$

where x_0 and y_0 are the x and y coordinates that define the position of the maximum field strength, B_0, and $\sigma_x(x)$ and $\sigma_y(y)$ are standard deviations that describe how the function falls on either side of the maximum field strength point in x and y respectively. In this case, electrons are deflected in the x-y plane, and the magnetic deflection field is applied in the z-direction. Figure 7-4 shows in-plane simulated electron trajectory ray paths for a beam of electrons travelling through an asymmetric Gaussian deflection field that is described by the above equation. Electrons have initial energies that range from 50 to 2,000 eV, with -25 to +25 mrad

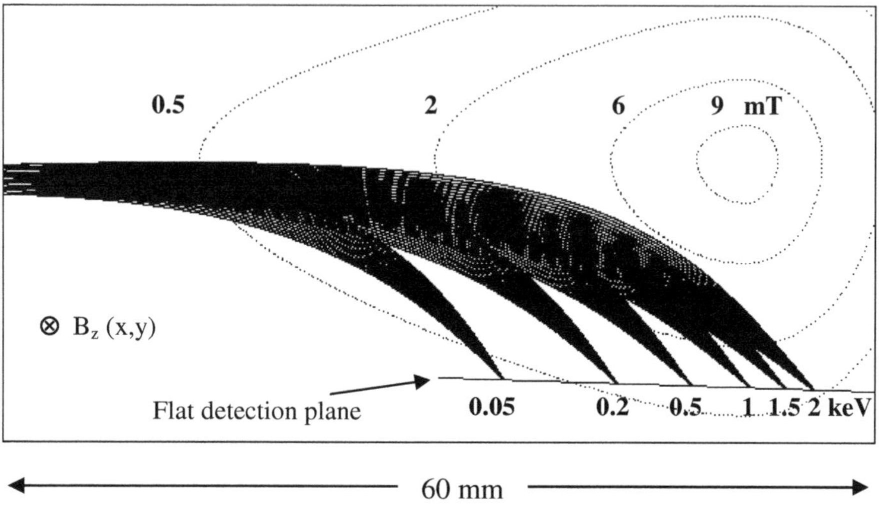

Fig. 7-4. Simulated trajectory paths of a beam of electrons having incident energies ranging from 0.05 to 2 keV with an angular spread of -25 to 25 mrad through an analytically generated asymmetric Gaussian magnetic deflection field. Dotted lines represent lines of constant magnetic field strength.

incident angles relative to the *x*-axis. The source in this case is located 40 mm from the left-hand side boundary. Dotted lines in Fig. 7-4 represent lines of constant magnetic field strength. Table 7-1 gives the field parameters required to generate the magnetic field distribution shown in Fig. 7-4, where the outer boundaries of the spectrometer are defined by $x_1 < x < x_2$ and $y_1 < y < y_2$ in the *x-y* plane.

Figure 7-5 shows that the focusing properties of the Gaussian magnetic deflection field distribution are predicted to be of first-order, where the final trace width is characterized by second-order spherical aberration. The simulated relative energy resolution varies from around 0.8% at 50 eV through to around 0.37% at 2,000 eV, which is comparable to the predicted energy resolution performance of the HFA for a similar energy range (Jacka, 2001). The transmittance is predicted to be approximately 0.063% for an angular spread of ±25 mrad in the polar and azimuthal directions, comparable to the transmittance of Jacka's HFA spectrometer proposal (for approximately the same energy resolution).

Table 7-1. Field parameters in Eq. (7.1) required to generate the magnetic field distribution shown in Fig. 7-4.

$B_0 = 9.822$ mT, $x_0 = 51$ mm, $y_0 = 19.5$ mm
$\sigma_x(x) = 7.5$ mm for $x_0 < x < x_2$,
$\sigma_x(x) = \sigma_{x0}\,[1 + 0.3\,(x_0 - x)/\sigma_{x0}]$ for $x_1 < x < x_0$, where $\sigma_{x0} = 11.5$ mm,
$\sigma_y(y) = 10.25$ mm for $y_1 < y < y_0$, and $\sigma_y(y) = 7.75$ mm for $y_0 < y < y_2$.

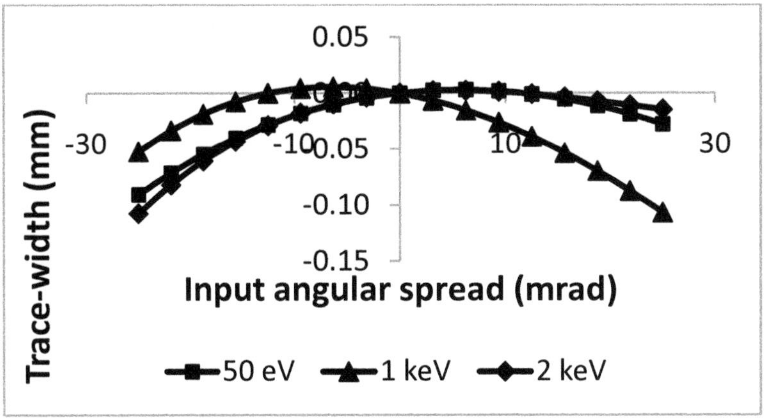

Fig. 7-5. Simulated trace width dependence on input angular spread for the asymmetric Gaussian field magnetic field distribution parallel spectrometer.

A wider range of parallel energy acquisition is predicted for suitably adjusting the shape of the field distribution. Consider the field parameters given in Table 7-2. In this case, simulated trajectory ray paths for energies ranging from 1 eV to 100 keV are deflected and focused on to a horizontal flat plate detector, as shown in Fig. 7-6. The energy range in this case, spans five orders of magnitude. The angular spread ranges from -25 to 25 mrad, and the source is located 20 mm from the left-hand side of the box.

Table 7-2. Field parameters in Eq. (7.1) required to generate the magnetic field distribution shown in Fig. 7-6.

$B_0 = 95$ mT, $x_0 = 76$ mm, $y_0 = 20$ mm
$\sigma_x(x) = 12.5$ mm for $x_0 < x < x_2$,
$\sigma_x(x) = \sigma_{x0}\,[1 + 0.2\,(x_0 - x)/\sigma_{x0}]$ for $x_1 < x < x_0$, where $\sigma_{x0} = 9.5$ mm,
$\sigma_y(y) = 8.75$ mm for $y_1 < y < y_0$, and $\sigma_y(y) = 7.25$ mm for $y_0 < y < y_2$.

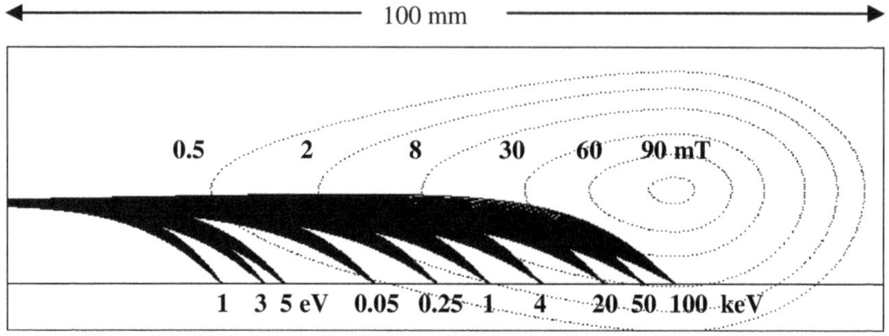

Fig. 7-6. Simulated trajectory paths through analytically generated asymmetric magnetic Gaussian magnetic deflection field for wide energy range parallel acquisition. Electron energies vary from 1 eV to 100 keV, and the angular spread varies from -25 to 25 mrad.

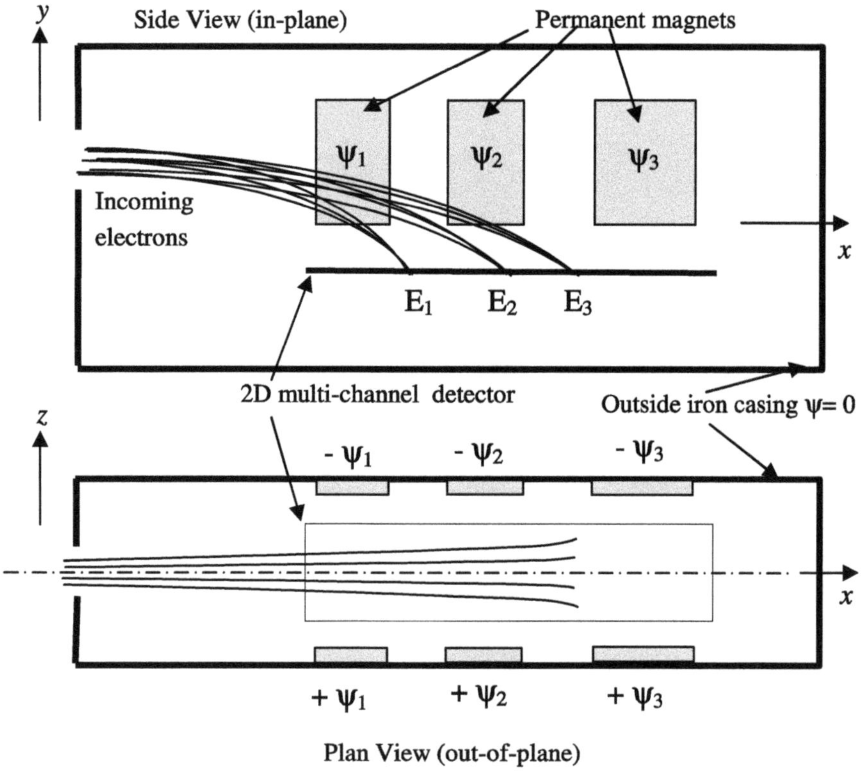

Fig. 7-7. Schematic diagram of a Gaussian field magnetic sector box spectrometer for parallel energy acquisition.

There are many ways to generate magnetic field distributions similar to the kind of asymmetric Gaussian field distribution described by Eq. 7.1. One approach is to use permanent magnets attached to the sides of an iron box, as shown in Fig. 7-7. The strengths of the permanent magnets here are represented by the scalar magnetic potentials, $\pm\psi_1$, $\pm\psi_2$, and $\pm\psi_3$. A finite element field solution, neglecting the effect of magnetic saturation for such a magnetic box spectrometer is shown in Fig. 7-8. Contours of constant magnetic scalar potential are indicated in the *x-y* plane along the surface of the magnets ($z = \pm 8$ mm), and the *x-z* plane midway through the magnets ($y = 20$ mm). This magnetic box spectrometer is designed for the central ray of the input electron beam

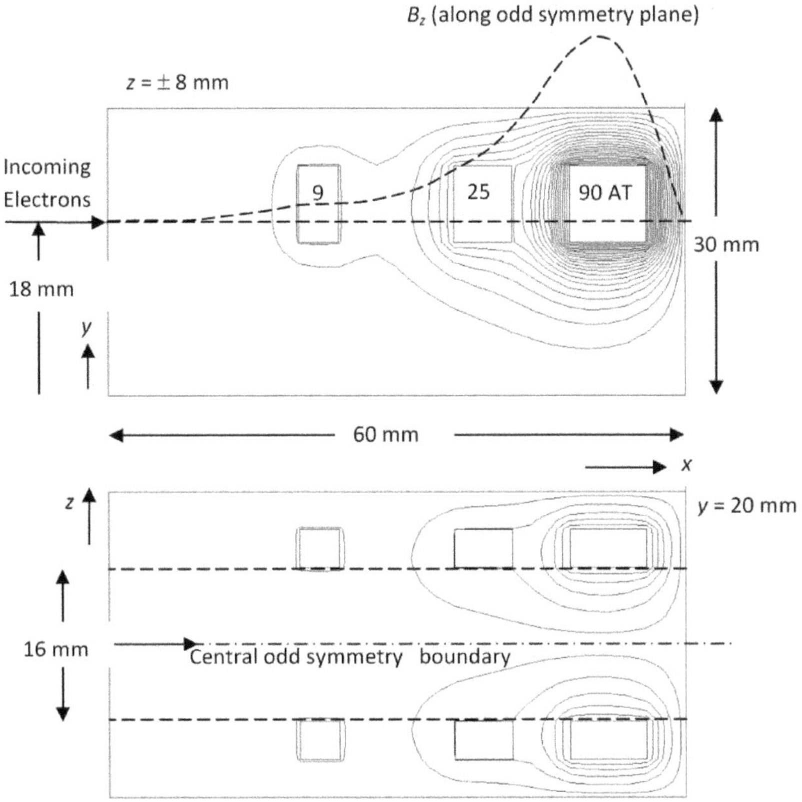

Fig. 7-8. Numerically solved magnetic scalar potential distribution for a Gaussian field magnetic sector box spectrometer for parallel energy acquisition.

entering along the mid-plane odd symmetry boundary ($z = 0$) at a height of y = 18 mm. For these conditions, simulated trajectory paths for electrons having energies of 0.05 to 2 keV with an angular spread of ±25 mrad are shown in Fig. 7-9(a), indicating that a magnetic box spectrometer of this type is expected to be able to deflect and focus electrons having a wide range of different energies on to a flat plate detector. The simulated relative energy resolution in this case lies between 0.42 to 0.62%, a little better than the corresponding analytically generated field distribution. Note that the detector plane, although straight, is slightly tilted. Figure 7-9(b) shows that with small adjustments in the field distribution, achieved through the use of a 4th pair of plates, the detection plane can be made approximately horizontal. The magnetic excitation strengths for the layout shown in Fig. 7-9(b) are increased by a factor of $\sqrt{20}$, so that the energy range spans from 1 to 30 keV.

Through the use of rare-earth permanent magnets, the kind of parallel energy magnetic sector spectrometer box depicted in Fig. 7-7 can be made relatively small. Figure 7-10(a) shows a photograph of such a prototype spectrometer, where round small cylindrical magnets are placed behind rectangular shaped iron plates. The number of magnets are adjusted to best fit the corresponding numerically simulated field distribution depicted in Fig. 7-9 (shown by the dotted contour lines). Figure 7-10(b) shows lines of constant magnetic field strength along the odd symmetry plane boundary ($z = 0$) for both the simulated values and experimentally measured ones, they show good agreement, and demonstrate that it is in principle feasible to build compact parallel energy magnetic spectrometer boxes. By using an array of such first-order focusing units in the azimuthal direction, the effective transmittance can be increased by over an order of magnitude, and is predicted to be around 0.8%. This is similar to the transmittance performance predicted for Read's PCMA proposal (at approximately the same energy resolution), which, by virtue of it having rotationally symmetric electrodes, has a simulated transmittance of around 0.9% (Read *et al.*, 2004).

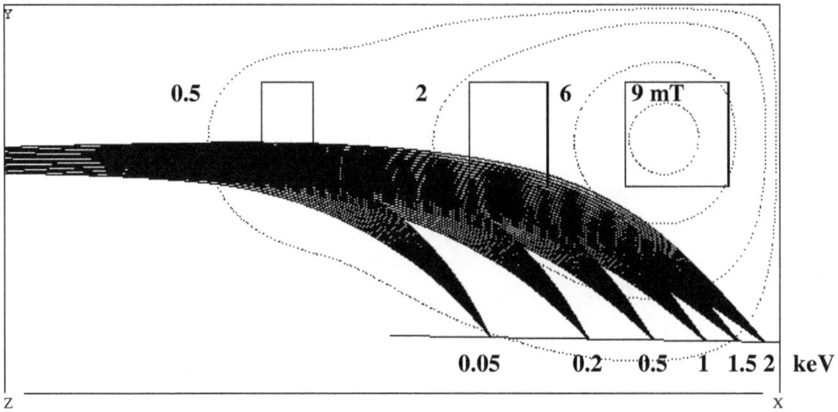

(a)

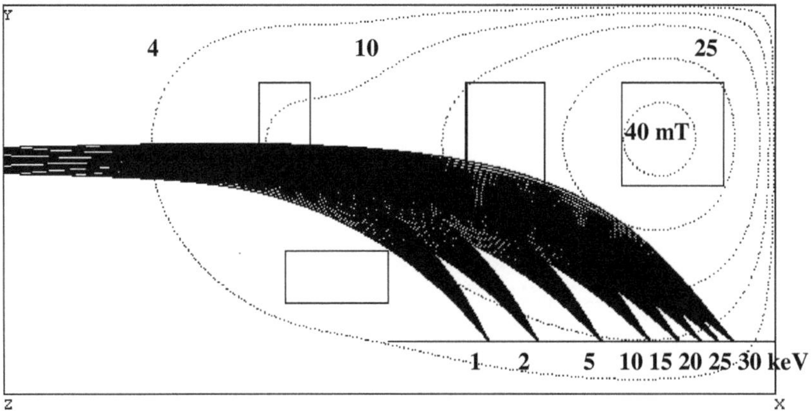

(b)

Fig. 7-9. Simulated trajectory paths through Gaussian field magnetic sector box spectrometer for an angular spread ranging from -25 to 25 mrad: (a) 50 eV to 2 keV range, (b) 1 to 30 keV range.

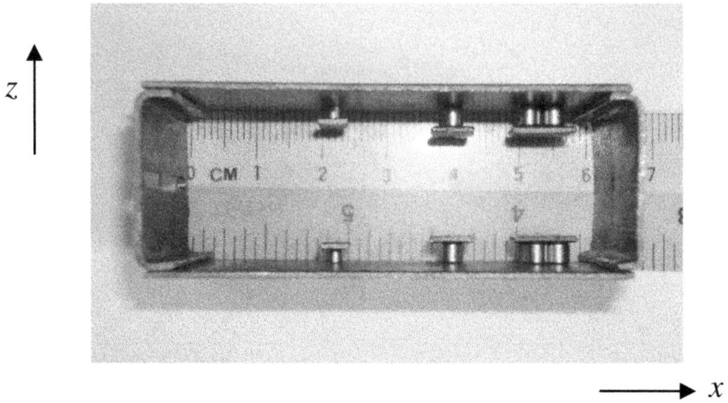

(a)

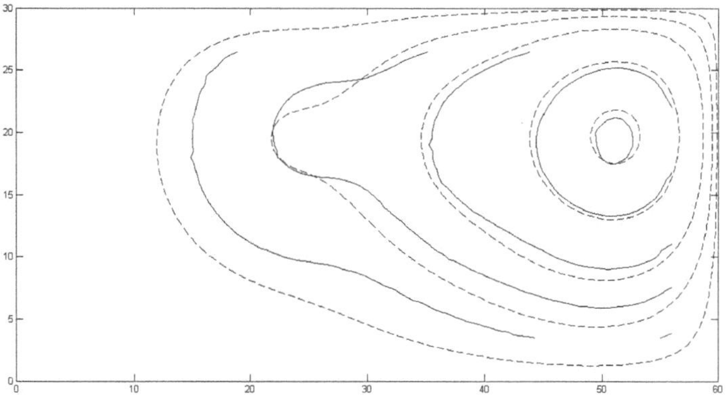

(b)

Fig. 7-10. Prototype Gaussian field magnetic sector spectrometer: (a) Plan view photograph, (b) Comparison of experimental and simulated contour lines of constant magnetic field strength at the mid-odd symmetry plane. Solid lines are from experiment while dotted lines are from simulation.

7.3 A round magnetic beam separator

Magnetic beam separators have been used in a variety of different applications, such as Low Electron Emission Microscopy (Tromp *et al.*, 1998; Bauer, 1998; Kolarik and Mankos, 1991), and are useful in situations where scattered or reflected electrons travel in the opposite direction to primary electrons, staying close to the column optical axis. In such situations, a magnetic deflector can be used to both deflect the primary beam on to the axis of the final lens, and at the same time, deflect scattered or reflected electrons away from it. Schematic diagrams for two magnetic deflectors of this type, one circular, the other square, are shown in Fig. 7-11(a). The magnetic field is directed in the z-direction, created between pairs of plates that are excited by magnetic scalar potentials, $\pm\psi$. Both deflectors are segmented, so that a pair of inner plates can be excited differently to a pair of outer plates. The plane of deflection (in-plane) occurs in the x-y plane, perpendicular to the x-z plane (out-of-plane). The purpose of segmenting the magnetic deflector is to achieve stigmatic focusing, that is, on exiting the deflector, both in-plane and out-of-plane electrons have the same exit angles and converge to (or appear to diverge from) the same point. The magnetic deflector therefore not only bends the primary beam, but also acts like a round lens.

The reason why segmenting the deflector is able to provide stigmatic focusing comes from the fact that the outer deflector plates pre-deflect the beam so that electrons enter the inner plates at an angle (non-normal to its edge), and as discussed in Chapter 2.4.4, this has the effect of focusing/defocusing the out-of-plane electrons. It turns out that the relative excitations of the inner and outer plates can be adjusted to provide the same focusing action on both in-plane and out-of-plane electrons upon their exit from the beam separator. This is illustrated for the in-plane in Fig. 7-11(b), where the trajectory paths for 3 primary rays are simulated for different entry angles. The rays converge towards the deflector centre at the beam separator entrance, and diverge by the same angles on exit, after having undergone a 90° deflection.

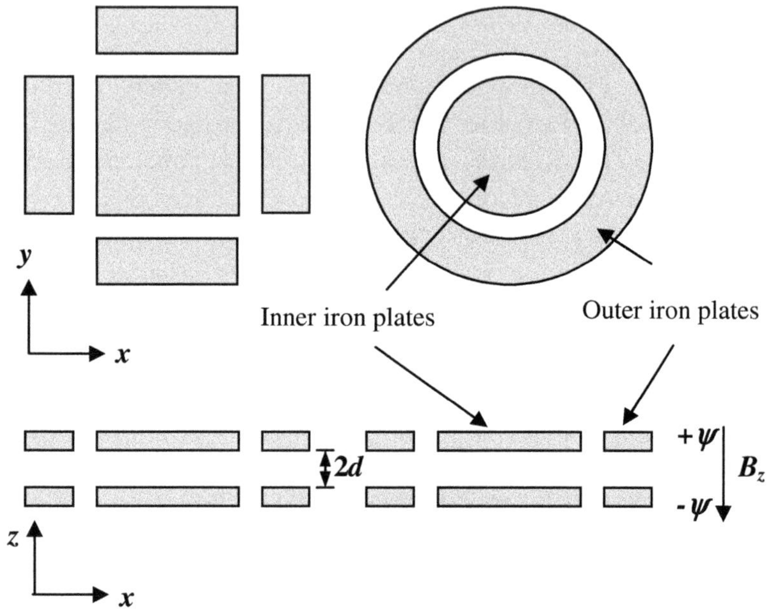

(a)

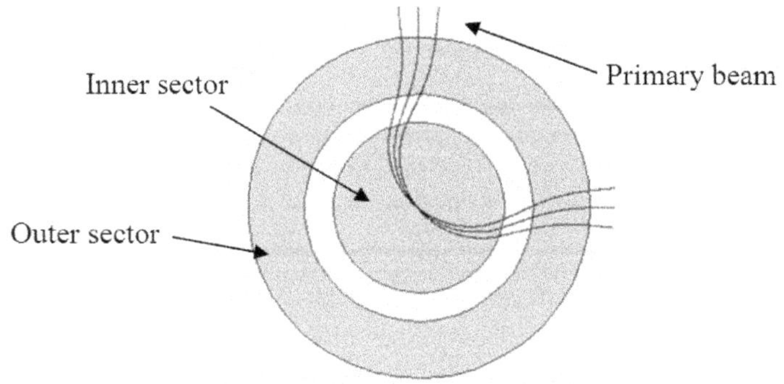

(b)

Fig. 7-11. Magnetic sector beam separators for stigmatic focusing: (a) Iron plate layout diagram for square and circular shapes, (b) Simulated stigmatic focusing of electron trajectory paths for a converging incoming primary beam in the circular layout.

Osterberg and Khursheed (2005) carried out detailed simulations for the circular magnetic beam separator model shown in Fig. 7-12. They found that for stigmatic focusing with minimum on-axis aberrations, the primary beam should enter the deflector converging towards its centre, and that the inner and outer plate excitations should be set to +70 AT and -9.11 AT respectively for a 6 keV primary beam. Simulated on-axis

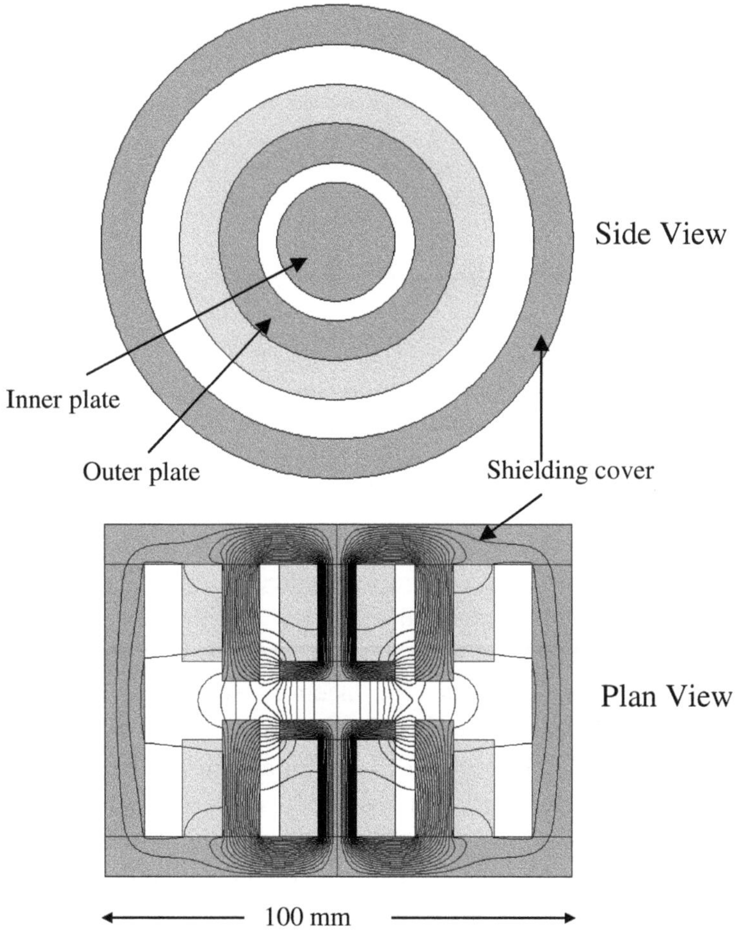

Fig. 7-12. Simulation model for circular magnetic sector beam separator. Light grey areas denote coil regions, while dark grey areas denote iron plates (Osterberg and Khursheed, 2005).

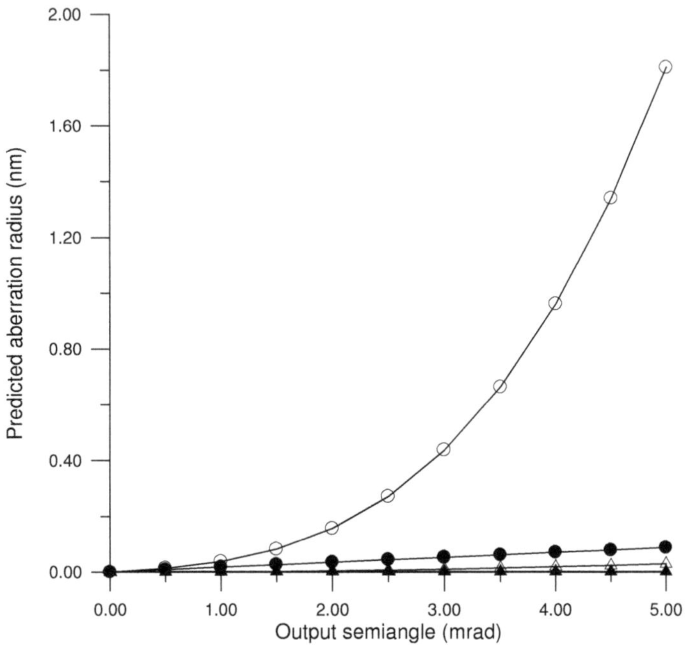

Figure 7-13. Predicted aberration limited resolutions as a function of the output semi-angle for a converging beam in the circular magnetic sector beam separator. The relative energy spread is taken as 10^{-4}. Aberrations have the following legend: △ in-plane spherical, ○ out-of-plane spherical, ▲ in-plane chromatic and ● out-of-plane chromatic.

aberrations for these conditions are depicted in Fig. 7-13, where the relative energy spread is assumed to be 10^{-4} and the semi-angle ranges from 0 to 5 mrad. Figure 7-13 indicates that the size of the virtual spot created by on-axis aberrations for electrons leaving the deflector is dominated by out-of-plane spherical aberration, and that the radius of the spot is predicted to be less than 2 nm for a semi-angle of 5 mrad. For most electron beam focusing systems, the beam separator can be placed at a point where the primary beam entrance semi-angles are well below this semi-angle. The size of beam separator on-axis aberrations on the final image are therefore expected to be relatively small, in the sub-nanometer range, negligible in comparison to the system's final lens aberrations.

For the SEM, a beam separator is useful for low landing energy mixed field immersion lenses, where the specimen is immersed in a magnetic field and a moderate to strong electric field (> 1 kV/mm) is used to slow down the primary beam just before it strikes the specimen surface (as described in Chapters 3 and 4). The electric field not only retards the primary beam, but it has the action of collimating and accelerating scattered electrons up into the SEM column. A Wien filter is sometimes used as a beam separator in this situation, and a parallel secondary electron energy spectrometer using this concept proposed by Kienle and Plies (2004) has already been described in Chap 5. The Wien filter as a beam separator however, provides a relatively small amount of dispersion, Kienle and Plies found it necessary to magnify the dispersion by the use of an additional spherical deflector analyzer as well as an electrostatic zoom lens. Khursheed and Osterberg (Khursheed and Osterberg 2006, and Osterberg and Khursheed, 2005) proposed an alternative approach, based upon the use of a stigmatic magnetic sector beam separator.

Khursheed and Osterberg propose that a magnetic sector beam separator be placed before the objective lens of the SEM to both bend the primary beam on to the lens axis, and disperse the scattered electrons in energy, using it as the first stage of a full range parallel energy spectrometer. A schematic diagram of their proposal is shown in Fig. 7-14. The SEM column, not shown, directs and focuses a primary beam into the centre of a circular magnetic sector deflector (of the type shown in Fig. 7-12). The SEM column has the usual gun, condenser lens, stigmator, aperture and scan coil components, and uses an auxiliary lens to focus the primary beam into the magnetic sector unit. The magnetic sector unit deflects the primary beam by 90°, operating in the stigmatic focusing condition, where it acts as a round lens. The primary beam then goes through a transfer lens which is set to have a relatively long focal length, so that the primary beam enters the bore of the objective lens nearly parallel. The specimen is biased (typically around -5 kV) and placed in a strong magnetic field (between say 0.3 to 1 Tesla). The subsequent scattered electrons from the specimen are focused back into the sector unit by the

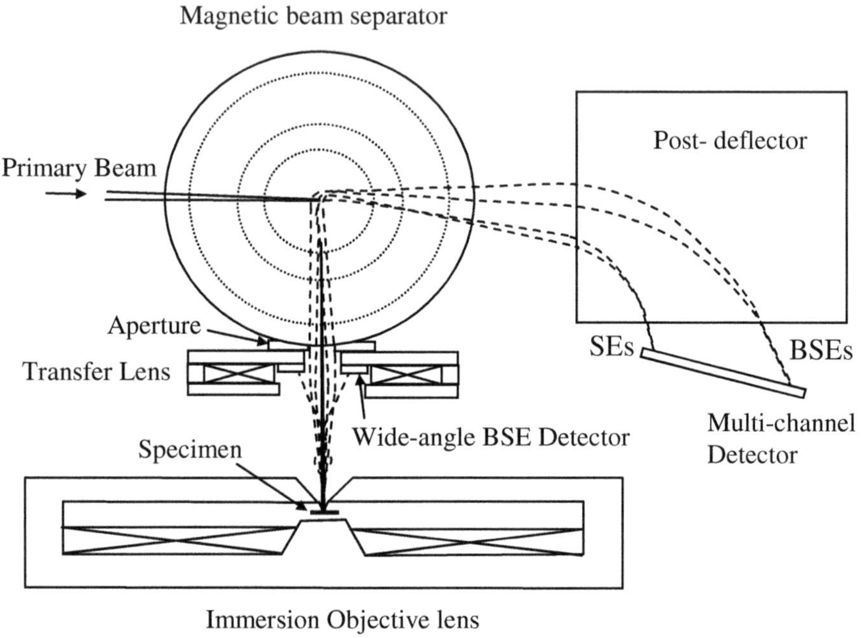

Fig. 7-14. Schematic diagram of magnetic beam separator spectrometer layout (Khursheed and Osterberg, 2006).

transfer lens, and deflected away from the SEM column. The wider-angle backscattered electrons are collected by a detector placed below the upper pole-piece of the transfer lens. The scattered electrons transmitted through the transfer lens and separator unit are then further deflected by a magnetic post-deflector unit, which directs and focuses them on to a multi-channel detector. Alternatively, in order to attain higher energy resolution, electrons can be further deflected on to an array of band-pass spectrometers. An aperture placed below the beam separator helps to restrict the angular dispersion of scattered electrons.

Khursheed and Hoang (2008b) carried out simulations to determine the trajectory paths of scattered electrons from the specimen to the detector plane for the kind of parallel energy beam separator spectrometer depicted in Fig. 7-14. They simulated the electric and magnetic field distributions of a mixed field immersion lens designed to retard a 10 keV primary beam to a landing energy of 5 keV (specimen biased to -5kV). The upper pole-piece of the objective lens has a 2 mm diameter hole through which the primary beam and scattered electrons pass. The magnetic gap in their lens model is 4 mm, and the distance of the specimen below the upper pole-piece of the objective lens is 2 mm. For these conditions, their simulations provide a peak magnetic field strength of 0.525 T. The primary beam is assumed to start 50 mm above the transfer lens, corresponding to the beam separator centre.

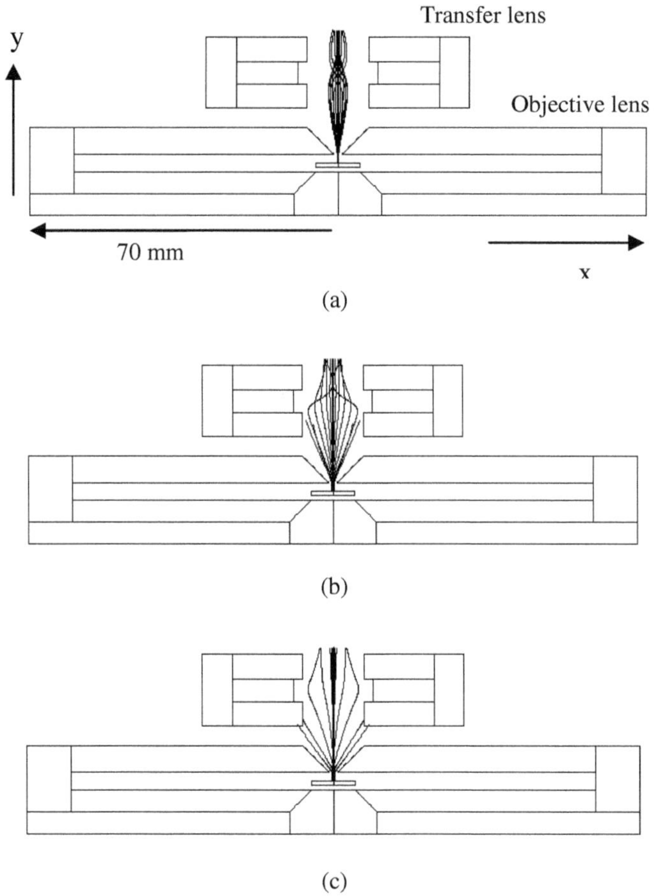

Fig. 7-15. Simulated trajectory paths through objective and transfer lenses for scattered electrons for emission angles ranging from 0 to 1.4 rad in 0.2 rad steps: (a) 500 eV, (b) 2 keV, (c) 5 keV (BSEs).

Figure 7-15 shows direct ray tracing of scattered electrons leaving the specimen with emission energies of 500 eV, 2 keV and 5 keV (BSE) over a wide range of emission angles relative to the optical axis (0 to 1.4 radians). The low energy secondary electrons (< 5 eV) were not plot since they remain close to the optical axis (within a radius of 0.5 mm). The trajectory paths shown in Fig. 7-15 illustrates how most scattered electrons are focused into the beam separator, while wider angle backscattered electrons are filtered through the transfer lens bore. The

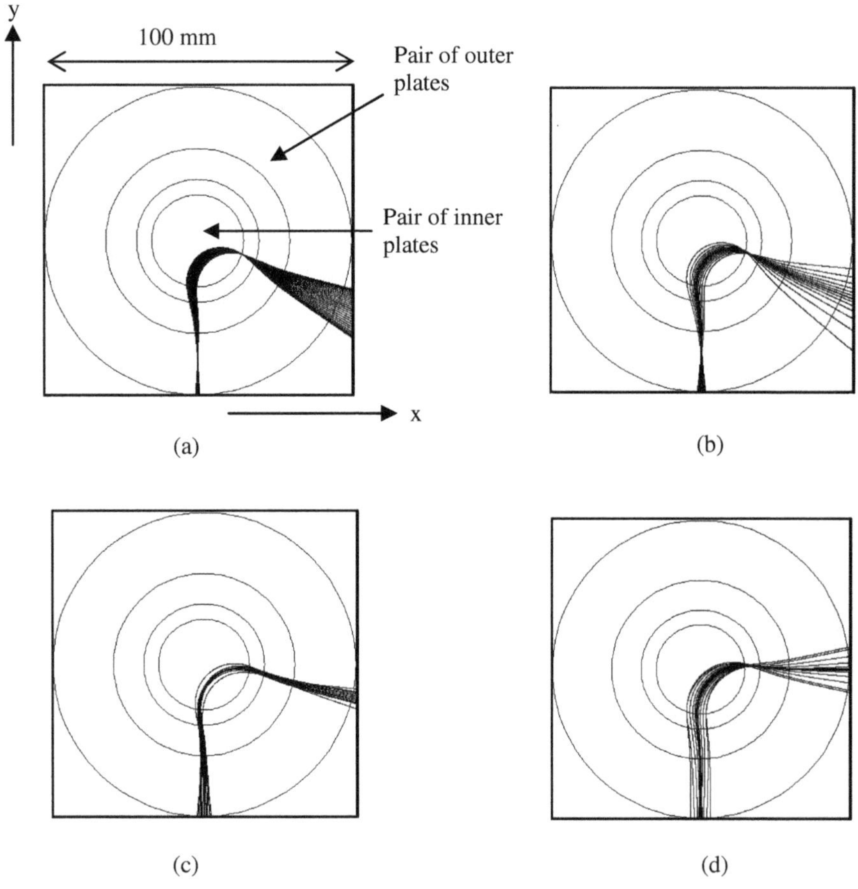

Fig. 7-16. Simulated in-plane (x-y) scattered electron trajectory paths through the beam separator. Emission angles plot in 0.1 rad steps: (a) 50 eV, 0–1.5 rad, (b) 500 eV, 0–0.6 rad, (c) 2 keV, 0–0.7 rad, (d) 5 keV, 0–0.9 rad.

excitations on the inner and outer plates necessary to provide the stigmatic focusing condition for a 10 keV primary beam deflected through 90 degrees are 85.85 AT and -12.14 AT respectively, more details on this are given by Khursheed and Osterberg (2006), and Osterberg and Khursheed (2005). The magnetic field distribution inside the beam separator is solved by using the finite-element method in combination with a Fourier-Bessel series (Hoang and Khursheed, 2008). The diameter of the beam separator in the x-y plane is 100 mm, and the gap between each pair of sector plates in the out-of-plane direction (z-direction) is 10 mm.

Ray tracing of the scattered electrons through the beam separator after having traveled through the transfer lens are shown in Figs. 7-16 and 7-17. These figures indicate that the angular spread effect is relatively small for the lower energy scattered electrons (represented by 50 eV in this instance), and the 2 keV electrons. The accelerating field at the specimen surface is responsible for reducing angular dispersion on the lower energy electrons, while the transfer lens focusing effect reduces it for the 2 keV electrons. For the backscattered electrons, the near-axis emission angle electrons clearly have a small amount of dispersion.

The aperture placed just below the beam separator helps restrict the in-plane angular dispersion of the scattered electrons over a wide range of emission energies. For a 2.7 mm wide aperture (full width), the emission angles of scattered electrons that enter the beam separator are limited to 0.4 radians for emission energies between 0.5 to 2 keV, 0.5 radians for 3 keV, and 0.7 radians for backscattered electrons (5 keV). Note that the aperture is not circular, but planar, that is, it only acts to restrict the angular dispersion in the plane of deflection.

In the out-of-plane direction, the beam separator dimensions largely determine the cut-off angle. For a 10 mm gap between each pair of sector plates, the out-of-plane cut-off angles on scattered electrons are: $\pi/2$ radians for 50 eV, 0.55 radians for 500 eV to 1 keV, 0.65 radians for 2 keV, 0.85 radians for 3 keV, and 0.95 radians for 5 keV (the

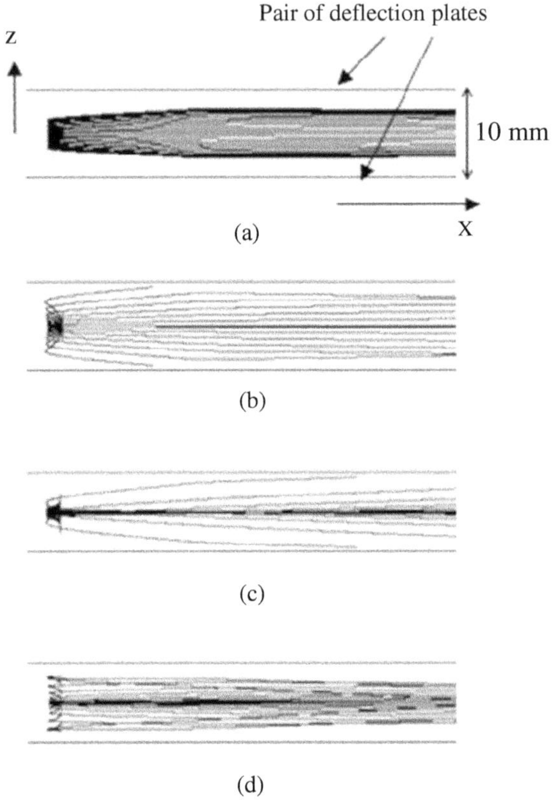

Fig. 7-17. Simulated out-of-plane (x-z) scattered electron trajectory paths through the beam separator. Emission angles plot in 0.1 rad steps: (a) 50 eV, 0–1.5 rad, (b) 500 eV, 0–0.6 rad, (c) 2 keV, 0–0.7 rad, (d) 5 keV, 0–0.9 rad.

backscattered electrons). These results are indicated by the trajectory paths shown in Fig. 7-17. Obviously, to increase transmission in the out-of-plane, the beam separator size can be increased, and there is no fundamental limit to transmission of the scattered electrons in the out-of-plane direction. For the in-plane direction however, the transmitted scattered electron angular dispersion must be selectively reduced in order to improve the attainable energy resolution at the detector plane.

After being dispersed by the beam separator, post-deflector units using either an electrostatic retarding field or magnetic sector field, or both, are required to focus the scattered electrons on to their respective detector planes. Accurate direct ray tracing of scattered electron trajectories through examples of these post-deflectors was achieved by using a combined 2D finite-element approach with 3D Fourier-Series expansions (Hoang and Khursheed, 2008). Figure 7-18 shows the ray paths of scattered trajectories which travel through the beam separator and are focused by a retarding magnetic sector unit for the secondary electrons, and a separate magnetic sector post-deflector for the backscattered electrons. Here the primary beam is assumed to be 6 keV, and scattered electrons are traced only from the beam separator entrance onwards, not taking the objective/transfer lenses into account. At the entrance of the beam separator, the scattered electrons are assumed to be diverging with a ±5 mrad cut-off angle. Figure 7-18 shows that this kind of post-deflector arrangement should be able to separate secondary electrons from backscattered electrons and achieve first-order focusing on to their respective detector planes.

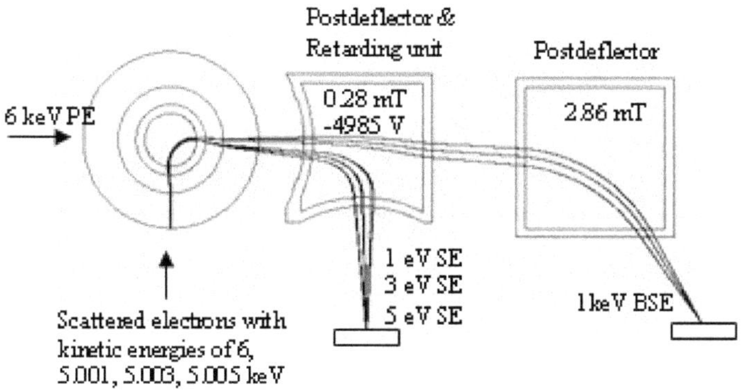

Fig. 7-18. Simulated scattered electron trajectory paths in the spectrometer for a 6 keV primary beam. Both SEs and BSEs enter the beam separator with angles of 0 and ±5 mrad.

Figure 7-19 shows a more complete simulation of scattered electron trajectory paths through the spectrometer. Here the beam separator is set to deflect a 10 keV primary beam through 90°, and Auger electrons, as well as secondary and backscattered electrons are considered. Although only ray paths through the beam separator and post-deflectors are shown, they are actually traced from the specimen with a wide-variety of emission energies and angles. The shapes, retarding voltages, and magnetic excitations on the post-deflectors are designed so that secondary, Auger and backscattered electrons are well separated and first-order focusing is achieved on their respective detector planes. The secondary electrons are mirrored back into the beam separator by the first post-deflector for a second pass. The transfer lens excitation is adjusted to minimize the angular dispersion for scattered electrons having energies from 1 to 2 keV.

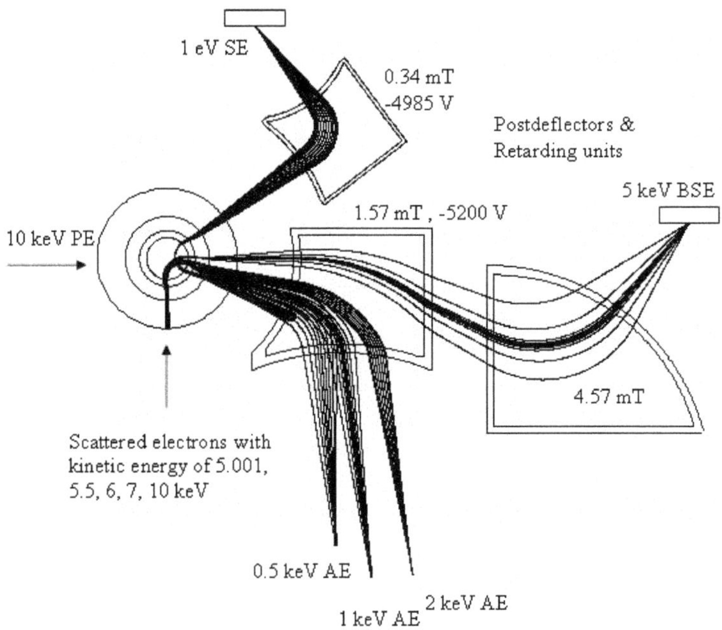

Fig. 7-19. Simulated scattered electron trajectory paths in the spectrometer for 1 eV SEs, 0.5–2 keV AEs, and 5 keV BSEs through the beam separator. The emission angles are plot in steps of 0.1 radians and range from 0 to 1.5 radians for SEs, 0 to 0.4 radians for AEs and 0.6 radians for BSEs.

In order to estimate the energy resolution achievable for a given degree of angular dispersion on the scattered electrons, detailed diagrams of simulated ray paths around the detection plane were examined. Some examples of these diagrams are shown in Fig. 7-20. By examining the relative dispersion created by different emission energies and angles at the detector plane, an energy resolution estimate was made. A cosine distribution of the emission polar angle is assumed (relative to the optical axis). The total transmission is estimated by averaging the respective transmissions for the in-plane and out-of-plane.

In the case of low energy secondary electrons, which have 100% transmission, an emission angle of 0.8 radians is used to estimate the energy resolution, corresponding approximately to the angular emission distribution peak (45°). For the Auger and backscattered electrons, cut-off angle trajectories lying around π/4 or below were used to estimate the energy resolution. The results are summarized in Table 7-3. They predict that the transmission of scattered electrons from the specimen to the detector plane for a given energy resolution will be relatively high. The transmission for 1 to 5 eV secondary electrons having an energy resolution of 148 meV to 193 meV is predicted to be 100%. The transmission varies between 20 to 40% for 0.5 to 3 keV Auger electrons whose energy resolution is between 4 to 7 eV, and is approximately 50% in the case of 5 keV elastic backscattered electrons having an energy resolution of 42 eV. The percentage energy resolution for the Auger and backscattered electrons for these conditions typically lies between 0.2 to 0.9%. It should be noted that the out-of-plane cut-off angles do not represent a fundamental limit to the spectrometer, since the beam separator size can always be scaled up to allow for 100% transmission in this direction, in which case, the total transmission value is likely to exceed 50%.

The predicted performance summarized in Table 7-3 for the magnetic beam separator spectrometer compares well with other types of electron energy spectrometers. It compares favorably for instance, with the Cylindrical Mirror Analyzer (CMA) operating at the second-order

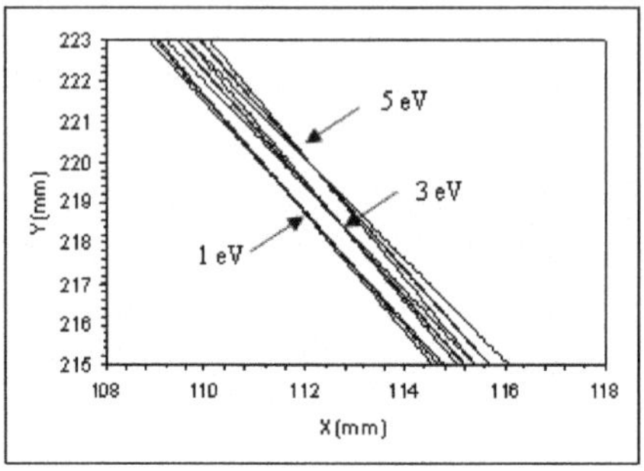

(a)

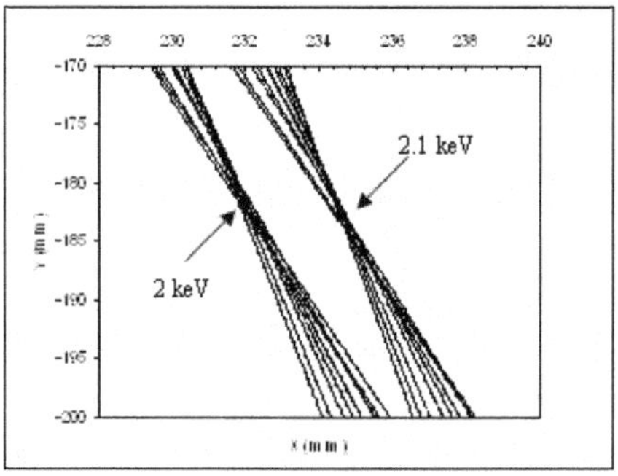

(b)

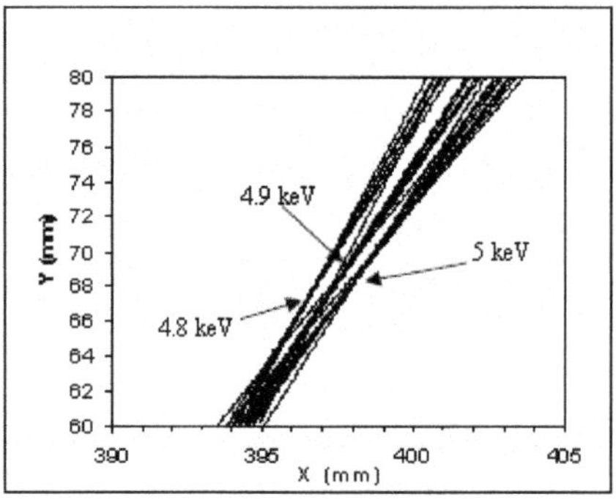

(c)

Fig. 7-20. Simulated trajectory paths around the detection plane for different emission conditions at the specimen: (a) 1, 3 and 5 eV SEs at 0, ±0.8 radians, (b) 2 and 2.1 keV AEs at 0 to 0.4 radians in 0.1 radian steps, (c) 4.8, 4.9 and 5 keV BSEs at 0 to 0.4 radians in 0.1 radian steps.

Table 7-3. Simulated energy resolution and transmission characteristics of spectrometer.

Emission energy (eV)	In-plane angle (radians)	Out-of-plane cut-off angle (radians)	Transmission (%)	ΔE (eV)	ΔE/E (%)
1 eV (SE)	π/2	π/2	100	0.148	14.8
3 eV (SE)	π/2	π/2	100	0.193	6.4
5 eV(SE)	π/2	π/2	100	0.175	3.5
0.5 keV (AE)	0.4	0.55	21.24	4.30	0.86
1 keV (AE)	0.4	0.55	21.24	6.88	0.69
2 keV (AE)	0.4	0.65	25.90	6.67	0.33
3 keV (AE)	0.5	0.85	39.50	6.93	0.23
5 keV (BSE)	0.6	0.95	49.02	41.93	0.83

focusing condition, where the CMA mid-entrance angle is 42.3° (see Chapter 2.4.2). Under this condition, the CMA for an entrance angular width of ±6° gives a relative theoretical energy resolution of around 0.328% and a transmission (over 2π sr) of around 16.84% transmission, assuming its input/output grids have 90% transparency. The present spectrometer for 2 keV energies (Auger electrons), has a predicted relative energy resolution of 0.33% with 25.9% transmission. However, a detailed comparison between the CMA and the present spectrometer is not meaningful, since the CMA does not provide a parallel energy spectrum. A better comparison would be with the parallel energy spectrometers proposed by Jacka *et al.* (1999), Read (2002) or Klienle and Plies (2004).

Both the hyperbolic field analyzer reported by Jacka and the parallel cylindrical mirror analyzer reported by Read require relatively small entrance angular spreads, typically in the 1 to 3° range, and depending on the mid-entrance angle, their transmissions will be smaller than 2%. For the widest energy reported by Read, 125 to 3,000 eV, an energy resolution of around 0.4% and relative transmission (over 2π sr) of around 2% is predicted (mid-entrance angle located at 27.04°). The widest energy range presented by Jacka is a few tens of eV to 2,000 eV, where an energy resolution of a few eV and a relative transmission of 0.05% is predicted (mid-entrance angle located at 20.76°). The second-order versions of the spectrometers reported by Read and Jacka cannot be compared to the present spectrometer since the detectable energy range is considerably smaller, no longer making them parallel in energy. The spectrometer presented by Kleine and Plies is similar to the present beam separator spectrometer in that it can obtain good separation between secondary and backscattered electrons. However, the Kleine and Plies cannot obtain the SE and BSE spectra at the same time, and the secondary electron energy range is limited to 20 eV, therefore, a detailed comparison to the Khursheed and Hoang magnetic beam separator spectrometer is also not meaningful. In the present context, it is sufficient to note that the magnetic beam separator spectrometer is predicted to have a relatively high transmission-energy resolution performance, comparable or better than the CMA operating at its optimum condition, while at the same time being capable of capturing the entire range of scattered electrons in parallel.

Osterberg and Khursheed (2008) carried out preliminary experiments on a small prototype magnetic beam separator which was fit on to the specimen stage of a conventional SEM stage. They used it together with a permanent magnet transmission lens to experimentally examine the on-axis aberrations imposed on the SEM primary beam by the beam separator. They found that these aberrations projected on to the specimen lie well below 2 nm, confirming previous simulations predictions (Osterberg and Khursheed, 2005).

In a separate experiment, preliminary spectral results from the prototype magnetic beam separator attachment were obtained by Osterberg *et al.* (2009). Figure 7-21 shows the experimental layout used. In this experiment, a 5 kV, 10 pA primary beam is deflected through 90° by the magnetic beam separator, and the specimen, located within the objective lens, is biased to around -3,650 V, yielding landing energies of 1,350 eV. After deflection, the scattered electrons are energy filtered by an electrostatic mirror, located below the separator. The mirror consists of

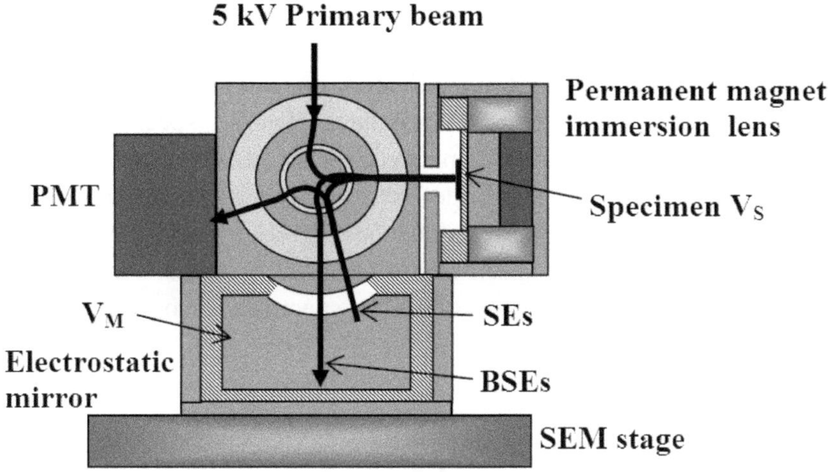

Fig. 7-21. Experimental layout used to acquire SE energy spectrum inside a conventional SEM chamber. V_S = -3,650 V and the beam separator inner sector excitation is 70.5 AT, while the outer sector excitation is -9AT.

an electrically insulted parallel plate pair, which is negatively biased to set up a retarding electric field. The mirror field sends secondary electrons back through the beam separator for a second pass, after which, they are collected by a photomultiplier tube (PMT).

Narrow-angle backscattered electrons (having higher energies) are transmitted through the mirror, and not monitored in this experiment. Moreover, due to the relatively small size of the prototype beam separator attachment, where only 2 mm separates the deflection plates, most backscattered electrons are lost by internal scattering in the out-of-plane direction. In order to obtain the BSE spectrum, a much larger prototype beam separator spectrometer needs to be made, requiring the construction of a custom made dedicated SEM specimen chamber.

The PMT current is monitored as a function of the applied mirror voltage V_M, effectively providing an integrated form of the secondary electron energy spectrum. The specimen voltage, V_S, is varied in one volt steps, and each spectrum takes 30 s to acquire. The output signal was averaged over 3 spectra, curve fitted, normalized, and differentiated. The results are shown in Fig. 7-22. They demonstrate that the spectrum shifts linearly, as expected for secondary electron detection. These preliminary results serve to show that the secondary electron spectrum can be obtained through the use of a magnetic beam separator spectrometer. The next step is to use a post-deflector, in order to project scattered electrons in parallel on to a multi-channel detector. Also, a much larger dedicated beam separator spectrometer needs to be made, similar to the design shown in Fig. 7-12, so that the BSE spectrum can be captured.

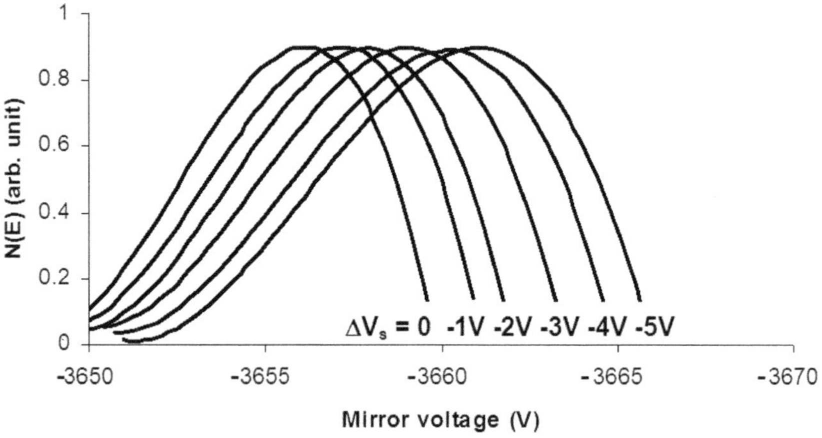

Fig. 7-22. Experimentally obtained secondary electron spectra as a function of the mirror voltage, where the specimen voltage is varied in one volt steps. The output signal has been curve fitted, normalized and differentiated in software form.

Chapter 8

Spectroscopic SEM proposals

The findings of this book can be summarized in the form of proposals for SEM redesign. Ways to combine energy spectrometers with focused scanning electron beams have of course, already been explored in Scanning Auger Microscopy (SAM), and it is natural to first see if any existing SAM can also function as a spectroscopic SEM. One such instrument, already described in Chapter 2, is the SAM design that places the objective lens inside a CMA spectrometer and captures scattered electrons on an off-axis multi-channel detector, as shown in Fig. 8-1. A commercial instrument based upon this design, made by Physical Electronics (PHI 700), limits the primary beam energy to be below 3 keV, which would make it suitable for low voltage SEM operation. In principle, the column/spectrometer can be redesigned to allow for higher primary beam energies, up to say 30 keV. As a low voltage spectroscopic SEM, this instrument would also benefit from the introduction of an aberration corrector unit, which would not only improve the image resolution to below 2 nm, but also increase the primary beam probe current by over an order of magnitude, greatly speeding up spectrum data acquisition times.

Figure 8-2(a) shows another possible spectroscopic SEM design, in this case, based upon the use of band-pass electrostatic spectrometers and flat plane detectors, as described in Chapter 6. The spectrometer might, in this case be a modified Fountain spectrometer or a Toroidal spectrometer. These band-pass electrostatic spectrometers can be fully rotationally symmetric, or segmented in the azimuthal angular direction, where each sector is separated by a high dielectric permittivity material, as shown in Fig. 8-2(b). Segmenting electric sectors in this way, means that they can function simultaneously with different pass energies, one set to capture SEs, and another designed to detect BSEs. A fully rotationally symmetric spectrometer in this type of column layout might also form the basis of a new SAM design. The Hoang (2010) modified Fountain analyzer or the Khursheed and Hoang (2008a) Toroidal analyzer are both predicted to have second-order focusing characteristics, comparable to or better than that of the CMA, while at the same time, operating in a limited form of parallel energy detection, like the HDA.

The spectroscopic SEM design based upon the use of electric band-pass analyzers should be able to provide high signal-to-noise SE signals, as shown by the experimental results presented in Chapter 6. This is suited for applications such as dopant concentration mapping of semiconductor specimens, where meV shifts in the SE spectrum need to be monitored. The technique of using two hemispherical extraction electrodes above the specimen, where the inner electrode and the specimen are biased to a negative voltage, say -10 V, greatly improves the signal-to-noise ratio of the output signal and makes it easier to track specimen voltage changes.

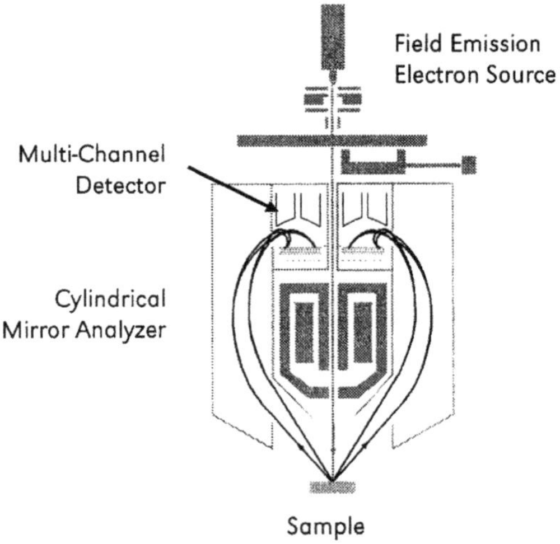

Fig. 8-1. Schematic diagram of a conventional objective lens inside a CMA. Courtesy of Physical Electronics.

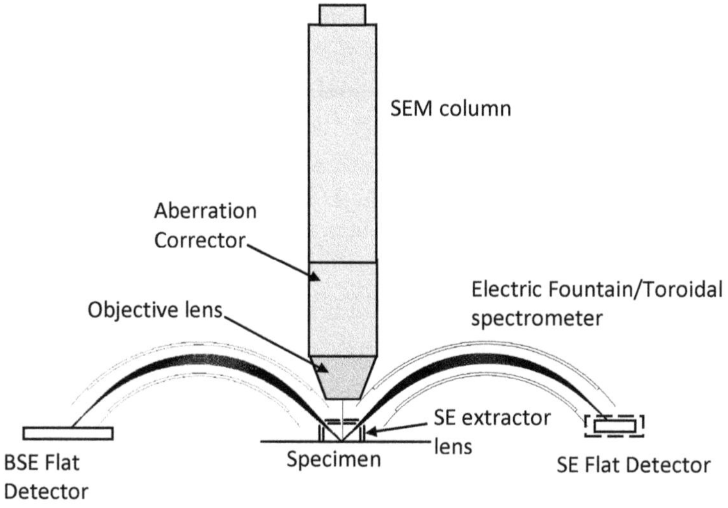

(a)

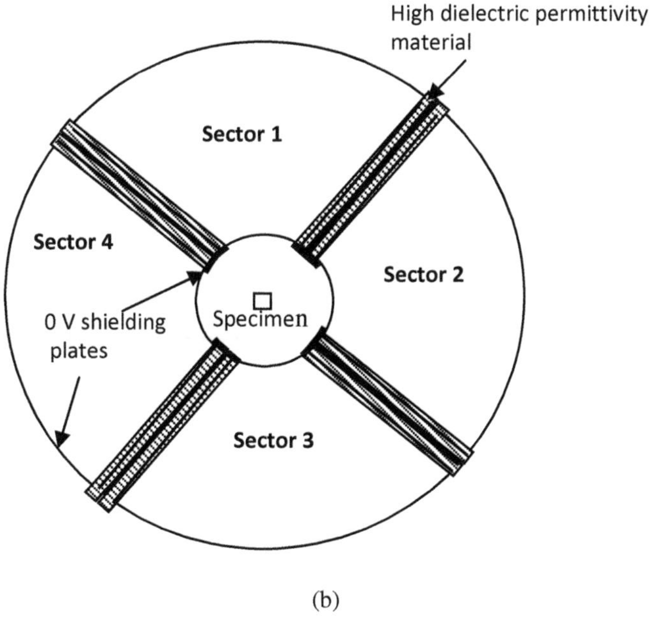

(b)

Fig. 8-2. Schematic layout of a conventional objective lens SEM combined with electric band-pass spectrometers. (a) Dual channel SE/BSE flat detectors, (b) Four channel sectors/detectors.

The SE signal is now transformed to have a sharp symmetric peak, and accurate open loop measurements of the specimen voltage can be made by monitoring peak position (or signal mean) changes. Experimental results in Chapter 6 demonstrate that the voltage resolution limited by shot noise for this method lies around 100 μV, more than an order of magnitude better than that obtained by previous voltage contrast spectrometers.

Experimental results in Chapter 6 also demonstrate that an electric band-pass analyzer can perform quantitative material contrast with high precision. By monitoring the BSE spectrum full-width at half-maximum height (FWHM), specimens that differ by an atomic number of one can be distinguished with a signal-to-noise of better than 20. Moreover, the FWHM falls monotonically with atomic number, providing the possibility of accurately estimating the atomic number of any specimen by capturing the BSE spectrum. This BSE method of elemental identification is a promising companion to the widely used EDS (Energy Dispersive Spectrometry) technique, commonly used for material analysis in SEMs. The BSE method can provide material contrast for the LVSEM mode of operation (beam energy between 0.5 to 1 keV, a probe current of a few pA and probe resolutions below 20 nm), overcoming standard EDS limitations of high primary beam voltages (typically between 10 to 20 keV), large probe currents (tens of nA), poor probe resolutions (several microns). Since a conventional objective lens design is used, an aberration corrector in this column can improve the image resolution when working either at low voltage or large working distances (or both), while at the same time, significantly increase the probe current. Larger probe currents for spectrometers are desirable, since they lead to smaller data-acquisition times.

Figure 8-3 shows how a parallel energy spectrometer can be combined with a conventional SEM objective lens column. The spectrometer is based upon the Gaussian field magnetic sector spectrometer described in Chapter 7. Several magnetic sector boxes of this type can be arranged in the azimuthal direction, and an electrostatic extractor lens can be used to increase the efficiency of SE collection. The SE and BSE spectra are

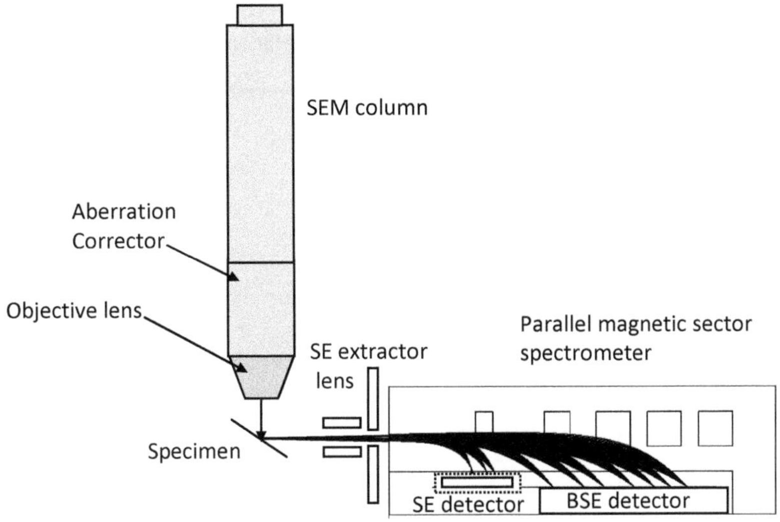

Fig. 8-3. Schematic layout of a conventional objective lens SEM combined with a Gaussian field magnetic sector parallel spectrometer

simultaneously captured on an array of detectors, or several multi-channel detectors. The advantage of capturing the energy spectrum in parallel is that the spectrometer does not have to be synchronized with the SEM scan coils, and a separate spectrum is directly captured for each pixel in the final image.

Figures 8-4 and 8-5 show schematic layouts of spectroscopic SEM designs that use a magnetic immersion objective lens. The trajectory paths of low energy SEs in both cases are collimated as they travel back up the column. In the first case, shown in Fig. 8-4, the output signal is related to the time-of-flight of scattered electrons, and is recorded by a high bandwidth MCP detection system located above a drift region. A beam blanker or pulsed source is used to pulse the primary beam. This type of spectroscopic SEM is inherently multi-channel, capable of capturing the entire spectral range of scattered electrons, and its time-sampling action makes it suitable to monitor fast changing periodic signals at the specimen surface. As demonstrated in Chapter 5, the

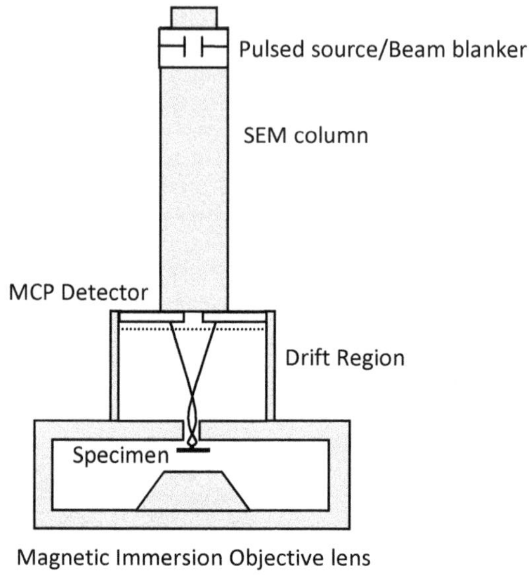

Fig. 8-4. Schematic layout for a time-of-flight spectrometer SEM design.

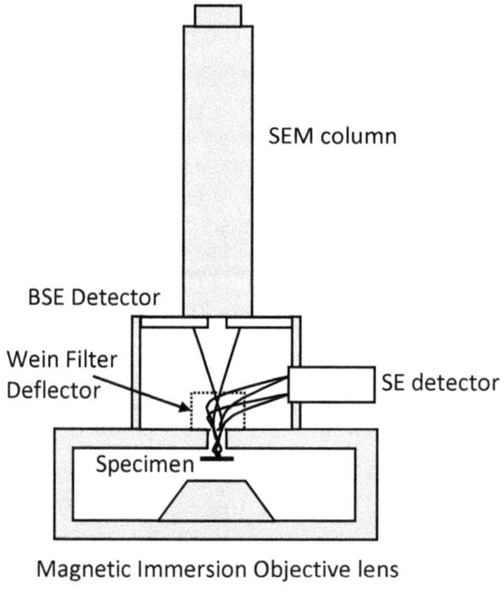

Fig. 8-5. Schematic layout of a magnetic immersion objective lens spectroscopic SEM.

time-of-flight spectrometer is capable of providing high signal-to-noise output SE signals, and the pulsing action of the blanker is only used when the spectrum is required. In normal imaging mode, a biased grid below the MCP functions to separate BSEs from SEs. The magnetic immersion objective lens provides high image resolution.

Figure 8-5 shows the schematic diagram of the same objective lens, but incorporating a band-pass or multi-channel spectrometer. A deflector, whether it be a Wein filter or an electrostatic deflection arrangement, is used to direct SEs off-axis to one side, where they are subsequently either detected in parallel, or go through a slit. Adverse effects on the image resolution are not created by the deflection field, since it is of relatively low strength, required only to deflect low energy secondaries away from the optical axis. However, if the spectrometer is used in band-pass mode, some form of image shift compensation is required as the deflection field strength is varied. In Chapter 5, SE energy spectra using an electrostatic deflector operating in band-pass mode with an add-on magnetic immersion lens is described, and experimental results confirm that this kind of spectroscopic SEM design is feasible.

Lastly, the schematic layout of an energy spectrometer combined with magnetic immersion retarding field objective lenses (mixed field immersion type) is shown Fig. 8-6. Objective lenses of this type, as analyzed in Chapters 3 and 4, provide the highest image resolution within the category of immersion lenses. They perform particularly well at low landing energies (≤ 1 keV). This spectroscopic SEM proposal is based upon the use of a magnetic beam separator, which bends the primary beam through a deflection angle, typically 90°, while at the same time dispersing scattered electrons, effectively acting as the first stage of a parallel energy spectrometer (Khursheed and Osterberg, 2006). The magnetic beam separator is designed to act like a low aberration round lens deflector for the primary beam (stigmatic focusing mode). This type of spectrometer is described in Chapter 7, in which preliminary experimental results for the SE spectrum are presented.

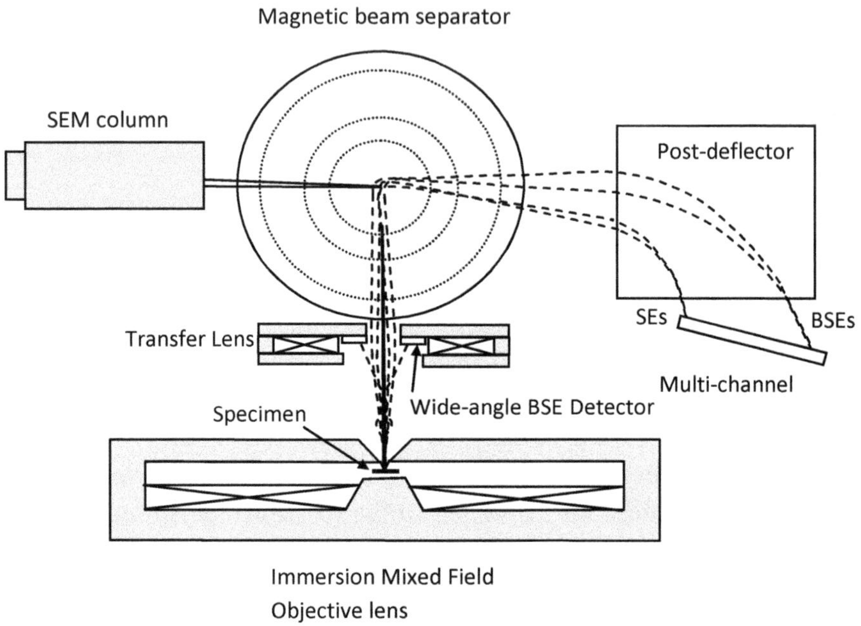

Fig. 8-6. Schematic layout of a mixed field immersion objective lens spectroscopic SEM based upon the use of a magnetic beam separator.

Appendix 1.0

Field Expansions

(i) Proof for

$$U(r,z) = U_0(z) - \frac{r^2}{4}\frac{\partial^2 U_0(z)}{\partial z^2} + \frac{r^4}{64}\frac{\partial^4 U_0(z)}{\partial z^4} - \frac{r^6}{2304}\frac{\partial^6 U_0(z)}{\partial z^6}\cdots.$$

The Laplace's equation for the electric field U in cylindrical coordinates (r,ϕ,z) is given by

$$\nabla^2 U = \frac{1}{r}\frac{\partial}{\partial r}\left(r\frac{\partial U}{\partial r}\right) + \frac{1}{r^2}\frac{\partial^2 U}{\partial \phi^2} + \frac{\partial^2 U}{\partial z^2} = 0. \tag{A1.0.1}$$

Since the electric field is axisymmetric, the field distribution is independent of ϕ and hence the Laplace's equation will become

$$\nabla^2 U = \frac{1}{r}\frac{\partial}{\partial r}\left(r\frac{\partial U}{\partial r}\right) + \frac{\partial^2 U}{\partial z^2} = 0. \tag{A1.0.2}$$

Since U is axisymmetric, $U(r,z) = U(-r,z)$. In this case, the distribution of the electric field can be expressed as a even function power series in r

$$U(r,z) = a_0(z) + a_2(z)r^2 + a_4(z)r^4 + \cdots \tag{A1.0.3}$$

$$\frac{\partial^2 U}{\partial z^2} = \frac{\partial^2 a_0(z)}{\partial z^2} + \frac{\partial^2 a_2(z)}{\partial z^2}r^2 + \frac{\partial^2 a_4(z)}{\partial z^2}r^4 + \cdots \tag{A1.0.4}$$

$$\frac{\partial U}{\partial r} = 2a_2(z)r + 4a_4(z)r^3 + \cdots \tag{A1.0.5}$$

$$r\frac{\partial U}{\partial r} = 2a_2(z)r^2 + 4a_4(z)r^4 + \cdots \tag{A1.0.6}$$

$$\frac{\partial}{\partial r}\left(r\frac{\partial U}{\partial r}\right) = 4a_2(z)r + 16a_4(z)r^3 + \cdots. \tag{A1.0.7}$$

Substituting Eq. (A1.0.4) and Eq. (A1.0.7) in Eq. (A1.0.2) we obtain,

$$\begin{aligned}\nabla^2 U = {} & \frac{\partial^2 a_0(z)}{\partial z^2} + \frac{\partial^2 a_2(z)}{\partial z^2} r^2 + \frac{\partial^2 a_4(z)}{\partial z^2} r^4 \\ & + 4a_2(z) + 16a_4(z)r^2 + \cdots = 0.\end{aligned} \tag{A1.0.8}$$

Substituting $r = 0$ in Eq. (A1.0.3), we obtain,

$$U(0,z) = U_0(z) = a_0(z) \tag{A1.0.9}$$

$$\frac{\partial^2 U(0,z)}{\partial z^2} = \frac{\partial^2 U_0(z)}{\partial z^2} \tag{A1.0.10}$$

$$\frac{\partial^4 U(0,z)}{\partial z^4} = \frac{\partial^4 U_0(z)}{\partial z^4}. \tag{A1.0.11}$$

Substituting the value of $a_0(z)$ in Eq. (A1.0.8) we obtain,

$$\begin{aligned}\nabla^2 U = {} & \left(\frac{\partial^2 U_0(z)}{\partial z^2} + 4a_2(z)\right) r^0 \\ & + \left(\frac{\partial^2 a_2(z)}{\partial z^2} + 16a_4(z)\right) r^2 + \frac{\partial^2 a_4(z)}{\partial z^2} r^4 + \cdots = 0.\end{aligned} \tag{A1.0.12}$$

Let $$f(r)=\left(\frac{\partial^2 U_0(z)}{\partial z^2}+4a_2(z)\right)r^0+\left(\frac{\partial^2 a_2(z)}{\partial z^2}+16a_4(z)\right)r^2+\frac{\partial^2 a_4(z)}{\partial z^2}r^4+\cdots$$

which is a Taylor series and hence its coefficients can be found using $f(0),\frac{\partial^2 f(0)}{\partial r^2},\frac{\partial^4 f(0)}{\partial r^4},\ldots$

$$f(r)=f(0)+\frac{\frac{\partial^2 f(0)}{\partial r^2}}{2!}r^2+\frac{\frac{\partial^4 f(0)}{\partial r^4}}{4!}r^4+\cdots.$$

Since the right hand side of the polynomial in Eq. (A1.0.12) is zero, then all the Taylor coefficient must be zero.

$$\begin{aligned}&\frac{\partial^2 U_0(z)}{\partial z^2}+4a_2(z)=0\\&a_2(z)=-\frac{1}{4}\frac{\partial^2 U_0(z)}{\partial z^2}\\&a_2(z)=\frac{(-1)^1}{2^2}\frac{\partial^2 U_0(z)}{\partial z^2}.\end{aligned}\tag{A1.0.13}$$

Similarly,

$$\begin{aligned}&\frac{\partial^2 a_2(z)}{\partial z^2}+16a_4(z)=0\\&\frac{-1}{4}\frac{\partial^4 U_0(z)}{\partial z^4}+16a_4(z)=0\\&a_4(z)=\frac{1}{64}\frac{\partial^4 U_0(z)}{\partial z^4}\\&a_4(z)=\frac{(-1)^2}{2^2\times 4^2}\frac{\partial^4 U_0(z)}{\partial z^4}.\end{aligned}\tag{A1.0.14}$$

Thus,

$$a_{2n}(z)=\frac{(-1)^n}{2^2\times4^2\times\cdots\times(2n)^2}\frac{\partial^{2n}U_0(z)}{\partial z^{2n}}. \tag{A1.0.15}$$

Substituting these values in Eq. (A1.0.3) we obtain,

$$U(r,z)=U_0(z)+\frac{(-1)^1}{2^2}\frac{\partial^2U_0(z)}{\partial z^2}r^2+\frac{(-1)^2}{2^2\times4^2}\frac{\partial^4U_0(z)}{\partial z^4}r^4$$
$$+\cdots+\frac{(-1)^n}{2^2\times4^2\times\cdots\times(2n)^2}\frac{\partial^{2n}U_0(z)}{\partial z^{2n}}r^{2n}$$

$$U(r,z)=U_0(z)-\frac{r^2}{4}\frac{\partial^2U_0(z)}{\partial z^2}+\frac{r^4}{64}\frac{\partial^4U_0(z)}{\partial z^4}-\frac{r^6}{2304}\frac{\partial^6U_0(z)}{\partial z^6}\cdots. \tag{A1.0.16}$$

(ii) Proof for $B_z(r,z)=B_0(z)-\frac{1}{4}\frac{\partial^2B_0(z)}{\partial z^2}r^2+\frac{1}{64}\frac{\partial^4B_0(z)}{\partial z^4}r^4\cdots$ and $B_r(r,z)=-\frac{1}{2}\frac{\partial B_0(z)}{\partial z}r+\frac{1}{16}\frac{\partial^3B_0(z)}{\partial z^3}r^3-\cdots$.

Let **B** be the magnetic flux density and **A** be the magnetic vector potential. We know that $\mathbf{B}=\nabla\times\mathbf{A}$. For an axisymmetric magnetic field, solving this is reduced to solving the component scalar equations in (r,θ,z).

$$\nabla\times A=\frac{1}{r}\begin{vmatrix}\hat{\mathbf{r}} & r\hat{\boldsymbol{\varphi}} & \hat{\mathbf{z}}\\ \frac{\partial}{\partial r} & \frac{\partial}{\partial\phi} & \frac{\partial}{\partial z}\\ A_r & rA_\phi & A_z\end{vmatrix} \tag{A1.0.17}$$

Therefore,

$$B_z = \frac{1}{r}\left[\frac{\partial(rA_\phi)}{\partial r} - \frac{\partial A_r}{\partial \phi}\right] \tag{A1.0.18}$$

$$B_r = \frac{1}{r}\left[\frac{\partial A_z}{\partial \phi} - \frac{\partial(rA_\phi)}{\partial z}\right] \tag{A1.0.19}$$

$$B_\phi = \frac{\partial A_r}{\partial z} - \frac{\partial A_z}{\partial r}. \tag{A1.0.20}$$

Due to rotational symmetry, all derivatives with respect to ϕ vanish in Eqs. (A1.0.18)–(A1.0.20). Therefore we obtain,

$$B_z = \frac{1}{r}\frac{\partial(rA_\phi)}{\partial r} \tag{A1.0.21}$$

$$B_r = -\frac{\partial A_\phi}{\partial z} \tag{A1.0.22}$$

$$B_\phi = \frac{\partial A_r}{\partial z} - \frac{\partial A_z}{\partial r}. \tag{A1.0.23}$$

Also, $B_\phi = 0$ (rotational symmetry of current carrying coil).

$$\begin{aligned} B_\phi &= \frac{\partial A_r}{\partial z} - \frac{\partial A_z}{\partial r} = 0 \\ \frac{\partial A_r}{\partial z} &= \frac{\partial A_z}{\partial r}. \end{aligned} \tag{A1.0.24}$$

Since B_z and B_r are independent of A_r and A_z in the above expressions, we can safely assume that:

$$A_z \equiv 0 \text{ and } A_r \equiv 0. \tag{A1.0.25}$$

We know that, $\nabla \times \mathbf{B} = 0$.

$$\nabla \times \mathbf{B} = \begin{vmatrix} \hat{\mathbf{r}} & \hat{\boldsymbol{\varphi}} & \hat{\mathbf{z}} \\ \dfrac{\partial}{\partial r} & \dfrac{1}{r}\dfrac{\partial}{\partial \phi} & \dfrac{\partial}{\partial z} \\ B_r & B_\phi & B_z \end{vmatrix} = 0 \tag{A1.0.26}$$

$$= \begin{vmatrix} \hat{\mathbf{r}} & \hat{\boldsymbol{\varphi}} & \hat{\mathbf{z}} \\ \dfrac{\partial}{\partial r} & \dfrac{1}{r}\dfrac{\partial}{\partial \phi} & \dfrac{\partial}{\partial z} \\ B_r & 0 & B_z \end{vmatrix} = 0$$

$$= \hat{\mathbf{r}}\left(\frac{1}{r}\frac{\partial B_z}{\partial \phi}\right) + \hat{\boldsymbol{\varphi}}\left(\frac{\partial B_r}{\partial z} - \frac{\partial B_z}{\partial r}\right) + \hat{\mathbf{z}}\left(-\frac{1}{r}\frac{\partial B_r}{\partial \phi}\right) = 0. \tag{A1.0.27}$$

Since the partial derivatives with respect to ϕ is zero, Eq. (A1.0.27) becomes

$$\frac{\partial B_r}{\partial z} - \frac{\partial B_z}{\partial r} = 0.$$

Substituting the values of B_z and B_r from Eqs. (A1.0.21) and (A1.0.22), we obtain

$$\begin{aligned} \frac{\partial}{\partial z}\left(-\frac{\partial A_\phi}{\partial z}\right) - \frac{\partial}{\partial r}\left[\frac{1}{r}\frac{\partial (rA_\phi)}{\partial r}\right] &= 0 \\ \frac{\partial^2 A_\phi}{\partial z^2} + \frac{\partial}{\partial r}\left[\frac{1}{r}\frac{\partial (rA_\phi)}{\partial r}\right] &= 0 \\ \frac{\partial^2 A_\phi}{\partial z^2} + \frac{\partial}{\partial r}\left[\frac{1}{r}\left(r\frac{\partial A_\phi}{\partial r} + A_\phi\right)\right] &= 0 \\ \frac{\partial^2 A_\phi}{\partial z^2} + \frac{\partial^2 A_\phi}{\partial r^2} + \frac{\partial}{\partial r}\left(\frac{A_\phi}{r}\right) &= 0. \end{aligned} \tag{A1.0.28}$$

The solution of the above equation can be given by the Taylor series,

$$A_\phi(r,z) = c_1(z)r + c_3(z)r^3 + c_5(z)r^5 + \cdots. \tag{A1.0.29}$$

From this, we can find the derivatives,

$$\frac{\partial^2 A_\phi}{\partial z^2} = \frac{\partial^2 c_1(z)}{\partial z^2} r + \frac{\partial^2 c_3(z)}{\partial z^2} r^3 + \frac{\partial^2 c_5(z)}{\partial z^2} r^5 + \cdots$$

$$\frac{\partial^2 A_\phi}{\partial r^2} = 6c_3(z)r + 20c_5(z)r^3 + \cdots$$

$$\frac{\partial}{\partial r}\left(\frac{A_\phi}{r}\right) = 2c_3(z)r + 4c_5(z)r^3 + \cdots.$$

Substituting these values in Eq. (A1.0.28) we obtain,

$$\left[\frac{\partial^2 c_1(z)}{\partial z^2} + 8c_3(z)\right]r + \left[\frac{\partial^2 c_3(z)}{\partial z^2} + 24c_5(z)\right]r^3 + \frac{\partial^2 c_5(z)}{\partial z^2} r^5 = 0. \tag{A1.0.30}$$

Since the right hand side of the above equation is zero, the Taylor coefficients must all be zero. Equating the Taylor coefficients to zero, we obtain

$$\frac{\partial^2 c_1(z)}{\partial z^2} + 8c_3(z) = 0$$

$$\frac{\partial^2 c_3(z)}{\partial z^2} = -\frac{1}{8}\frac{\partial^4 c_1(z)}{\partial z^4}$$

$$\frac{\partial^2 c_3(z)}{\partial z^2} + 24c_5(z) = 0$$

$$c_5(z) = -\frac{1}{24}\left(-\frac{1}{8}\frac{\partial^4 c_1(z)}{\partial z^4}\right) = \frac{\frac{\partial^4 c_1(z)}{\partial z^4}}{2\times 4^2 \times 6}$$

$$A_\varphi(r,z)=c_1(z)r-\frac{1}{24}\frac{\partial^2 c_1(z)}{\partial z^2}r^3+\frac{\frac{\partial^4 c_1(z)}{\partial z^4}}{2\times 4^2\times 6}r^5+\cdots. \quad \text{(A1.0.31)}$$

Since $B_\phi(r,z)=0$, $\mathbf{B}=B_r\hat{\mathbf{r}}+B_z\hat{\mathbf{z}}$.

$$B_z(r,z)=\frac{1}{r}\frac{\partial(rA_\phi)}{\partial r}$$

$$=\frac{1}{r}\left(2c_1(z)r-\frac{4\frac{\partial^2 c_1(z)}{\partial z^2}r^3}{2\times 4}+\frac{6\frac{\partial^4 c_1(z)}{\partial z^4}r^5}{2\times 4^2\times 6}+\cdots\right)$$

$$=2c_1(z)-\frac{\frac{\partial^2 c_1(z)}{\partial z^2}r^2}{2}+\frac{\frac{\partial^4 c_1(z)}{\partial z^4}r^4}{2\times 4^2}+\cdots$$

$$B_z(0,z)=2c_1(z)=B_0(z) \quad \text{(A1.0.31a)}$$

$$c_1(z)=\frac{B_0(z)}{2}. \quad \text{(A1.0.31b)}$$

Therefore,

$$B_z(r,z)=B_0(z)-\frac{\frac{\partial^2 B_0(z)}{\partial z^2}r^2}{2^2}+\frac{\frac{\partial^4 B_0(z)}{\partial z^4}r^4}{2^2\times 4^2}+\cdots$$

$$B_z(r,z)=B_0(z)-\frac{1}{4}\frac{\partial^2 B_0(z)}{\partial z^2}r^2+\frac{1}{64}\frac{\partial^4 B_0(z)}{\partial z^4}r^4\cdots. \quad \text{(A1.0.32)}$$

$$B_r(r,z) = -\frac{\partial A_\phi}{\partial z}$$

$$= -\frac{\partial c_1(z)}{\partial z} r + \frac{1}{2\times 4}\frac{\partial^3 c_1(z)}{\partial z^3} r^3 - \frac{\dfrac{\partial^5 c_1(z)}{\partial z^5} r^5}{2\times 4^2\times 6} + \cdots$$

$$= -\frac{\dfrac{\partial B_0(z)}{\partial z}}{2} r + \frac{\dfrac{\partial^3 B_0(z)}{\partial z^3}}{2^2\times 4} r^3 - \frac{\dfrac{\partial^5 B_0(z)}{\partial z^5}}{2^2\times 4^2\times 6} r^5 + \cdots$$

$$B_r(r,z) = -\frac{1}{2}\frac{\partial B_0(z)}{\partial z} r + \frac{1}{16}\frac{\partial^3 B_0(z)}{\partial z^3} r^3 - \cdots. \qquad \text{(A1.0.33)}$$

Appendix 1.1

Derivation of the Paraxial Equation

We know that,

$$\mathbf{F} = \mathbf{F}_e + \mathbf{F}_m = -e\mathbf{E} - e(\mathbf{v} \times \mathbf{B}). \tag{A1.1.1}$$

For rotationally symmetric field distributions we have $E_\phi = 0$ and $B_\phi = 0$. Thus,

$$E_z = -\frac{\partial \mathbf{U}(r,z)}{\partial z} \qquad E_r = -\frac{\partial \mathbf{U}(r,z)}{\partial r}. \tag{A1.1.2}$$

Using Eq. (A1.0.16), we can find the values of Eq. (A1.1.2) as

$$E_z = -\frac{\partial \mathbf{U}(r,z)}{\partial z} = -\frac{dU_0(z)}{dz} + \frac{1}{4}\frac{d^3U_0(z)}{dz^3}r^2 + \cdots \tag{A1.1.3}$$

$$E_r = -\frac{\partial \mathbf{U}(r,z)}{\partial r} = \frac{1}{2}\frac{d^2U_0(z)}{dz^2}r - \frac{1}{16}\frac{d^4U_0(z)}{dz^4}r^3 + \cdots \tag{A1.1.4}$$

The vector equation of motion cylindrical in coordinates is given by

$$\mathbf{F} = m\left[(\ddot{r} - r\dot{\phi}^2)\hat{\mathbf{r}} + \left(\frac{1}{r}\frac{\mathrm{d}}{\mathrm{dt}}(r^2\dot{\phi})\right)\hat{\boldsymbol{\varphi}} + \ddot{z}\hat{\mathbf{z}}\right]$$

$$F_r = m(\ddot{r} - r\dot{\phi}^2) \tag{A1.1.5}$$

$$F_\phi = m\frac{1}{r}\frac{\mathrm{d}}{\mathrm{dt}}(r^2\dot{\phi})$$

$$F_z = m\ddot{z}.$$

The vector equation for velocity in cylindrical coordinates can be written as

$$\begin{aligned}\mathbf{v} &= \dot{r}\hat{\mathbf{r}} + r\dot{\phi}\hat{\boldsymbol{\varphi}} + \dot{z}\hat{\mathbf{z}} \\ v_r &= \dot{r} \\ v_\phi &= r\dot{\phi} \\ v_z &= \dot{z}\end{aligned} \tag{A1.1.6}$$

$$\begin{aligned}\mathbf{v}\times\mathbf{B} &= \begin{vmatrix} \hat{\mathbf{r}} & \hat{\boldsymbol{\varphi}} & \hat{\mathbf{z}} \\ \dot{r} & r\dot{\phi} & \dot{z} \\ -\dfrac{\partial A}{\partial z} & 0 & \dfrac{\partial A}{\partial r} \end{vmatrix} \\ &= \left(r\dot{\phi}\frac{\partial A}{\partial r} \right)\hat{\mathbf{r}} - \left(\dot{r}\frac{\partial A}{\partial r} + \frac{\partial A}{\partial z}\dot{z} \right)\hat{\boldsymbol{\varphi}} + \left(\frac{\partial A}{\partial z} r\dot{\phi} \right)\hat{\mathbf{z}}.\end{aligned} \tag{A1.1.7}$$

Substituting Eqs. (A1.1.2) and (A1.1.7) in Eq. (A1.1.1), we obtain

$$\begin{aligned}F_r &= e\frac{\partial U}{\partial r} - e\dot{\phi}\frac{\partial (rA)}{\partial r} \\ F_\phi &= \frac{e}{r}\frac{\mathrm{d}(rA)}{\mathrm{dt}} \\ F_z &= e\frac{\partial U}{\partial z} - er\dot{\phi}\frac{\partial A}{\partial z}.\end{aligned} \tag{A1.1.8}$$

Comparing Eqs. (A1.1.5) and (A1.1.8), we obtain,

$$\begin{aligned}m(\ddot{r} - r\dot{\phi}^2) &= e\frac{\partial U}{\partial r} - e\dot{\phi}\frac{\partial (rA)}{\partial r} \\ m\frac{1}{r}\frac{\mathrm{d}}{\mathrm{dt}}(r^2\dot{\phi}) &= \frac{e}{r}\frac{\mathrm{d}(rA)}{\mathrm{dt}} \\ m\ddot{z} &= e\frac{\partial U}{\partial z} - er\dot{\phi}\frac{\partial A}{\partial z}.\end{aligned} \tag{A1.1.9}$$

Integrating the middle equation in the system of Eq. (A1.1.9) we obtain,

$$\frac{m}{e}r^2\dot{\phi} = rA + C. \tag{A1.1.9a}$$

From this,

$$\dot{\phi} = \frac{rA + C}{r^2}\frac{e}{m}. \tag{A1.1.10}$$

Substituting Eq. (A1.1.10) in the first and last equation in the system of Eq. (A1.1.9),

$$m\left[\ddot{r} - r\left(\frac{rA+C}{r^2}\frac{e}{m}\right)^2\right] = e\frac{\partial U}{\partial r} - e\frac{rA+C}{r^2}\frac{e}{m}\frac{\partial (rA)}{\partial r}$$

$$\frac{m}{e}\ddot{r} = r\left(\frac{rA+C}{r^2}\right)^2\frac{e}{m} + \frac{\partial U}{\partial r} - \frac{rA+C}{r^2}\frac{e}{m}\frac{\partial (rA)}{\partial r}$$

$$= \frac{\partial U}{\partial r} + \frac{e}{m}r\left(\frac{rA+C}{r^2}\right)^2 - \frac{e}{m}\frac{rA+C}{r^2}\frac{\partial (rA)}{\partial r}$$

$$= \frac{\partial U}{\partial r} - \frac{e}{m}\left(\frac{\partial (rA)}{\partial r}\frac{rA+C}{r^2} - r\left(\frac{rA+C}{r^2}\right)^2\right)$$

$$= \frac{\partial U}{\partial r} - \frac{e}{m}\left(\frac{\partial (rA)}{\partial r}\frac{rA+C}{r^2} - \frac{rA+C}{r^2}\frac{rA+C}{r}\right)$$

$$= \frac{\partial U}{\partial r} - \frac{e}{m}\left(\frac{\partial (rA)}{\partial r}\frac{rA+C}{r^2} - \frac{rA+C}{r^2}\left(A + \frac{C}{r}\right)\right)$$

$$= \frac{\partial U}{\partial r} - \frac{e}{m}\left(\frac{\partial (rA)}{\partial r}\frac{rA+C}{r^2} - A\frac{rA+C}{r^2} - \frac{C}{r}\frac{rA+C}{r^2}\right)$$

$$= \frac{\partial U}{\partial r} - \frac{e}{m}\left(\left(\frac{\partial (rA)}{\partial r} - A\right)\frac{rA+C}{r^2} - \frac{C}{r}\frac{rA+C}{r^2}\right)$$

$$= \frac{\partial U}{\partial r} - \frac{e}{m}\left(\left(r\frac{\partial A}{\partial r}\right)\frac{rA+C}{r^2} - \frac{C}{r}\frac{rA+C}{r^2}\right)$$

$$=\frac{\partial U}{\partial r}-\frac{e}{m}\left(\frac{\partial A}{\partial r}\frac{rA+C}{r}-\frac{C}{r^2}\frac{rA+C}{r}\right)$$

$$=\frac{\partial U}{\partial r}-\frac{e}{m}\left(\left(A+\frac{C}{r}\right)\left(\frac{\partial A}{\partial r}-\frac{C}{r^2}\right)\right)$$

$$=\frac{\partial U}{\partial r}-\frac{e}{m}\left(A+\frac{C}{r}\right)\left(\frac{\partial A}{\partial r}+\frac{\partial}{\partial r}\left(\frac{C}{r}\right)\right)$$

$$=\frac{\partial U}{\partial r}-\frac{e}{m}\left(A+\frac{C}{r}\right)\frac{\partial}{\partial r}\left(A+\frac{C}{r}\right)$$

$$=\frac{\partial U}{\partial r}-\frac{e}{2m}\frac{\partial}{\partial r}\left(A+\frac{C}{r}\right)^2$$

$$\frac{m\ddot{r}}{e}=\frac{\partial}{\partial r}\left[U-\frac{e}{2m}\left(A+\frac{C}{r}\right)^2\right] \qquad \text{(A1.1.11)}$$

$$\frac{m\ddot{z}}{e}=\frac{\partial U}{\partial z}-r\frac{rA+C}{r^2}\frac{e}{m}\frac{\partial A}{\partial z}$$

$$=\frac{\partial U}{\partial z}-\left(A+\frac{C}{r}\right)\frac{e}{m}\frac{\partial A}{\partial z}$$

$$=\frac{\partial U}{\partial z}-\frac{e}{m}\frac{1}{2}\frac{\partial}{\partial z}\left(A+\frac{C}{r}\right)^2.$$

$$\frac{m\ddot{z}}{e}=\frac{\partial}{\partial z}\left[U-\frac{e}{2m}\left(A+\frac{C}{r}\right)^2\right]. \qquad \text{(A1.1.12)}$$

Let $\Phi(r,z)=U-\frac{e}{2m}\left(A+\frac{C}{r}\right)^2$ (called Störmer Potential). (A1.1.12a)

Therefore, Eqs. (A1.1.11), (A1.1.12) and (A1.1.9a) become,

$$\begin{aligned} \frac{m}{e}\ddot{r} &= \frac{\partial \Phi}{\partial r} \\ \frac{m}{e}\ddot{z} &= \frac{\partial \Phi}{\partial z} \\ \frac{m}{e}\dot{\phi} &= \frac{A}{r} + \frac{C}{r^2}. \end{aligned} \tag{A1.1.13}$$

$$\mathbf{F} = \frac{\mathrm{d}}{\mathrm{dt}}(m\mathbf{v}) = -e\mathbf{E} - e(\mathbf{v}\times\mathbf{B}). \tag{A1.1.14}$$

Taking scalar dot product by $\mathbf{v}$ on both sides, we obtain,

$$\begin{aligned} \mathbf{v}\cdot\frac{\mathrm{d}}{\mathrm{dt}}(m\mathbf{v}) &= -e\mathbf{E}\cdot\mathbf{v} - e\mathbf{v}\cdot(\mathbf{v}\times\mathbf{B}) \\ \frac{\mathrm{d}}{\mathrm{dt}}\left(\frac{mv^2}{2}\right) &= -e\mathbf{E}\cdot\mathbf{v} \\ &= -e(E_x v_x + E_y v_y + E_z v_z) \\ &= e\left(\frac{\partial U}{\partial x}\frac{\mathrm{dx}}{\mathrm{dt}} + \frac{\partial U}{\partial y}\frac{\mathrm{dy}}{\mathrm{dt}} + \frac{\partial U}{\partial z}\frac{\mathrm{dz}}{\mathrm{dt}}\right) \\ &= e\frac{\mathrm{d}U}{\mathrm{dt}}. \end{aligned}$$

Integrating the above result on both sides, we obtain,

$$\frac{1}{2}mv^2 = eU + \text{constant}. \tag{A1.1.15}$$

Equation (A1.1.15) is the energy equation of an electron in an electromagnetic field. In cylindrical coordinates, it can be expanded as

$$\frac{1}{2}m(\dot{r}^2 + r^2\dot{\phi}^2 + \dot{z}^2) = e\mathrm{U(r,z)} \tag{A1.1.16}$$

$$\begin{aligned}
\dot{z}^2 &= 2\frac{\mathrm{e}}{\mathrm{m}}U(r,z) - \dot{r}^2 - r^2\dot{\phi}^2 \\
&= 2\frac{\mathrm{e}}{\mathrm{m}}U(r,z) - \left(\frac{\mathrm{d}r}{\mathrm{d}t}\right)^2 - r^2\left(\frac{\mathrm{d}\phi}{\mathrm{d}t}\right)^2 \\
&= 2\frac{\mathrm{e}}{\mathrm{m}}U(r,z) - \left(\frac{\mathrm{d}z}{\mathrm{d}z}\frac{\mathrm{d}r}{\mathrm{d}t}\right)^2 - r^2\left(\frac{\mathrm{d}z}{\mathrm{d}z}\frac{\mathrm{d}\phi}{\mathrm{d}t}\right)^2 \\
&= 2\frac{\mathrm{e}}{\mathrm{m}}U(r,z) - \left(\frac{dr}{dz}\dot{z}\right)^2 - r^2\left(\frac{d\phi}{dz}\dot{z}\right)^2 \\
&= 2\frac{\mathrm{e}}{\mathrm{m}}U(r,z) - \dot{z}^2\left[\left(\frac{dr}{dz}\right)^2 + r^2\left(\frac{d\phi}{dz}\right)^2\right].
\end{aligned}$$

$$\dot{z}^2 = \frac{2e}{m}\frac{U(r,z)}{\left(1 + \left(\frac{dr}{dz}\right)^2 + r^2\left(\frac{d\phi}{dz}\right)^2\right)}. \tag{A1.1.17}$$

$$\begin{aligned}
\ddot{r} &= \frac{\mathrm{d}}{\mathrm{d}t}\left(\frac{\mathrm{d}z}{\mathrm{d}z}\frac{\mathrm{d}r}{\mathrm{d}t}\right) = \frac{\mathrm{d}}{\mathrm{d}t}\left(\frac{\mathrm{d}z}{\mathrm{d}t}\frac{\mathrm{d}r}{\mathrm{d}z}\right) = \frac{\mathrm{d}}{\mathrm{d}t}(\dot{z}\frac{dr}{dz}) \\
&= \dot{z}\frac{\mathrm{d}}{\mathrm{d}t}\left(\frac{\mathrm{d}r}{\mathrm{d}z}\right) + \frac{\mathrm{d}r}{\mathrm{d}z}\ddot{z} = \dot{z}\frac{\mathrm{d}}{\mathrm{d}t}\left(\frac{\mathrm{d}z}{\mathrm{d}z}\frac{\mathrm{d}r}{\mathrm{d}z}\right) + \ddot{z}\frac{dr}{dz} \\
&= \dot{z}\frac{\mathrm{d}z}{\mathrm{d}t}\frac{\mathrm{d}}{\mathrm{d}z}\left(\frac{\mathrm{d}r}{\mathrm{d}z}\right) + \ddot{z}\frac{dr}{dz}.
\end{aligned}$$

$$\ddot{r} = \dot{z}^2\frac{d^2r}{dz} + \ddot{z}\frac{dr}{dz}. \tag{A1.1.18}$$

Substituting Eq. (A1.1.18) in Eq. (A1.1.13), we obtain,

$$
\begin{aligned}
\frac{m}{e}\ddot{r} &= \frac{\partial \Phi}{\partial r} \\
\frac{m}{e}(\dot{z}^2 \frac{d^2 r}{dz^2} + \ddot{z}\frac{dr}{dz}) &= \frac{\partial \Phi}{\partial r} \\
\frac{m}{e}\dot{z}^2 \frac{d^2 r}{dz^2} + \frac{m}{e}\ddot{z}\frac{dr}{dz} &= \frac{\partial \Phi}{\partial r} \\
\frac{m}{e}\dot{z}^2 \frac{d^2 r}{dz^2} + \frac{\partial \Phi}{\partial z}\frac{dr}{dz} &= \frac{\partial \Phi}{\partial r} \\
\frac{m}{e}\dot{z}^2 \frac{d^2 r}{dz^2} &= \frac{\partial \Phi}{\partial r} - \frac{\partial \Phi}{\partial z}\frac{dr}{dz}.
\end{aligned}
\tag{A1.1.19}
$$

Substituting Eq. (A1.1.17) in the above equation, we obtain,

$$
\begin{aligned}
\frac{m}{e}\frac{2e}{m}\frac{U(r,z)}{\left(1+\left(\frac{dr}{dz}\right)^2 + r^2\left(\frac{d\phi}{dz}\right)^2\right)}\frac{d^2 r}{dz^2} &= \frac{\partial \Phi}{\partial r} - \frac{\partial \Phi}{\partial z}\frac{dr}{dz} \\
\frac{2U(r,z)}{\left(1+\left(\frac{dr}{dz}\right)^2 + r^2\left(\frac{d\phi}{dz}\right)^2\right)}\frac{d^2 r}{dz^2} &= \frac{\partial \Phi}{\partial r} - \frac{\partial \Phi}{\partial z}\frac{dr}{dz}.
\end{aligned}
\tag{A1.1.20}
$$

Since $\dot{\phi} = 0$, from Eq. (A1.1.13), we will obtain $C = -Ar$. Substituting this in Eq. (A1.1.12a), we obtain

$$
\Phi(r,z) = U(r,z). \tag{A1.1.21}
$$

Using Eq. (A1.1.21), we can write Eq. (A1.1.20) as

$$\frac{2\Phi(r,z)}{\left(1+\left(\frac{dr}{dz}\right)^2+r^2\left(\frac{d\phi}{dz}\right)^2\right)}\frac{d^2r}{dz^2}=\frac{\partial\Phi(r,z)}{\partial r}-\frac{\partial\Phi(r,z)}{\partial z}\frac{dr}{dz}$$

$$\frac{\mathrm{d}^2r}{\mathrm{d}z^2}=\frac{1}{2\Phi(r,z)}\left[1+\left(\frac{\mathrm{d}r}{\mathrm{d}z}\right)^2+r^2\left(\frac{\mathrm{d}\phi}{\mathrm{d}z}\right)^2\right]\left[\frac{\partial\Phi(r,z)}{\partial r}-\frac{\partial\Phi(r,z)}{\partial z}\frac{\mathrm{d}r}{\mathrm{d}z}\right].$$

Since $\dot{\phi}=0$, $\frac{\mathrm{d}\phi}{\mathrm{d}t}=\frac{\mathrm{d}\iota}{\mathrm{d}z}\frac{\mathrm{d}z}{\mathrm{d}t}=0\Rightarrow\frac{\mathrm{d}\phi}{\mathrm{d}z}=0$ (because $\dot{z}\neq 0$). Therefore, the above equation becomes,

$$\frac{\mathrm{d}^2r}{\mathrm{d}z^2}-\frac{1+\left(\frac{\mathrm{d}r}{\mathrm{d}z}\right)^2}{2\Phi(r,z)}\left[\frac{\partial\Phi(r,z)}{\partial r}-\frac{\partial\Phi(r,z)}{\partial z}\frac{\mathrm{d}r}{\mathrm{d}z}\right]=0. \tag{A1.1.22}$$

Equation (A1.1.22) is the general equation for the electron path in a rotationally symmetric field.

For a paraxial electron path (near-axis path), to determine the first order optical properties, such as the beam's focal position or focal length, it is sufficient to make the following assumptions:

$$U(r,z)=U_0(z)-\frac{r^2}{4}\frac{d^2U_0(z)}{dz^2} \tag{A1.1.22a}$$

$$E_z=-\frac{\partial\mathbf{U}(r,z)}{\partial z}=-\frac{dU_0(z)}{dz} \tag{A1.1.23}$$

$$E_r=-\frac{\partial\mathbf{U}(r,z)}{\partial r}=\frac{r}{2}\frac{d^2U_0(z)}{dz^2}. \tag{A1.1.24}$$

Because we are considering a paraxial electron, $U(r,z)=U_0(z)$ and $\frac{\partial U_0(z)}{\partial z}=\frac{dU_0(z)}{dz}$.

Using Eqs. (A1.0.31) and (A1.0.31a) we have,

$$A=A_\phi(r,z)=c_1(z)r=\frac{B_0(z)}{2}r. \tag{A1.1.25}$$

Also Eqs. (A1.0.32) and (A1.0.33) will become,

$$B_z=B_0(z) \tag{A1.1.26}$$

$$B_r=-\frac{\partial B_0(z)}{\partial z}\frac{r}{2}. \tag{A1.1.27}$$

With these values, the Störmer potential becomes,

$$\Phi(r,z)=U_0(z)-\frac{r^2}{4}\frac{d^2U_0(z)}{dz^2}-\frac{e}{2m}\left(\frac{B_0(z)}{2}r+\frac{C}{r}\right)^2$$

$$=U_0(z)-\frac{r^2}{4}\frac{d^2U_0(z)}{dz^2}-\frac{e}{2m}\left(\frac{(B_0(z))^2}{4}r^2+\frac{C^2}{r^2}+B_0(z)C\right)$$

$$\frac{\partial\Phi(r,z)}{\partial r}=-\frac{r}{2}\frac{d^2U_0(z)}{dz^2}-\frac{e}{2m}\left[\frac{(B_0(z))^2}{4}2r-\frac{2C^2}{r^3}\right]$$

$$=-\frac{r}{2}\frac{d^2U_0(z)}{dz^2}-\frac{e}{m}r\left[\frac{(B_0(z))^2}{4}-\frac{C^2}{r^4}\right]$$

$$=-r\left(\frac{\frac{d^2U_0(z)}{dz^2}}{2}+\frac{e}{m}\left[\frac{(B_0(z))^2}{4}-\frac{C^2}{r^4}\right]\right).$$

From Eq. (A1.1.13),

$$\frac{m\ddot{r}}{e}=\frac{\partial\Phi(r,z)}{\partial r}=-r\left(\frac{\frac{d^2U_0(z)}{dz^2}}{2}+\frac{e}{m}\left[\frac{(B_0(z))^2}{4}-\frac{C^2}{r^4}\right]\right). \tag{A1.1.28}$$

With the assumption $C=0$, Eq. (A1.1.28) becomes,

$$\frac{m\ddot{r}}{e}=\frac{\partial\Phi(r,z)}{\partial r}=-r\left(\frac{\frac{d^2U_0(z)}{dz^2}}{2}+\frac{e}{m}\frac{(B_0(z))^2}{4}\right). \tag{A1.1.28a}$$

The assumption of paraxial electron gives

$$v\approx\frac{\mathrm{d}z}{\mathrm{d}t}. \tag{A1.1.29}$$

Therefore, from Eq. (A1.1.15), we obtain,

$$v^2=\left(\frac{\mathrm{d}z}{\mathrm{d}t}\right)^2=2\frac{e}{m}U_0(z). \tag{A1.1.30}$$

$$\dot{r} = \frac{\mathrm{d}r}{\mathrm{d}t} = \frac{\mathrm{d}r}{\mathrm{d}z}\frac{\mathrm{d}z}{\mathrm{d}t} = v\frac{\mathrm{d}r}{\mathrm{d}z}$$

$$\ddot{r} = \frac{\mathrm{d}\dot{r}}{\mathrm{d}t} = \frac{\mathrm{d}\dot{r}}{\mathrm{d}z}\frac{\mathrm{d}z}{\mathrm{d}t} = v\frac{\mathrm{d}}{\mathrm{d}z}\left(v\frac{\mathrm{d}r}{\mathrm{d}z}\right)$$

$$= \sqrt{\frac{2e}{m}U_0(z)}\left(\sqrt{\frac{2e}{m}U_0(z)}\frac{\mathrm{d}^2 r}{\mathrm{d}z^2} + \frac{\mathrm{d}r}{\mathrm{d}z}\sqrt{\frac{2e}{m}}\frac{\frac{dU_0(z)}{dz}}{2\sqrt{U_0(z)}}\right) \tag{A1.1.31}$$

$$= \frac{2e}{m}\left(U_0(z)\frac{\mathrm{d}^2 r}{\mathrm{d}z^2} + \frac{\frac{dU_0(z)}{dz}}{2}\frac{\mathrm{d}r}{\mathrm{d}z}\right).$$

Using the above value in Eq. (A1.1.28a), we obtain,

$$\frac{m}{e}\frac{2e}{m}\left(U_0(z)\frac{\mathrm{d}^2 r}{\mathrm{d}z^2} + \frac{\frac{dU_0(z)}{dz}}{2}\frac{\mathrm{d}r}{\mathrm{d}z}\right) = -r\left(\frac{\frac{d^2U_0(z)}{dz^2}}{2} + \frac{e}{m}\frac{(B_0(z))^2}{4}\right)$$

$$\left(U_0(z)\frac{\mathrm{d}^2 r}{\mathrm{d}z^2} + \frac{\frac{dU_0(z)}{dz}}{2}\frac{\mathrm{d}r}{\mathrm{d}z}\right) + r\left(\frac{\frac{d^2U_0(z)}{dz^2}}{4} + \frac{e}{8m}(B_0(z))^2\right) = 0.$$

Rewritten as,

$$\left(\frac{\mathrm{d}^2 r}{\mathrm{d}z^2} + \frac{1}{2U_0(z)}\frac{dU_0(z)}{dz}\frac{\mathrm{d}r}{\mathrm{d}z}\right) + \left(\frac{1}{4U_0(z)}\frac{d^2U_0(z)}{dz^2} + \frac{e}{8mU_0(z)}(B_0(z))^2\right)r = 0. \tag{A1.1.32}$$

This the paxial equation of motion in electric/magnetic rotationally symmetric field distributions.

Appendix 1.2

Spherical Aberration

Spherical aberration coefficient can be obtained by including terms up to the third order in the series expansion of electric and magnetic field. Thus, if we use

$$U(r,z)=U-\frac{r^2}{4}U''+\frac{r^4}{64}U^{IV}$$

$$A(r,z)=A_\phi=\frac{1}{2}rB-\frac{1}{16}r^3B''$$

the Störmer Potential will become (with $C=0$),

$$\Phi(r,z)=U-\frac{r^2}{2}\left(\phi''+\frac{e}{2m}B^2\right)+\frac{r^4}{64}\left(U^{IV}+\frac{2e}{m}BB''\right).$$

By using the above value of Störmer Potential in Eq. (A1.1.22), we obtain the trajectory equation for a mixed lens which produces a combined electric and magnetic field as

$$\begin{aligned} r''+\frac{1}{2}\frac{U'}{U}r'+\frac{1}{4U}\left(U''+\frac{e}{2m}B^2\right)r=\frac{1}{32U}\left(U^{IV}+\frac{2e}{m}BB''\right)r^3 \\ -\frac{1}{16U^2}\left(U''+\frac{e}{2m}B^2\right)^2r^3+\frac{1}{8U}\left(U'''+\frac{e}{m}BB'\right)r^2r' \\ -\frac{1}{8U^2}\left(U''+\frac{e}{2m}B^2\right)U'r^2r'-\frac{1}{4U}\left(U''+\frac{e}{2m}B^2\right)rr'^2-\frac{1}{2}\frac{U'}{U}r'^3. \end{aligned} \quad \text{(A1.2.1)}$$

Let us write $r = r_0 + \varepsilon$ where r_0 represents the Gaussian trajectory originating in a point object in some chosen direction. The perturbation ε is sufficiently small for powers greater than unity to be negligible. It is possible to find the equation of this trajectory by using the two independent solutions $g(z)$ and $h(z)$ which was given by Eq. (A1.5).

The equation for perturbation is then given by

$$\varepsilon'' + \frac{1}{2}\frac{U'}{U}\varepsilon' + \frac{1}{4U}\left(U'' + \frac{e}{2m}B^2\right)\varepsilon = \frac{1}{32U}\left(U^{IV} + \frac{2e}{m}BB''\right)r_0^3$$
$$-\frac{1}{16U^2}\left(U'' + \frac{e}{2m}B^2\right)^2 r_0^3 + \frac{1}{8U}\left(U''' + \frac{e}{m}BB'\right)r_0^2 r_0'$$
$$-\frac{1}{8U^2}\left(U'' + \frac{e}{2m}B^2\right)U' r_0^2 r_0' - \frac{1}{4U}\left(U'' + \frac{e}{2m}B^2\right)r_0 r_0'^2 - \frac{1}{2}\frac{U'}{U}r_0'^3. \tag{A1.2.2}$$

This is a linear second order differential equation with a term independent of ε on the right hand side. We can solve this equation using "variation of parameters" method.

From Eq. (A1.5), we know that, $r(z) = Cg(z) + Dh(z)$, where $g(z)$ and $h(z)$ are two independent solutions which satisfy the paraxial equation. Let the solution to the inhomogeneous Eq. (A1.2.2) be given by

$$\varepsilon = G_1(z)h(z) + G_2(z)g(z) \tag{A1.2.3}$$

in which G_1 and G_2 are functions of z. We need to determine G_1 and G_2 satisfying Eq. (A1.2.3).

$$\varepsilon' = G_1'h + G_1h' + G_2'g + G_2g'. \tag{A1.2.4}$$

If we choose,

$$G_1'h + G_2'g = 0 \tag{A1.2.4a}$$

then,

$$\varepsilon' = G_1 h' + G_2 g' \tag{A1.2.5}$$

and

$$\varepsilon'' = G_1' h' + G_1 h'' + G_2' g' + G_2 g''. \tag{A1.2.6}$$

If we substitute Eqs. (A1.2.3), (A1.2.5) and (A1.2.6) in Eq. (A1.2.2), we obtain

$$\begin{aligned}
&G_1' h' + G_1 h'' + G_2' g' + G_2 g'' + \frac{1}{2}\frac{U'}{U}(G_1 h' + G_2 g') \\
&+\frac{1}{4U}\left(U'' + \frac{e}{2m}B^2\right)(G_1 h + G_2 g) = \frac{1}{32U}\left(U^{IV} + \frac{2e}{m}BB''\right)r_0^3 \\
&-\frac{1}{16U^2}\left(U'' + \frac{e}{2m}B^2\right)^2 r_0^3 + \frac{1}{8U}\left(U''' + \frac{e}{m}BB'\right)r_0^2 r_0' \\
&-\frac{1}{8U^2}\left(U'' + \frac{e}{2m}B^2\right)U' r_0^2 r_0' - \frac{1}{4U}\left(U'' + \frac{e}{2m}B^2\right)r_0 r_0'^2 - \frac{1}{2}\frac{U'}{U}r_0'^3 \\
&\Rightarrow G_1\left[h'' + \frac{1}{2}\frac{U'}{U}h' + \frac{1}{4U}\left(U'' + \frac{e}{2m}B^2\right)h\right] + \\
&G_2\left[g'' + \frac{1}{2}\frac{U'}{U}g' + \frac{1}{4U}\left(U'' + \frac{e}{2m}B^2\right)g\right] + G_1' h' + G_2' g' = f(z).
\end{aligned} \tag{A1.2.7}$$

Since $h(z)$ and $g(z)$ are solutions of the paraxial ray equation, the coefficient of G_1 and G_2 in Eq. (A1.2.7) are zero and we obtain

$$G_1' h' + G_2' g' = f(z). \tag{A1.2.8}$$

Equations (A1.2.4a) and (A1.2.8) are solved simultaneously.

$$\begin{bmatrix} h & g \\ h' & g' \end{bmatrix} \begin{bmatrix} G_1' \\ G_2' \end{bmatrix} = \begin{bmatrix} 0 \\ f(z) \end{bmatrix}$$

$$G_1' = \frac{\begin{vmatrix} 0 & g \\ f(z) & g' \end{vmatrix}}{\begin{vmatrix} h & g \\ h' & g' \end{vmatrix}} = \frac{-gf(z)}{hg' - gh'} = \frac{gf(z)}{gh' - hg'}$$

and

$$G_2' = \frac{\begin{vmatrix} h & 0 \\ h' & f(z) \end{vmatrix}}{hg' - gh'} = \frac{hf(z)}{hg' - gh'}$$

$$\therefore G_1' = \frac{gf(z)}{gh' - hg'} \text{ and } G_2' = \frac{hf(z)}{hg' - gh'}.$$

$$G_1 = \int_{z_0}^{z_i} \frac{gf(z)}{gh' - hg'} dz + G_3 \text{ and } G_2 = \int_{z_0}^{z_i} \frac{hf(z)}{hg' - gh'} dz + G_4 \quad \text{(A1.2.9)}$$

where G_3 and G_4 are constants of integration.

$$\varepsilon = G_1 h + G_2 g$$

$$\varepsilon = h \int_{z_0}^{z_i} \frac{gf(z)}{gh' - hg'} dz + G_3 h + g \int_{z_0}^{z_i} \frac{hf(z)}{hg' - gh'} dz + G_4 g. \quad \text{(A1.2.10)}$$

At the object coordinate $z = z_0$, the trajectories of the ray with and without spherical aberration coincide and they have the same slope. $\therefore \varepsilon(z_0) = 0$, $\varepsilon'(z_0) = 0$. Also $h(z_0) = 0$ and therefore $G_4 = 0$ and $G_3 = 0$.

$$\varepsilon(z) = h(z)\int_{z_0}^{z_i} \frac{f(z)g(z)}{g(z)h'(z) - h(z)g'(z)} dz$$
$$+ g(z)\int_{z_0}^{z_i} \frac{f(z)h(z)}{h(z)g'(z) - g(z)h'(z)} dz \qquad \text{(A1.2.11)}$$

using the following simultaneous equations:

$$h'' + \frac{1}{2}\frac{U'}{U}h' + \frac{1}{4U}\left(U'' + \frac{e}{2m}B^2\right)h - \{\frac{1}{32U}\left(U^{IV} + \frac{2e}{m}BB''\right)h^3$$
$$-\frac{1}{16U^2}\left(U'' + \frac{e}{2m}B^2\right)^2 h^3 + \frac{1}{8U}\left(U''' + \frac{e}{m}BB'\right)h^2h'$$
$$-\frac{1}{8U^2}\left(U'' + \frac{e}{2m}B^2\right)U'h^2h' - \frac{1}{4U}\left(U'' + \frac{e}{2m}B^2\right)hh'^2$$
$$-\frac{1}{2}\frac{U'}{U}h'^3\} = 0$$

$$g'' + \frac{1}{2}\frac{U'}{U}g' + \frac{1}{4U}\left(U'' + \frac{e}{2m}B^2\right)g - \{\frac{1}{32U}\left(U^{IV} + \frac{2e}{m}BB''\right)g^3$$
$$-\frac{1}{16U^2}\left(U'' + \frac{e}{2m}B^2\right)^2 g^3 + \frac{1}{8U}\left(U''' + \frac{e}{m}BB'\right)g^2g'$$
$$-\frac{1}{8U^2}\left(U'' + \frac{e}{2m}B^2\right)U'g^2g' - \frac{1}{4U}\left(U'' + \frac{e}{2m}B^2\right)gg'^2$$
$$-\frac{1}{2}\frac{U'}{U}g'^3\} = 0$$

$$U^{1/2}h''+\frac{1}{2}\frac{U'}{U^{1/2}}h'+\frac{1}{4U^{1/2}}\left(U''+\frac{e}{2m}B^2\right)h-U^{1/2}$$

$$\{\frac{1}{32U}\left(U^{IV}+\frac{2e}{m}BB''\right)h^3-\frac{1}{16U^2}\left(U''+\frac{e}{2m}B^2\right)^2h^3$$

$$+\frac{1}{8U}\left(U'''+\frac{e}{m}BB'\right)h^2h'-\frac{1}{8U^2}\left(U''+\frac{e}{2m}B^2\right)U'h^2h'$$

$$-\frac{1}{4U}\left(U''+\frac{e}{2m}B^2\right)hh'^2-\frac{1}{2}\frac{U'}{U}h'^3\}=0$$

$$\{1\}$$

$$U^{1/2}g''+\frac{1}{2}\frac{U'}{U^{1/2}}g'+\frac{1}{4U^{1/2}}\left(U''+\frac{e}{2m}B^2\right)g-U^{1/2}$$

$$\{\frac{1}{32U}\left(U^{IV}+\frac{2e}{m}BB''\right)g^3-\frac{1}{16U^2}\left(U''+\frac{e}{2m}B^2\right)^2g^3$$

$$+\frac{1}{8U}\left(U'''+\frac{e}{m}BB'\right)g^2g'-\frac{1}{8U^2}\left(U''+\frac{e}{2m}B^2\right)U'g^2g'$$

$$-\frac{1}{4U}\left(U''+\frac{e}{2m}B^2\right)gg'^2-\frac{1}{2}\frac{U'}{U}g'^3\}=0$$

$$\{2\}$$

$\{1\}\Rightarrow$

$$\frac{d}{dz}(U^{1/2}h')+\frac{1}{4U^{1/2}}\left(U''+\frac{e}{2m}B^2\right)h-U^{1/2}\{\frac{1}{32U}\left(U^{IV}+\frac{2e}{m}BB''\right)h^3$$

$$-\frac{1}{16U^2}\left(U''+\frac{e}{2m}B^2\right)^2h^3+\frac{1}{8U}\left(U'''+\frac{e}{m}BB'\right)h^2h'-\frac{1}{8U^2}$$

$$\left(U''+\frac{e}{2m}B^2\right)U'h^2h'-\frac{1}{4U}\left(U''+\frac{e}{2m}B^2\right)hh'^2-\frac{1}{2}\frac{U'}{U}h'^3\}=0$$

$$\{3\}$$

$\{2\} \Rightarrow$

$$\frac{d}{dz}(U^{1/2}q')+\frac{1}{4U^{1/2}}\left(U''+\frac{e}{2m}B^2\right)g-U^{1/2}\{\frac{1}{32U}\left(U^{IV}+\frac{2e}{m}BB''\right)g^3$$

$$-\frac{1}{16U^2}\left(U''+\frac{e}{2m}B^2\right)^2 g^3+\frac{1}{8U}\left(U'''+\frac{e}{m}BB'\right)g^2g'-\frac{1}{8U^2}$$

$$\left(U''+\frac{e}{2m}B^2\right)U'g^2g'-\frac{1}{4U}\left(U''+\frac{e}{2m}B^2\right)gg'^2-\frac{1}{2}\frac{U'}{U}g'^3\}=0 \qquad \{4\}$$

$$g\times\{3\}-h\times\{4\}\Rightarrow g\frac{d}{dz}(U^{1/2}h')-h\frac{d}{dz}(U^{1/2}g')=0$$

$$\Rightarrow\frac{d}{dz}\left[U^{1/2}(h'g-g'h)\right]=0$$

$$U^{1/2}(h'g-g'h)=\text{constant}. \qquad \text{(A1.2.12)}$$

For $z=z_0$, $h(z_0)=0, h'(z_0)=1$, $g(z_0)=1$ and $U=U_0$

Therefore, constant = $U_0^{1/2}$. Thus Eq. (A1.2.12) becomes

$$(h'g-g'h)=\sqrt{\frac{U_0}{U}}. \qquad \text{(A1.2.13)}$$

We already know that $h(z_i)=0$. Substituting $f(z)$ in Eq. (A1.2.11) we obtain, the radius of the aberration disc,

$$\varepsilon(z_i)=|w|=g(z_i)\int_{z_0}^{z_i}\frac{h(z)}{h(z)g'(z)-h'(z)g(z)}\{\frac{1}{32U}\left(U^{IV}+\frac{2e}{m}BB''\right)r_0^3$$

$$-\frac{1}{16U^2}\left(U''+\frac{e}{2m}B^2\right)^2 r_0^3+\frac{1}{8U}\left(U'''+\frac{e}{m}BB'\right)r_0^2r_0'-\frac{1}{8U^2}$$

$$\left(U''+\frac{e}{2m}B^2\right)U'r_0^2r_0'-\frac{1}{4U}\left(U''+\frac{e}{2m}B^2\right)r_0r_0'^2-\frac{1}{2}\frac{U'}{U}r_0'^3\}dz.$$

(A1.2.14)

Using Eq. (A1.2.13), we obtain

$$\varepsilon(z_i) = |w| = -\frac{g(z_i)}{\sqrt{U_0}} \int_{z_0}^{z_i} hU^{1/2} \{ \frac{1}{32U}\left(U^{IV} + \frac{2e}{m}BB''\right) r_0^3$$

$$-\frac{1}{16U^2}\left(U'' + \frac{e}{2m}B^2\right)^2 r_0^3 + \frac{1}{8U}\left(U''' + \frac{e}{m}BB'\right) r_0^2 r_0' - \frac{1}{8U^2}$$

$$\left(U'' + \frac{e}{2m}B^2\right) U' r_0^2 r_0' - \frac{1}{4U}\left(U'' + \frac{e}{2m}B^2\right) r_0 r_0'^2 - \frac{1}{2}\frac{U'}{U} r_0'^3 \} dz.$$

(A1.2.15)

If M is the transverse linear magnification which is given by $g(z_i)$, then

$$\varepsilon(z_i) = |w| = -\frac{M}{\sqrt{U_0}} \int_{z_0}^{z_i} hU^{1/2} \{ \frac{1}{32U}\left(U^{IV} + \frac{2e}{m}BB''\right) r_0^3 - \frac{1}{16U^2}$$

$$\left(U'' + \frac{e}{2m}B^2\right)^2 r_0^3 + \frac{1}{8U}\left(U''' + \frac{e}{m}BB'\right) r_0^2 r_0' - \frac{1}{8U^2}\left(U'' + \frac{e}{2m}B^2\right)$$

$$U' r_0^2 r_0' - \frac{1}{4U}\left(U'' + \frac{e}{2m}B^2\right) r_0 r_0'^2 - \frac{1}{2}\frac{U'}{U} r_0'^3 \} dz.$$

(A1.2.16)

r_0 can be given by $r_0 = \alpha h(z)$ where α is the inclination. Substituting the value of r_0 in the above equation, we obtain,

$$\varepsilon(z_i) = |w| = -\frac{\alpha^3 M}{\sqrt{U_0}} \int_{z_0}^{z_i} h^4 U^{1/2} \{ \frac{1}{32U}\left(U^{IV} + \frac{2e}{m}BB''\right) - \frac{1}{16U^2}$$

$$\left(U'' + \frac{e}{2m}B^2\right)^2 + \frac{1}{8U}\left(U''' + \frac{e}{m}BB'\right)\frac{h'}{h} - \frac{1}{8U^2}\left(U'' + \frac{e}{2m}B^2\right) U' \frac{h'}{h}$$

$$-\frac{1}{4U}\left(U'' + \frac{e}{2m}B^2\right)\left(\frac{h'}{h}\right)^2 - \frac{1}{2}\frac{U'}{U}\left(\frac{h'}{h}\right)^3 \} dz.$$

(A1.2.17)

The aberration constant C_s can be defined by the relation

$$\varepsilon(z_i) = |w| = MC_s\alpha^3 . \qquad \text{(A1.2.18)}$$

Comparing Eqs. (A1.2.17) and (A1.2.18) we have,

$$C_s = -\frac{1}{\sqrt{U_0}}\int_{z_0}^{z_i} h^4 U^{1/2}\{\frac{1}{32U}\left(U^{IV} + \frac{2e}{m}BB''\right) - \frac{1}{16U^2}\left(U'' + \frac{e}{2m}B^2\right)^2$$
$$+\frac{1}{8U}\left(U''' + \frac{e}{m}BB'\right)\frac{h'}{h} - \frac{1}{8U^2}\left(U'' + \frac{e}{2m}B^2\right)U'\frac{h'}{h}$$
$$-\frac{1}{4U}\left(U'' + \frac{e}{2m}B^2\right)\left(\frac{h'}{h}\right)^2 - \frac{1}{2}\frac{U'}{U}\left(\frac{h'}{h}\right)^3\}dz.$$

(A1.2.19)

Now,

$$\int_{z_0}^{z_i} h^3 h' \frac{U'''}{U^{1/2}} dz = \int_{z_0}^{z_i} h^3 h' \frac{1}{U^{1/2}} dU'' = \left\{h^3 h' \frac{U''}{U^{1/2}}\right\}_{z_0}^{z_i} - \int_{z_0}^{z_i} U'' d\left(\frac{h^3 h'}{U^{1/2}}\right).$$

(A1.2.20)

We have chosen h such that it vanishes at $z = z_0$ and at $z = z_i$. Therefore, the term $\left\{h^3 h' \frac{U''}{U^{1/2}}\right\}_{z_0}^{z_i}$ vanishes and we obtain

$$\int_{z_0}^{z_i} h^3 h' \frac{U'''}{U^{1/2}} dz = -\int_{z_0}^{z_i} U'' d\left(\frac{h^3 h'}{U^{1/2}}\right)$$
$$= -\int_{z_0}^{z_i} U''\left(\frac{U^{1/2} d(h^3 h') - h^3 h' d(U)^{1/2}}{U}\right)$$
$$= -\int_{z_0}^{z_i} U''\left(\frac{h^3 h'' + h' 3h^2 h'}{U^{1/2}} - \frac{1}{2}\frac{h^3 h'}{U^{3/2}} U'\right) dz.$$

(A1.2.21)

We use the paraxial relation

$$h'' = \frac{-U'}{2U}h' - \frac{U''}{4U}h - \frac{eB^2}{8mU}h \qquad \text{(A1.2.22)}$$

to eliminate h'' and we finally obtain Eq. (A1.2.21)

$$= -\int_{z_0}^{z_i} U''\left\{\frac{h^3}{U^{1/2}}\left(\frac{-U'}{2U}h' - \frac{U''}{4U}h - \frac{eB^2}{8mU}h\right) + \frac{3h^2h'^2}{U^{1/2}} - \frac{1}{2}\frac{h^3h'}{U^{3/2}}U'\right\}dz$$

$$= -\int_{z_0}^{z_i}\{3h^2h'^2\frac{U''}{U^{1/2}} - h^4\frac{U''^2}{4U^{3/2}} - \frac{1}{2}\frac{h^3h'U'U''}{U^{3/2}} - \frac{1}{2}\frac{h^3h'U'U''}{U^{3/2}}$$

$$-h^4\frac{eB^2U''}{8mU^{3/2}}\}dz.$$

Thus,

$$\int_{z_0}^{z_i} h^3h'\frac{U'''}{U^{1/2}}dz$$

$$= -\int_{z_0}^{z_i}\left\{3h^2h'^2\frac{U''}{U^{1/2}} - h^4\frac{U''^2}{4U^{3/2}} - \frac{h^3h'U'U''}{U^{3/2}} - h^4\frac{eB^2U''}{8mU^{3/2}}\right\}dz. \qquad \text{(A1.2.23)}$$

Similarly,

$$\int_{z_0}^{z_i}\frac{h^4U^{IV}}{U^{1/2}}dz = \int_{z_0}^{z_i}\frac{h^4}{U^{1/2}}d(U''') = \left[\frac{h^4U'''}{U^{1/2}}\right]_{z_0}^{z_i} - \int_{z_0}^{z_i}U'''d\left(\frac{h^4}{U^{1/2}}\right)$$

$$= -\int_{z_0}^{z_i}U'''\left\{\frac{U^{1/2}4h^3h' - h^4\left(\frac{1}{2}U^{-1/2}U'\right)}{U}\right\}dz$$

$$= -\int_{z_0}^{z_i}\left[4h^3h'\frac{U'''}{U^{1/2}}dz - \frac{h^4U'U'''}{2U^{3/2}}\right]dz.$$

Substituting the value of $h^3 h' \dfrac{U'''}{U^{1/2}} dz$ from Eq. (A1.2.23), we obtain,

$$= \int_{z_0}^{z_i} \left\{ 12h^2 h'^2 \frac{U''}{U^{1/2}} - \frac{h^4 U''^2}{U^{3/2}} - 4h^3 h' \frac{U'U''}{U^{3/2}} - 4h^4 \frac{eB^2 U''}{8mU^{3/2}} \right\} dz$$

$$+ \frac{1}{2} \int_{z_0}^{z_i} h^4 \frac{U'U'''}{U^{3/2}} dz. \tag{A1.2.24}$$

But, $\displaystyle\int_{z_0}^{z_i} h^4 \frac{U'U'''}{U^{3/2}} dz = \int_{z_0}^{z_i} \frac{h^4 U'}{U^{3/2}} d(U'')$

$$= \left[\frac{h^4 U'U''}{U^{3/2}} \right]_{z_0}^{z_i} - \int_{z_0}^{z_i} U'' d\left(\frac{h^4 U'}{U^{3/2}} \right) \quad \text{(the first term here will vanish)}$$

$$= - \int_{z_0}^{z_i} U'' \left\{ \frac{U^{3/2} d\left(h^4 U'\right) - \left(h^4 U'\right) \frac{3}{2} U^{1/2} U'}{U^3} \right\} dz$$

$$= - \int_{z_0}^{z_i} U'' \left\{ \frac{U^{3/2}}{U^3} \left(h^4 U'' + U' 4h^3 h' \right) - \frac{3}{2} \frac{h^4 U^{1/2} U'^2}{U^3} \right\} dz$$

$$= - \int_{z_0}^{z_i} \frac{h^4 U''^2}{U^{3/2}} dz - 4 \int_{z_0}^{z_i} \frac{h^3 h' U'U''}{U^{3/2}} dz + \frac{3}{2} \int_{z_0}^{z_i} \frac{h^4 U'^2 U''}{U^{5/2}} dz.$$

Therefore Eq. (A1.2.24) becomes

$$\int_{z_0}^{z_i} U'' \{ \frac{12h^2 h'^2}{U^{1/2}} - \frac{h^4 U''}{U^{3/2}} - 4\frac{h^3 h' U'}{U^{3/2}} - h^4 \frac{eB^2}{2mU^{3/2}} - \frac{1}{2} \frac{h^4 U''}{U^{3/2}} - 2\frac{h^3 h' U'}{U^{3/2}}$$

$$+ \frac{3}{4} \frac{h^4 U'^2}{U^{5/2}} \} dz.$$

Thus, $\displaystyle\int_{z_0}^{z_i} \frac{h^4 U^{IV}}{U^{1/2}} dz =$

$$\int_{z_0}^{z_i} U'' \left(\frac{12h^2 h'^2}{U^{1/2}} - \frac{3}{2} \frac{h^4 U''}{U^{3/2}} - \frac{6h^3 h' U'}{U^{3/2}} - \frac{eB^2 h^4}{2mU^{3/2}} + \frac{3}{4} \frac{h^4 U'^2}{U^{5/2}} \right) dz$$

$$\tag{A1.2.25}$$

and

$$\int_{z_0}^{z_i} \frac{h^4 BB''}{U^{1/2}} dz = \int_{z_0}^{z_i} \frac{h^4 B}{U^{1/2}} d(B')$$

$$= \left[\frac{h^4 B}{U^{1/2}} B'\right]_{z_0}^{z_i} - \int_{z_0}^{z_i} B' d\left(\frac{h^4 B}{U^{1/2}}\right) = - \int_{z_0}^{z_i} B' d\left(\frac{h^4 B}{U^{1/2}}\right)$$

$$= - \int_{z_0}^{z_i} B' \left\{ \frac{U^{1/2} d\left(h^4 B\right) - h^4 B \frac{1}{2} U^{-1/2} U'}{U} \right\} dz$$

$$\int_{z_0}^{z_i} \frac{h^4 BB''}{U^{1/2}} dz = - \int_{z_0}^{z_i} B' \left(\frac{4h^3 h' B}{U^{1/2}} + h^4 \frac{B'}{U^{1/2}} - \frac{h^4 BU'}{2U^{3/2}} \right) dz. \quad \text{(A1.2.26)}$$

Substituting Eqs. (A1.2.23), (A1.2.25) and (A1.2.26) in (A.2.2.19), we obtain

$$C_s = \frac{1}{U_0^{1/2}} \int_{z_0}^{z_i} h^4 \{ \frac{5}{64} \frac{U''^2}{U^{3/2}} + \frac{3}{16} \frac{U''U'}{U^{3/2}} \frac{h'}{h} + \frac{U''}{4U^{1/2}} \left(\frac{h'}{h}\right)^2 + \frac{eB^2}{16m} \frac{U''}{U^{3/2}}$$
$$- \frac{3}{128} \frac{U''U'^2}{U^{5/2}} \frac{eBB'}{8mU^{1/2}} \frac{h'}{h} + \frac{eB'^2}{16mU^{1/2}} - \frac{eBB'U'}{32mU^{3/2}} + \frac{e^2 B^4}{64m^2 U^{3/2}}$$
$$+ \frac{eB^2 U'}{16mU^{3/2}} \frac{h'}{h} + \frac{eB^2}{8mU^{1/2}} \left(\frac{h'}{h}\right)^2 + \frac{U'}{2U^{1/2}} \left(\frac{h'}{h}\right)^3 \} dz.$$
$$\text{(A1.2.27)}$$

Now,

$$\int_{z_0}^{z_i} hh'^3 \frac{U'}{2U^{1/2}} dz = \int_{z_0}^{z_i} hh'^2 \frac{U'}{2U^{1/2}} dh$$

$$= - \int_{z_0}^{z_i} \{ \frac{h^2 h'^2 U''}{2U^{1/2}} - \frac{h^2 h'^2 U'^2}{4U^{3/2}} + h^2 h' h'' \frac{U'}{U^{1/2}} + \frac{hh'^3 U'}{2U^{1/2}} \} dz.$$

Using Eq. (A1.2.22) to eliminate h'',

$$\int_{z_0}^{z_i} hh'^3 \frac{\phi'}{2U^{1/2}} dz$$

$$= -\int_{z_0}^{z_i} \{\frac{h^2h'^2U''}{2U^{1/2}} - \frac{h^2h'^2U'^2}{4U^{3/2}} + h^2h'\frac{U'}{U^{1/2}}[-\frac{U'}{2U}h' - \frac{U''}{4U}h - \frac{eB^2}{8mU}h] + \frac{hh'^3U'}{2U^{1/2}}\}dz$$

$$= -\int_{z_0}^{z_i} \{\frac{h^2h'^2U''}{2U^{1/2}} - \frac{h^2h'^2U'^2}{4U^{3/2}} - \frac{h^2h'^2U'^2}{2U^{3/2}} - \frac{h^3h'U''U'}{4U^{3/2}} - \frac{h^3h'eB^2U'}{8mU^{3/2}}$$

$$+ \frac{hh'^3U'}{2U^{1/2}}\}dz$$

$$= \int_{z_0}^{z_i} [h^2h'^2(\frac{3}{8}\frac{U'^2}{U^{3/2}} - \frac{U''}{4U^{1/2}}) + h^3h'(\frac{U''U'}{8U^{3/2}} + \frac{eB^2U'}{16mU^{3/2}})]dz. \qquad \text{(A1.2.28)}$$

If we substitute Eq. (A1.2.28) in Eq. (A1.2.27), we obtain

$$C_s = \frac{1}{U_0^{1/2}} \int_{z_0}^{z_i} [h^4(\frac{5U''^2}{64U^{3/2}} + \frac{eU''B^2}{16mU^{3/2}} + \frac{e^2B^4}{64m^2U^{3/2}} - \frac{3U'^2U''}{128U^{5/2}}$$

$$+ \frac{eB'^2}{16mU^{1/2}} - \frac{eBB'U'}{32mU^{3/2}}) + h^3h'(\frac{5U'U''}{16U^{3/2}} + \frac{eB^2U'}{8mU^{3/2}} + \frac{eBB'}{8mU^{1/2}})$$

$$+ h^2h'^2(\frac{3U'^2}{8U^{3/2}} + \frac{eB^2}{8mU^{1/2}})]dz.$$

$$\text{(A1.2.29)}$$

If we write down Scherzer's result (Scherzer, 1936) in the notation as used here, we obtain

$$C_3 = \frac{1}{U_0^{1/2}} \int_{z_0}^{z_i} [h^4(\frac{5U''^2}{64U^{3/2}} + \frac{5U'^4}{384U^{7/2}} + \frac{eB'^2}{16mU^{1/2}} + \frac{3e^2B^4}{128m^2U^{3/2}}$$

$$+ \frac{35eU'^2B^2}{265mU^{5/2}} - \frac{3eU'BB'}{16mU^{3/2}}) + h^3h'(\frac{7U'^3}{24U^{5/2}} - \frac{eU'B^2}{8mU^{3/2}})$$

$$+ h^2h'^2(\frac{-3U'^2}{32U^{3/2}} - \frac{eB^2}{16mU^{1/2}})]dz.$$

$$\text{(A1.2.30)}$$

The difference between the terms with $h^2h'^2$ in Eq. (2.2.29) and in Eq. (2.2.30) is

$$h^2h'^2[(\frac{3U'^2}{8U^{3/2}}+\frac{eB^2}{8mU^{1/2}})-(-\frac{3U'^2}{32U^{3/2}}-\frac{eB^2}{16mU^{1/2}})]$$

$$=h^2h'^2(\frac{15U'^2}{32U^{3/2}}+\frac{3eB^2}{16mU^{1/2}}) \qquad \text{(A1.2.31)}$$

and we try to express this difference under the integral sign in terms with h^3h' and h^4.

$$\int_{z_0}^{z_i} h^2h'^2(\frac{15U'^2}{32U^{3/2}}+\frac{3eB^2}{16mU^{1/2}})dz=\int_{z_0}^{z_i} h'h^2(\frac{15U'^2}{32U^{3/2}}+\frac{3eB^2}{16mU^{1/2}})d(h).$$

(A1.2.32)

Carrying out the partial integration and substituting the value for h'' from Eq. (A1.2.22) and Eq. (A.2.2.32) this becomes

$$=\int_{z_0}^{z_i}\{h^3h'(\frac{5U'^3}{16U^{5/2}}+\frac{eB^2U'}{16mU^{3/2}}-\frac{5U'U''}{16U^{3/2}}-\frac{eBB'}{8mU^{1/2}})+h^4(\frac{5U'^2U''}{128U^{5/2}}$$
$$+\frac{5eU'^2B^2}{256mU^{5/2}}+\frac{eB^2U''}{64mU^{3/2}}+\frac{e^2B^4}{128m^2U^{3/2}})\}dz.$$

(A1.2.33)

When we use the results of Eqs. (A1.2.31) and (A1.2.33) in Eq. (A1.2.29), we obtain,

$$C_s=\frac{1}{U_0^{1/2}}\int_{z_0}^{z_i}\{h^4(\frac{5U''^2}{64U^{3/2}}+\frac{eU''B^2}{16mU^{3/2}}+\frac{e^2B^4}{64m^2U^{3/2}}-\frac{3U'^2U''}{128U^{5/2}}$$
$$+\frac{eB'^2}{16mU^{1/2}}-\frac{eBB'U'}{32mU^{3/2}})+h^3h'(\frac{5U'U''}{16U^{3/2}}+\frac{eBB'}{8mU^{1/2}}+\frac{eB^2U'}{8mU^{3/2}})$$
$$+h^2h'^2[(\frac{15U'^2}{32U^{3/2}}+\frac{3eB^2}{16mU^{1/2}})+(-\frac{3U'^2}{32U^{3/2}}-\frac{eB^2}{16mU^{1/2}})]\}dz$$

$$
\begin{aligned}
&= \frac{1}{U_0^{1/2}} \int_{z_0}^{z_i} \{h^4 (\frac{5U''^2}{64U^{3/2}} + \frac{5eU''B^2}{64mU^{3/2}} + \frac{3e^2B^4}{128m^2U^{3/2}} + \frac{U'^2U''}{64U^{5/2}} + \frac{eB'^2}{16mU^{1/2}} \\
&\quad - \frac{eBB'U'}{32mU^{3/2}} + \frac{5eU'^2B^2}{256mU^{5/2}}) + h^3h'(\frac{3eB^2U'}{16mU^{3/2}} + \frac{5U'^3}{16U^{5/2}}) \\
&\quad + h^2h'^2(-\frac{3U'^2}{32U^{3/2}} - \frac{eB^2}{16mU^{1/2}})\}dz. \qquad \text{(A1.2.34)}
\end{aligned}
$$

The difference between the terms with h^3h' as in Eq. (A1.2.34) and in Eq. (A1.2.30) is:

$$
\begin{aligned}
&h^3h'([\frac{5U'^3}{16U^{5/2}} + \frac{3eB^2U'}{16mU^{3/2}}) - (\frac{7U'^3}{24U^{5/2}} - \frac{eU'B^2}{8mU^{3/2}})] \\
&\quad = h^3h'(\frac{U'^3}{48U^{5/2}} + \frac{5eB^2U'}{16mU^{3/2}}). \qquad \text{(A1.2.35)}
\end{aligned}
$$

Therefore,

$$
\int_{z_0}^{z_i} h^3h'(\frac{U'^3}{48U^{5/2}} + \frac{5eB^2U'}{16mU^{3/2}})dz = \int_{z_0}^{z_i} h^3(\frac{U'^3}{48U^{5/2}} + \frac{5eB^2U'}{16mU^{3/2}})d(h). \qquad \text{(A1.2.36)}
$$

Carrying out the partial integration and simplifying we obtain,

$$
\begin{aligned}
&\int_{z_0}^{z_i} h^3h'(\frac{U'^3}{48U^{5/2}} + \frac{5eB^2U'}{16mU^{3/2}})dz \\
&= -\int_{z_0}^{z_i} \{h^4(\frac{3}{64}\frac{U'^2U''}{U^{5/2}} - \frac{5}{384}\frac{U'^4}{U^{7/2}} + \frac{5}{64}\frac{eB^2U''}{mU^{3/2}} + \frac{5}{32}\frac{eBB'U'}{mU^{3/2}} - \frac{15}{128}\frac{eB^2U'^2}{mU^{5/2}})\}dz. \qquad \text{(A1.2.37)}
\end{aligned}
$$

Using the results from Eqs. (A1.2.35) and (A1.2.37) in Eq. (A1.2.34), we obtain

$$C_s = \frac{1}{U_0^{1/2}} \int_{z_0}^{z_i} [h^4 (\frac{5U''^2}{64U^{3/2}} + \frac{5U'^4}{384U^{7/2}} + \frac{eB'^2}{16mU^{1/2}} + \frac{3e^2B^4}{128m^2U^{3/2}}$$

$$+ \frac{35eU'^2B^2}{265mU^{5/2}} - \frac{3eU'BB'}{16mU^{3/2}}) + h^3h'(\frac{7U'^3}{24U^{5/2}} - \frac{eU'B^2}{8mU^{3/2}})$$

$$+ h^2h'^2(\frac{-3U'^2}{32U^{3/2}} - \frac{eB^2}{16mU^{1/2}})]dz$$

(A1.2.38)

which is identical to Eq. (A1.2.30).

The final equation can be written as

$$C_s = \frac{1}{16U_0^{1/2}} \int_{z_0}^{z_i} h^4U^{1/2} (\frac{5}{4}\left(\frac{U''}{U}\right)^2 + \frac{5}{24}\left(\frac{U'}{U}\right)^4 + \frac{14}{3}\left(\frac{U'}{U}\right)^3 \frac{h'}{h}$$

$$-\frac{3}{2}\left(\frac{U'}{U}\right)^2 \frac{h'^2}{h^2} + \frac{eB'^2}{mU} + \frac{3e^2B^4}{8m^2U^2} - \frac{eB^2}{mU}\frac{h'^2}{h^2} + \frac{35eU'^2B^2}{16mU^3} - \frac{3eU'BB'}{mU}$$

$$-\frac{2eB^2U'}{mU^2}\frac{h'}{h})dz.$$

(A1.2.39)

Appendix 1.3

Chromatic Aberration

We shall start with the paraxial ray equation.

$$r'' + \frac{1}{2}\frac{U'(z)}{U(z)}r' + \frac{1}{4U(z)}[U''(z) + \frac{e}{2m}B(z)^2]r = 0. \qquad \text{(A1.3.1)}$$

Here $U(z)$ is the potential at any point z along the axis. Let us assume that the potential varies by ΔU so that U must be replaced by $U + \Delta U$. Similarly we assume that B changes to $B + \Delta B$ because of the variation of magnetic field. We shall also assume that we operate in the linear region of the $B - H$ curve below saturation such that the shape of the field does not vary. In other words,

$$\frac{\Delta B}{\Delta B_{\max}} = \frac{B}{B_{\max}} \text{ (or) } \Delta B = B\left(\frac{\Delta B_{\max}}{B_{\max}}\right). \qquad \text{(A1.3.2)}$$

The variation in potential and magnetic field are assumed to be relatively small so that $\Delta B_{\max}/B_{\max} << 1$ and $\Delta U/U << 1$. We shall, therefore, neglect all higher orders of ΔB and ΔU. Equation (A1.3.1) then changes to

$$r'' + \frac{1}{2}\frac{(U + \Delta U)'}{(U + \Delta U)}r' + \frac{1}{4(U + \Delta U)}\left[(U + \Delta U)'' + \frac{e}{2m}(B + \Delta B)^2\right]r = 0$$

$$(U + \Delta U)r'' + \frac{1}{2}U'r' + \frac{1}{4}\left[U'' + \frac{e}{2m}(B^2 + 2B\Delta B)\right]r = 0$$

$$(U + \Delta U)r'' + \frac{U'r'}{2} + \frac{1}{4}\left[U'' + \frac{eB^2}{2m} + \frac{eB^2}{m}\frac{\Delta B_{\max}}{B_{\max}}\right]r = 0.$$

(A1.3.3)

Equation (A1.3.1) has two independent solutions. Let $h(z)$ be a solution with initial conditions $h(z)=0$ at $z=z_0$. This is an ideal ray with no aberrations, it will cross the axis at $z=z_i$ in the Gaussian plane. Let $r(z)$ be a solution to Eq. (A1.3.3), which crosses the axis farther away from z_i, and at the Gaussian plane, let the radius be $\varepsilon=\Delta r$, caused by chromatic aberration. Thus we have

$$r(z)=h(z)+\varepsilon(z). \tag{A1.3.4}$$

We substitute Eq. (A1.3.4) in Eq. (A1.3.3)

$$(U+\Delta U)(h+\varepsilon)''+\frac{1}{2}U'(h+\varepsilon)'$$
$$+\frac{1}{4}\left(U''+\frac{e}{2m}B^2+\frac{e}{m}B^2\frac{\Delta B_{\max}}{B_{\max}}\right)(h+\varepsilon)=0$$

$$(U+\Delta U)(h''+\varepsilon'')+\frac{U'}{2}(h'+\varepsilon')$$
$$+\frac{1}{4}\left(U''+\frac{e}{2m}B^2+\frac{e}{m}B^2\frac{\Delta B_{\max}}{B_{\max}}\right)(h+\varepsilon)=0$$

$$Uh''+U\varepsilon''+\Delta Uh''+\Delta U\varepsilon''+\frac{U'h'}{2}+\frac{U'\varepsilon'}{2}+\frac{1}{4}\left(U''+\frac{eB^2}{2m}\right)h$$
$$+\frac{1}{4}\left(U''+\frac{eB^2}{2m}\right)\varepsilon+\frac{eB^2}{4m}\frac{\Delta B_{\max}}{B_{\max}}h+\frac{eB^2}{4m}\frac{\Delta B_{\max}}{B_{\max}}\varepsilon=0$$

and neglecting all products of ε, ΔU, and ΔB that are higher than first-order, we obtain,

$$\varepsilon''+\frac{U'\varepsilon'}{2U}+\frac{1}{4U}\left(U''+\frac{eB^2}{2m}\right)\varepsilon=-\frac{\Delta U}{U}h''-\frac{e}{4m}\frac{B^2}{U}\frac{\Delta B_{\max}}{B_{\max}}h. \tag{A1.3.5}$$

From Eq. (A1.3.1), we know that, $h'' + \frac{1}{2}\frac{U'}{U}h' + \frac{1}{4U}\left(U'' + \frac{eB^2}{2m}\right)h = 0.$

The difference between Eqs. (A1.3.5) and (A1.3.1) is the inhomogeneous part on the RHS of Eq. (A1.3.5) which we shall call $f(z)$.

$$f(z) = -\frac{\Delta U}{U}h'' - \frac{e}{4m}\frac{B^2}{U}\frac{\Delta B_{\max}}{B_{\max}}h.$$

Let $q(z)$ be the other independent solution of Eq. (A1.3.1). Therefore, $c_1 h + c_2 q$ is also a solution of Eq. (A1.3.1) in which c_1 and c_2 are constants. We now assume that the solutions to the inhomogeneous Eq. (A1.3.5) can also be written as

$$\varepsilon = G_1(z)h(z) + G_2(z)q(z) \tag{A1.3.6}$$

in which G_1 and G_2 are functions of z. We shall attempt to determine G_1 and G_2 so that Eq. (A1.3.5) is satisfied.

We know that,

$$\frac{d\varepsilon}{dz} = \frac{dG_1}{dz}h + G_1\frac{dh}{dz} + \frac{dG_2}{dz}q + G_2\frac{dq}{dz}. \tag{A1.3.7}$$

If we choose,

$$\frac{dG_1}{dz}h + \frac{dG_2}{dz}q = 0 \tag{A1.3.8}$$

then,

$$\frac{d\varepsilon}{dz} = G_1\frac{dh}{dz} + G_2\frac{dq}{dz} \tag{A1.3.9}$$

and

$$\frac{d^2\varepsilon}{dz^2} = \frac{dG_1}{dz}\frac{dh}{dz} + G_1\frac{d^2h}{dz^2} + \frac{dG_2}{dz}\frac{dq}{dz} + G_2\frac{d^2q}{dz^2}. \qquad \text{(A1.3.10)}$$

If we substitute Eq. (A1.3.10) in Eq. (A1.3.5), we obtain

$$\frac{dG_1}{dz}\frac{dp}{dz} + G_1\frac{d^2p}{dz^2} + \frac{dG_2}{dz}\frac{dq}{dz} + G_2\frac{d^2q}{dz^2} + \frac{1}{2}\frac{U'}{U}\left(G_1\frac{dh}{dz} + G_2\frac{dq}{dz}\right)$$

$$+\frac{1}{4}\left(U'' + \frac{e}{2m}B^2\right)\left(G_1h + G_2q\right) = f(z)$$

$$\Rightarrow G_1\{\frac{d^2h}{dz^2} + \frac{1}{2}\frac{U'}{U}\frac{dh}{dz} + \frac{1}{4}\left(U'' + \frac{e}{2m}B^2\right)h\} + G_2\{\frac{d^2q}{dz^2} + \frac{1}{2}\frac{U'}{U}\frac{dq}{dz}$$

$$+\frac{1}{4}\left(U'' + \frac{e}{2m}B^2\right)q\} + \frac{dG_1}{dz}\frac{dh}{dz} + \frac{dG_2}{dz}\frac{dq}{dz} = f(z).$$

(A1.3.11)

Since $h(z)$ and $q(z)$ are solutions of the paraxial ray Eq. (A1.3.1), the coefficient of G_1 and G_2 in Eq. (A1.3.11) are zero and we obtain

$$\frac{dG_1}{dz}\frac{dh}{dz} + \frac{dG_2}{dz}\frac{dq}{dz} = f(z). \qquad \text{(A1.3.12)}$$

Equations (A1.3.12) and (A1.3.8) are solved simultaneously:

$$\begin{bmatrix} h & q \\ \frac{dh}{dz} & \frac{dq}{dz} \end{bmatrix} \begin{bmatrix} \frac{dG_1}{dz} \\ \frac{dG_2}{dz} \end{bmatrix} = \begin{bmatrix} 0 \\ f(z) \end{bmatrix}$$

$$\frac{dG_1}{dz} = \frac{\begin{bmatrix} 0 & q \\ f(z) & \frac{dq}{dz} \end{bmatrix}}{\begin{bmatrix} h & q \\ \frac{dh}{dz} & \frac{dq}{dz} \end{bmatrix}} = \frac{-qf(z)}{h\frac{dq}{dz} - q\frac{dh}{dz}} = \frac{qf(z)}{q\frac{dh}{dz} - h\frac{dq}{dz}}$$

and

$$\frac{dG_2}{dz} = \frac{\begin{bmatrix} h & 0 \\ \frac{dh}{dz} & f(z) \end{bmatrix}}{hq' - qh'} = \frac{hf(z)}{hq' - qh'}$$

$$\therefore \frac{dG_1}{dz} = \frac{qf(z)}{qh' - hq'} \text{ and } \frac{dG_2}{dz} = \frac{hf(z)}{hq' - qh'} \qquad \text{(A1.3.13)}$$

$$G_1 = \int_{z_0}^{z} \frac{qf(z)}{qh' - hq'} dz + G_3 \text{ and } G_2 = \int_{z_0}^{z} \frac{hf(z)}{hq' - qh'} dz + G_4$$

where G_3 and G_4 are constants of integration.

$$\varepsilon = G_1 h + G_2 q$$

$$\varepsilon = h \int_{z_0}^{z} \frac{qf(z)}{qh' - hq'} dz + G_3 h + q \int_{z_0}^{z} \frac{hf(z)}{hq' - qh'} dz + G_4 q. \qquad \text{(A1.3.14)}$$

At the object coordinate $z = z_0$, the trajectories of the ray with and without chromatic aberration coincide and they have the same slope.

$\therefore \varepsilon(z_0) = 0, \varepsilon'(z_0) = 0$. Also $h(z_0) = 0$ and therefore $G_4 = 0$ and $G_3 = 0$.

$$\varepsilon(z) = h(z) \int_{z_0}^{z} \frac{f(z)q(z)}{q(z)h'(z) - h(z)q'(z)} dz + q(z) \int_{z_0}^{z} \frac{f(z)h(z)}{h(z)q'(z) - q(z)h'(z)} dz \qquad \text{(A1.3.15)}$$

using the following simultaneous equations:

$$h''+\frac{1}{2}\frac{U'}{U}h'+\frac{1}{4U}\left(U''+\frac{e}{2m}B^2\right)h=0 \qquad \text{(A1.3.16)}$$

$$q''+\frac{1}{2}\frac{U'}{U}q'+\frac{1}{4U}\left(U''+\frac{e}{2m}B^2\right)q=0 \qquad \text{(A1.3.17)}$$

$$U^{1/2}h''+\frac{1}{2}\frac{U'}{U^{1/2}}h'+\frac{1}{4U^{1/2}}\left(U''+\frac{e}{4m}B^2\right)h=0 \qquad \{1\}$$

$$U^{1/2}q''+\frac{1}{2}\frac{U'}{U^{1/2}}q'+\frac{1}{4U^{1/2}}\left(U''+\frac{e}{4m}B^2\right)q=0 \qquad \{2\}$$

$$\{1\}\Rightarrow\frac{d}{dz}(U^{1/2}h')+\frac{1}{4U^{1/2}}\left(U''+\frac{e}{2m}B^2\right)h=0 \qquad \{3\}$$

$$\{2\}\Rightarrow\frac{d}{dz}(U^{1/2}q')+\frac{1}{4U^{1/2}}\left(U''+\frac{e}{2m}B^2\right)q=0 \qquad \{4\}$$

$$q\times\{3\}-h\times\{4\}\Rightarrow q\frac{d}{dz}(U^{1/2}h')-h\frac{d}{dz}(U^{1/2}q')=0$$

$$\Rightarrow\frac{d}{dz}\left[U^{1/2}\left(h'q-q'h\right)\right]=0$$

$$U^{1/2}\left(h'q-q'h\right)=\text{constant.} \qquad \text{(A1.3.18)}$$

For $z=z_i$, $h(z_i)=0$ and $U=U_i$,

therefore, constant = $h'(z_i)q(z_i)U_i^{1/2}$. Thus Eq. (A1.3.18) becomes

$$\left(hq'-qh'\right)U^{1/2}=-h'(z_i)q(z_i)U_i^{1/2}. \qquad \text{(A1.3.19)}$$

Let us now return to Eq. (A1.3.15) where we substitute for $f(z)$ from the RHS of Eq. (A1.3.14).

$$\varepsilon(z) = h(z)\int_{z_0}^{z}\frac{q(z)}{q(z)h'(z)-h(z)q'(z)}\left(\frac{-\Delta U}{U}h'' - \frac{e}{4m}\frac{B^2}{U}\frac{\Delta B_{\max}}{B_{\max}}h\right)dz$$
$$+q(z)\int_{z_0}^{z}\frac{h(z)}{h(z)q'(z)-h'(z)q(z)}\left(\frac{-\Delta U}{U}h'' - \frac{e}{4m}\frac{B^2}{U}\frac{\Delta B_{\max}}{B_{\max}}h\right)dz. \tag{A1.3.20}$$

But we already know that $h(z_i) = 0$.

$$\therefore\ \varepsilon(z_i) = q(z_i)\int_{z_0}^{z_i}\frac{h(z)}{h(z)q'(z)-h'(z)q(z)}\{\frac{-\Delta U}{U}h''$$
$$-\frac{e}{4m}\frac{B^2}{U}\frac{\Delta B_{\max}}{B_{\max}}h\}dz \tag{A1.3.21}$$

and using Eq. (A1.3.19) and putting $q(z_i)$ inside the integral sign since it is a constant, we obtain

$$\varepsilon(z_i) = \int_{z_0}^{z_i}\left(\Delta U h'' + \frac{e}{4m}\frac{\Delta B_{\max}}{B_{\max}}B^2 h\right)\frac{hU^{1/2}}{UU_i^{1/2}h'(z_i)}dz. \tag{A1.3.22}$$

We refer the aberration to the object side, as C_c.

$$\text{We write, } (C_c)_{\text{at object}} = \frac{\varepsilon(z_i)}{M} \tag{A1.3.23}$$

$$(C_c)_{\text{at object}} = \left(Mh'(z_i)U_i^{1/2}\right)^{-1}\int_{z_0}^{z_i}\left(\Delta U h h'' + \frac{e}{4m}\frac{\Delta B_{\max}}{B_{\max}}B^2h^2\right)\frac{dz}{U^{1/2}}. \tag{A1.3.24}$$

From the law of Helmholtz-Lagrange, we know that for small inclinations

$$(U_0)^{1/2}h'(z_0) = -Mh'(z_i)(U_i)^{1/2} \tag{A1.3.25}$$

$$C_c = \frac{-1}{(U_0)^{1/2} h'(z_0)} \int_{z_0}^{z_i} \left(\Delta U h h'' + \frac{e}{4m} \frac{\Delta B_{\max}}{B_{\max}} B^2 h^2 \right) \frac{dz}{U^{1/2}}. \quad \text{(A1.3.26)}$$

We substitute for h'' from Eq. (A1.3.17) and obtain

$$C_c = \frac{-1}{(U_0)^{1/2} h'(z_0)} \int_{z_0}^{z_i} \left(\frac{-\Delta U}{2} \frac{U'}{U} h h' - \frac{\Delta U}{4U} \left(U'' + \frac{eB^2}{2m} \right) h^2 + \frac{e}{4m} \frac{\Delta B_{\max}}{B_{\max}} B^2 h^2 \right)$$

(A1.3.27)

$$\int_{z_0}^{z_i} \frac{U'}{U^{3/2}} h h' dz = \int_{z_0}^{z_i} \frac{U'}{U^{3/2}} h d(h) = \left[\frac{U'}{U^{3/2}} h^2 \right]_{z_0}^{z_i} - \int_{z_0}^{z_i} h d\left(\frac{U'h}{U^{3/2}} \right)$$

$$= 0 - \int_{z_0}^{z_i} h \left\{ \frac{U^{3/2} (U'h' + hU'') - U'h \frac{3}{2} U^{1/2} U'}{U^3} \right\} dz$$

$$= - \int_{z_0}^{z_i} \frac{U'h'h + h^2 U''}{U^{3/2}} dz + \int_{z_0}^{z_i} \frac{3}{2} \frac{h^2 U'^2}{U^{5/2}} dz.$$

Since $h(z_i) = h(z_0) = 0$,

$$\int_{z_0}^{z_i} \frac{U'}{U^{3/2}} h h' dz = -\frac{1}{2} \int_{z_0}^{z_i} \frac{U'' h^2 dz}{U^{3/2}} + \frac{3}{4} \int_{z_0}^{z_i} \frac{U'^2 h^2}{U^{5/2}} dz. \quad \text{(A1.3.28)}$$

Substituting the value of Eq. (A1.3.29) in Eq. (A1.3.28), we obtain

$$C_c = \frac{\Delta U}{2(U_0)^{1/2} h'(z_0)} \left[\frac{3}{4} \int_{z_0}^{z_i} \frac{U'^2 h^2}{U^{5/2}} dz - \frac{1}{2} \int_{z_0}^{z_i} \frac{U'' h^2}{U^{3/2}} dz \right]$$

$$+ \frac{\Delta U}{4(U_0)^{1/2} h'(z_0)} \int_{z_0}^{z_i} \left(U'' + \frac{eB^2}{2m} \right) \frac{h^2}{U^{3/2}} dz$$

$$- \frac{e}{4m U_0^{1/2} h'(z_0)} \frac{\Delta B_{\max}}{B_{\max}} \int_{z_0}^{z_i} \frac{B^2 h^2}{U^{1/2}} dz$$

$$C_c = \frac{3\Delta U}{8U_0^{1/2}h'(z_0)}\int_{z_0}^{z_i}\frac{U'^2h^2}{U^{5/2}}dz + \frac{\Delta Ue}{8mU_0^{1/2}h'(z_0)}\int_{z_0}^{z_i}\frac{B^2h^2}{U^{3/2}}dz$$
$$-\frac{e}{4mU_0^{1/2}h'(z_0)}\frac{\Delta B_{\max}}{B_{\max}}\int_{z_0}^{z_i}\frac{B^2h^2}{U^{1/2}}dz. \qquad (A1.3.29)$$

The above equation can be separated into two parts, one caused by a variation of potential ΔU, and the other by a variation of the magnetic field ΔB,

$$C_{ce} = \frac{1}{8}\frac{\Delta U}{h'(z_0)U_0^{1/2}}\int_{z_0}^{z_i}\left[3\left(\frac{U'}{U}\right)^2 + \frac{e}{m}\frac{B^2}{U}\right]\frac{h^2dz}{U^{1/2}} \qquad (A1.3.30)$$

and

$$C_{cm} = \frac{-e}{4mh'(z_0)U_0^{1/2}}\frac{\Delta B_{\max}}{B_{\max}}\int_{z_0}^{z_i}B^2\frac{h^2dz}{U^{1/2}} \qquad (A1.3.31)$$

where C_{ce} and C_{cm} are the electric and magnetic chromatic aberrations referred to the object respectively.

Appendix 2

Multipole Lenses

Multipole elements consist of electric or magnetic "poles" placed symmetrically, with alternating polarity, around an axis. The number of poles in a multipole, N, is always an even number. This is due to symmetry, since it is not possible to arrange an odd number of poles around the axis in such a way that all neighboring poles have opposite polarity. Symmetry is also the reason why multipoles, with the exception of the dipole (two poles of opposite polarity), do not have an electric/magnetic potential or radial fields on the axis. Multipoles are named according to how many poles are used. The most common are: Dipole $N = 2$; Quadrupole $N = 4$; Hexapole $N = 6$; Octupole $N = 8$; Decapole $N = 10$; Dodecapole $N = 12$. Two examples, a quadrupole and a hexapole, are illustrated in Fig. A2-1.

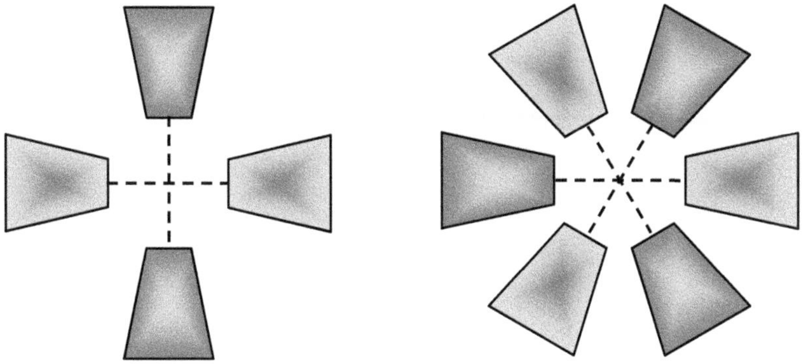

Fig. A2-1. Schematic layout of a quadrupole (left) and a hexapole (right).

Multipoles have M-fold symmetry, where M is defined to mean the number of symmetry planes, which equal the number of poles divided by 2: Dipole $M = 1$; Quadrupoles $M = 2$; Hexapoles $M = 3$; and Octupoles $M = 4$ etc. The use of this convention is more obvious in relation to the equations that govern multipole field distributions. This text is concerned mainly with derivation of equations using scalar potentials.

The following analytical approximation to the field distribution of multipoles is based upon using a simple two-dimensional potential distribution model. Consider an infinitely long circular "wedge", as shown in Fig. A2-2. We want to find the potential distribution $\Phi(r,\theta)$ in the two-dimensional domain contained within the dashed lines, which represents the field distribution between odd symmetry lines in the multipole. The two side walls extend radially at zero potential (making the potential along the cylindrical axis zero also), while the outer "cap", at radius a, has a potential distribution given by $\Phi(a,\theta) = h(\theta)$.

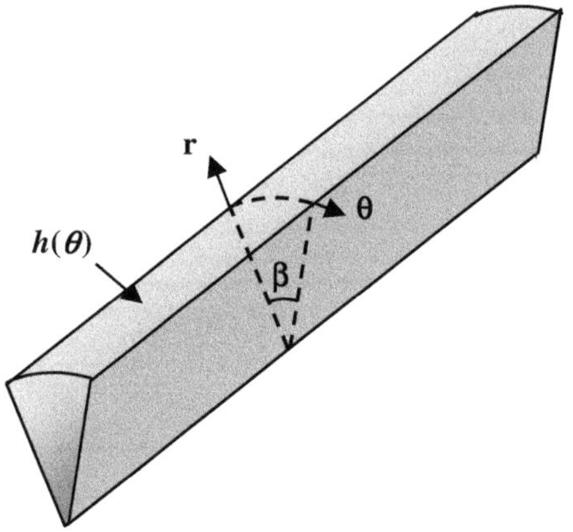

Fig. A2-2. Schematic diagram showing the circular coordinate system, and the geometry of the wedge.

The potential distribution $\Phi(r,\theta)$ within the wedge domain is governed by the circular Laplace's equation

$$\frac{\partial^2\Phi}{\partial r^2}+\frac{1}{r}\frac{\partial\Phi}{\partial r}+\frac{1}{r^2}\frac{\partial^2\Phi}{\partial\theta^2}=0. \tag{A2.1}$$

Using the method of separation of variables, we write

$$\Phi(r,\theta)=R(r)\Theta(\theta). \tag{A2.2}$$

Inserting this expression into (A2.1) and rearranging the terms gives

$$r^2\frac{R_{rr}}{R}+r\frac{R_r}{R}=-\frac{\Theta_{\theta\theta}}{\Theta}=m^2. \tag{A2.3}$$

The expression in (A2.3) gives us two separate equations:

$$r^2R_{rr}+rR_r-m^2R=0 \tag{A2.4}$$

with boundary conditions: $R(0) = 0$ and $R(a,0) = h(\theta)$ and

$$\Theta_{\theta\theta}+m^2\Theta=0 \tag{A2.5}$$

with boundary conditions: $\Theta(0) = 0$ and $\Theta(\beta)=0$.

We solve Eq. (A2.5) for Θ first. It has a standard solution with sin and cos terms, but the boundary conditions forces the solution to be

$$\Theta(\theta)=\sin\left(\frac{n\pi}{\beta}\theta\right) \tag{A2.6}$$

where n is an integer. This means that

$$m = \frac{n\pi}{\beta}. \tag{A2.7}$$

The Eq. (A2.4) for R is of Euler type and we set

$$R(r) = r^{\alpha}. \tag{A2.8}$$

Inserting into (A2.4) yields

$$\alpha(\alpha-1)r^{\alpha} + \alpha r^{\alpha} - m^2 r^{\alpha} = 0$$
$$\Leftrightarrow$$
$$\alpha^2 - m^2 = 0 \Rightarrow \alpha = \pm m \tag{A2.9}$$

but R must be zero at $r = 0$ so we cannot use the negative exponent.

Using Eqs. (A2.6) and (A2.8), the solution now becomes

$$\Phi(r,\theta) = \sum_{n=1} A_n r^{n\pi/\beta} \sin\left(\frac{n\pi}{\beta}\theta\right). \tag{A2.10}$$

A_n is found by imposing the boundary condition at $r = a$ in this Fourier Series. For simplicity, assume that $\Phi(a, \theta) = h(\theta) = V$ is a constant. In practice, the multipole has a θ dependence that is defined by the shape of the pole surfaces. Imposing this boundary condition into Eq. (A2.10) gives

$$\Phi(a,\theta) = V = \sum_{n=1} A_n a^{n\pi/\beta} \sin\left(\frac{n\pi}{\beta}\theta\right). \tag{A2.11}$$

Using the standard formula for Fourier coefficients

$$A_n = \frac{2Va^{-n\pi/\beta}}{\beta}\int_0^{\beta}\sin\left(\frac{n\pi}{\beta}\theta\right)d\theta = \frac{2Va^{-n\pi/\beta}}{\beta}\frac{\beta}{n\pi}\left[1-(-1)^n\right]$$

$$\Leftrightarrow$$

$$A_{2n+1} = \frac{4Va^{-(2n+1)\pi/\beta}}{(2n+1)\pi}. \tag{A2.12}$$

Since $A_{2n} = 0$,

the final expression becomes

$$\Phi(r,\theta) = \frac{4V}{\pi}\sum_{n=1}\frac{r^{(2n+1)\pi/\beta}}{(2n+1)}\sin\left(\frac{(2n+1)\pi}{\beta}\theta\right). \tag{A2.13}$$

From the potential distribution (A2.13) it is clear that the exponent of r grows if β is decreased.

For the case of multipoles, the wedge angle β is always a certain fraction of π. In fact, for multipoles of multiplicity M we have

$$\beta = \frac{\pi}{M} \tag{A2.14}$$

so that

$$\Phi(r,\theta) = \frac{4V}{\pi}\sum_{n=1}\frac{r^{(2n+1)M}}{(2n+1)}\sin\left((2n+1)M\theta\right). \tag{A2.15}$$

The above theory can be applied to different multipole geometries to obtain an approximate analytical expression for their cross-sectional scalar potential distribution. Consider the simple case of a quadrupole, where $M = 2$, so $\beta = \pi/2$. The potential distribution is

$$\Phi_{QUAD}(r,\theta) = \frac{4V}{\pi}\sum_{n=0} \frac{r^{2(2n+1)}}{(2n+1)} \sin(2(2n+1)\theta)$$
$$= \frac{4V}{\pi}\left(r^2 \sin 2\theta + \frac{r^6}{3} \sin 6\theta + \ldots\right). \quad \text{(A2.16)}$$

In terms of x-y coordinates

$$\Phi_{QUAD}(x,y) = \frac{4V}{\pi}\left(2xy + \frac{2}{3}xy\left(x^4 - \frac{10}{3}x^2y^2 + y^4\right) + \ldots\right)$$
$$= \frac{8V}{\pi}xy\left(1 + \left(\frac{x^4}{3} - \frac{10}{9}x^2y^2 + \frac{y^4}{3}\right)\right) \quad \text{(A2.17)}$$

Near the axis, only the first term of the series is significant. Below are listed simplified expressions for quadrupole, hexapole, and the octupole, expressed both in rectilinear and polar coordinates (with the term $4V/\pi$ omitted):

$$\Phi_{QUAD} = r^2 \sin 2\varphi = 2r^2 \sin\varphi \cos\varphi = 2xy \quad \text{(A2.18a)}$$

$$\Phi_{QUAD} = r^2 \cos 2\varphi = r^2\left(\cos^2\varphi - \sin^2\varphi\right) = x^2 - y^2. \quad \text{(A2.18b)}$$

$$\Phi_{HEX} = r^3 \sin 3\varphi = r^3\left(\sin 2\varphi \cos\varphi + \sin\varphi \cos 2\varphi\right)$$
$$= r^3\left(2\sin\varphi\cos^2\varphi + \sin\varphi\cos^2\varphi - \sin^3\varphi\right) = 3x^2y - y^3.$$
(A2.19a)

$$\Phi_{HEX} = r^3 \cos 3\varphi = r^3\left(\cos 2\varphi \cos\varphi - \sin 2\varphi \sin\varphi\right)$$
$$= r^3\left(\cos^3\varphi - \sin^2\varphi\cos\varphi - 2\sin^2\varphi\cos\varphi\right) = x^3 - 3xy^2.$$
(A2.19b)

$$\Phi_{OCTU} = r^4 \sin 4\varphi = 2r^4 \sin 2\varphi \cos 2\varphi$$
$$= 4r^4 \sin\varphi\cos\varphi\left(\cos^2\varphi - \sin^2\varphi\right) = 4x^3y - 4xy^3. \tag{A2.20a}$$

$$\Phi_{OCTU} = r^4 \cos 4\varphi = r^4\left(\cos^2 2\varphi - \sin^2 2\varphi\right)$$
$$= r^4\left[\left(\cos^2\varphi - \sin^2\varphi\right)^2 - \left(2\sin\varphi\cos\varphi\right)^2\right] = x^4 - 6x^2y^2 + y^4. \tag{A2.20b}$$

The expressions in parts (a) and (b) of the above equations differ only in that they represent potential distributions rotated relative to each other by an angle $\pi/(2M)$.

Electric and magnetic fields are calculated by taking the gradient of the potential

$$\vec{E} = -\vec{\nabla}\phi \tag{A2.21}$$
$$\vec{B} = -\mu_0\vec{\nabla}\psi. \tag{A2.22}$$

Below are x- and y-derivatives of the quadru-, hexa-, and octupole potentials (with the term $4V/\pi$ omitted)

$$F_{QUAD,X} = 2y, 2x \tag{A2.23a}$$
$$F_{QUAD,Y} = 2x, -2y \tag{A2.23b}$$
$$F_{HEX,X} = 6xy, 3x^2 - 3y^2 \tag{A2.24a}$$
$$F_{HEX,Y} = 3x^2 - 3y^2, -6xy \tag{A2.24b}$$
$$F_{OCTU,X} = 12x^2y - 4y^3, 4x^3 - 12xy^2 \tag{A2.25a}$$
$$F_{OCTU,Y} = 4x^3 - 12xy^2, -12x^2y + 4y^3. \tag{A2.25b}$$

Figure A2-3 shows plots of the potential distributions for Eqs. (A2.18)–(A2.20).

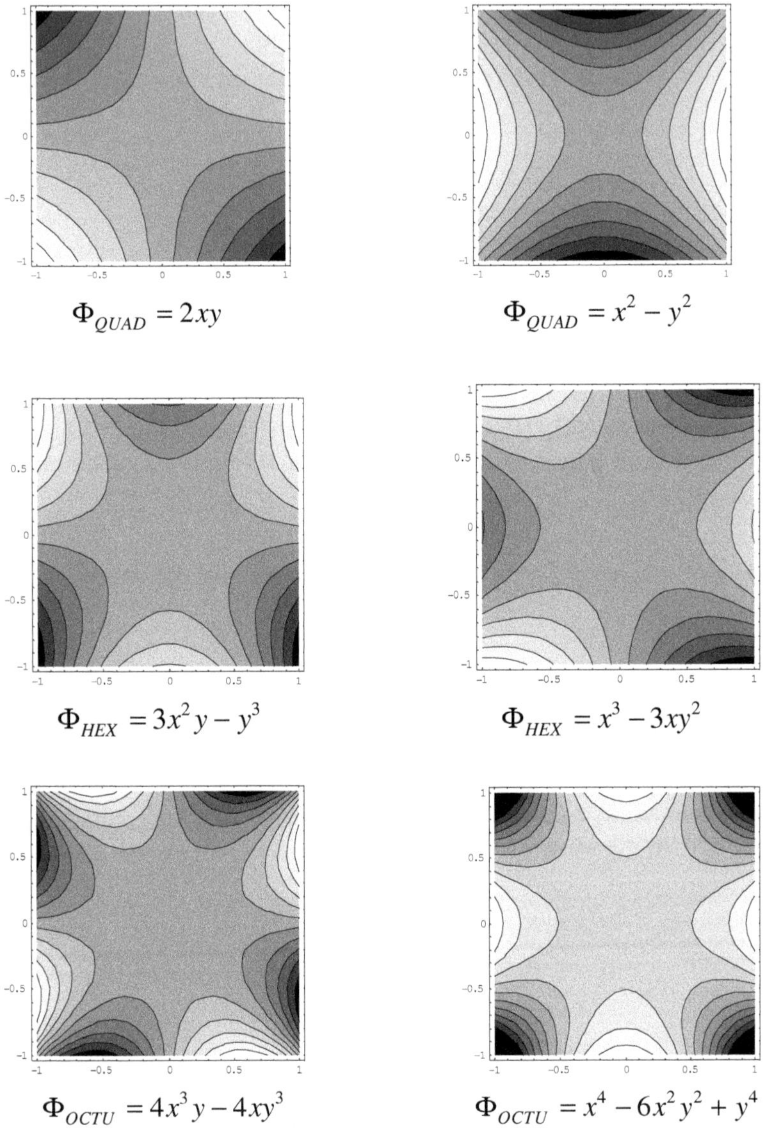

Fig. A2-3. Graphs showing equipotential *x*-*y* plots for equations describing quadrupole, hexapole, and octupoles.

The potential distributions in Figs. A2-4 and A2-5 show the potential distribution for some selected multipoles. The grey area, indicating a weak field, stretches further out from the symmetry centre as the number of poles grow. In addition, the transition between grey and black/white takes place over a shorter distance. This indicates that the dependence on r increases as the multiplicity M grows, in agreement with Eq. (A2.15).

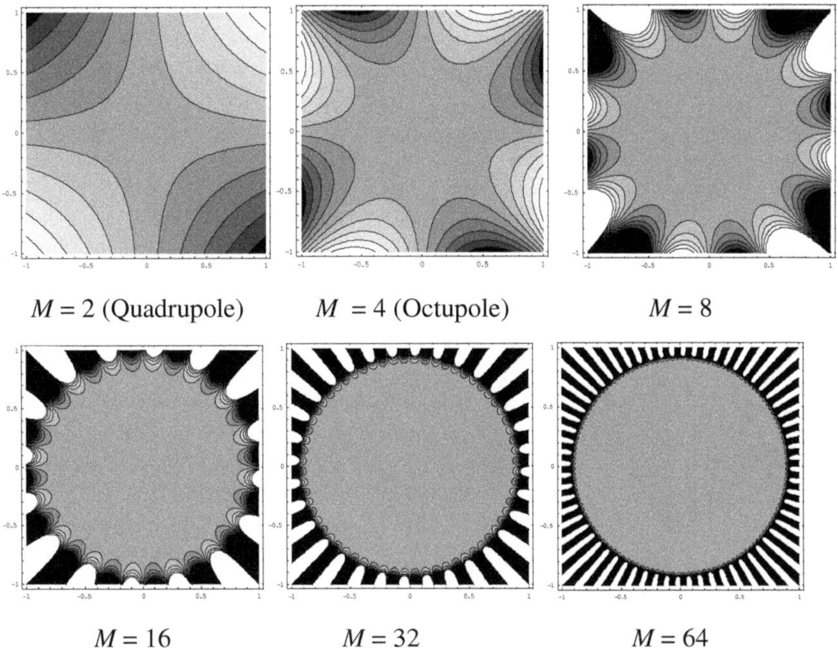

Fig. A2-4. Equipotential plots for some multipoles. The gray areas in the centre of these plots indicate a weak-field region, where potentials and fields are close to zero. Black and white areas represent opposite polarities.

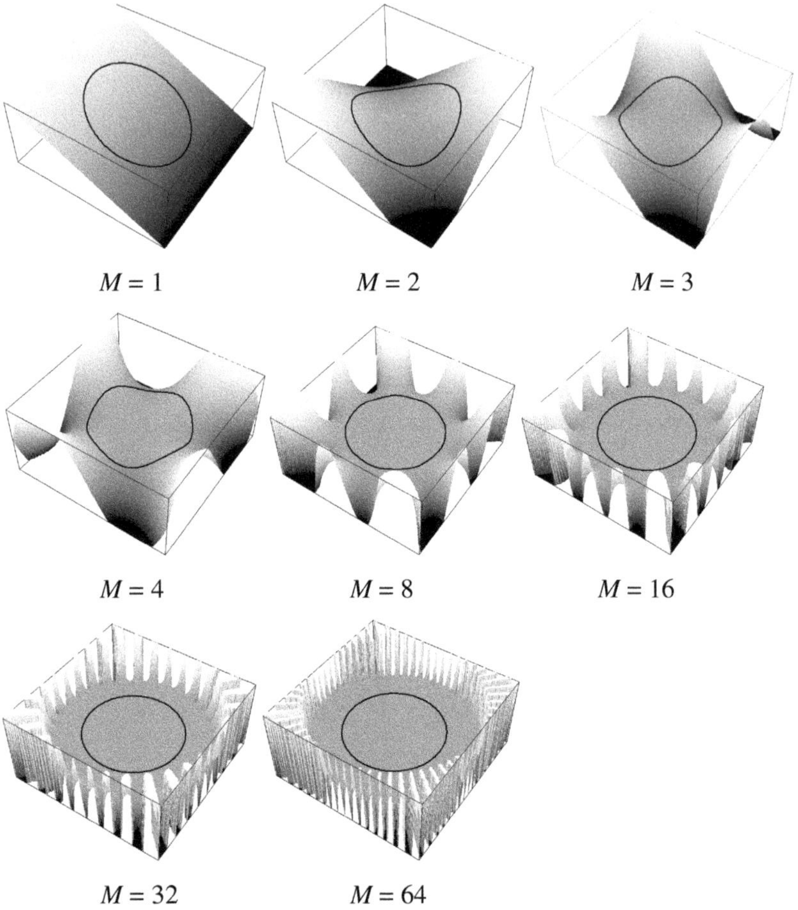

Fig. A2-5. Schematic 3D views of multipole potential distributions. The black closed loops indicate the potential for equidistant points from the symmetry centre. On a plane surface it forms a circle, because all points have the same potential.

Bibliography

Balasubramanyam, M., Munro, E. and Taylor, J. (1995). Secondary electron detection in the scanning electron microscope, *Nucl. Instrum. and Meth. in Phys. Res. A* **363**, pp. 270–275.

Balk, L. J., Feuerbaum, H. P., Kubalek, E. and Menzel, E. (1976). Quantitative voltage contrast at high frequencies in the SEM, *Scanning Electron Microscopy*, Vol. IV (Illinois Institute of Technology Research Institute, Chicago), pp. 615–624.

Banford, A. P. (1966a). *The Transport of Charged Particle Beams* (E. and F. N. Spon Limited, London), pp. 92–120.

Banford, A. P. (1966b). *The Transport of Charged Particle Beams* (E. and F. N. Spon Limited, London), pp. 84–90.

Banford, A. P. (1966c). *The Transport of Charged Particle Beams* (E. and F. N. Spon Limited, London), pp. 63–84.

Barth, J. E. and Kruit, P. (1996). Addition of different contributions to the charged particle probe size, *Optik* **101**(3), pp. 101–109.

Batiné, M., Begert, D. and Kubalek, E. (1995). Pulsed electron beam generation via laser stimulation, *Nucl. Instrum. and Meth. in Phys. Res. A* **363**, pp. 43–47.

Bauer, E. (1998). LEEM basics, *Surf. Rev. Lett.* **5**, p. 1275.

Beck, S., Plies, E. and Schiebel, B. (1995). Low-voltage probe forming columns for electrons, *Nucl. Instrum. and Meth. in Phys. Res. A* **363**, pp. 31–32.

Beha, H. and Clauberg, R. (1993). Picosecond photoemission probing, in J. T. L. Thong (ed.), *Electron Beam Testing Technology* (Plenum Press, New York), p. 298.

Belhaj, M., Jbara, O. and Fakhfakh, S. (2008). Quantitative voltage contrast method for electron irradiated insulators in SEM, *J. Phys. D: Appl. Phys.* **41**, p. 175304.

Bell, D. C. (2009). Contrast mechanisms and image formation in helium ion microscopy, *Microsc. Microanal.* **15**, pp. 147–153.

Benis, E. P. and Zouros, T. M. J. (2008). The hemispherical deflector analyser revisited II. Electron-optical properties, *J. Electron Spectrosc. Relat. Phenom.* **163**, pp. 28–39.

Boyes, E. D. (2002). Analytical potential of EDS at low voltages, *Mikrochim Acta* **138**, pp. 225–234.

Chung, M. S. and Everhart, T. E. (1974). Simple calculation of low-energy secondary electrons emitted from metals under bombardment, *J. Appl. Phys.* **45**, pp. 707–709.

Cosslett, V. E. (1972). Probe size and probe current in the STEM, *Optik* **36**, pp. 85–92.

Crewe, A. V. (1997). The scanning electron microscope, in J. Orloff (1st edn.), *Handbook of Charged Particle Optics* (CRC Press, New York), pp. 409–410.

Delong, A. (1993). Electron sources for electron microscopes, *Microscopy and Analysis*, **3**, pp. 19–21.

Dinnis, A. R. (1993). Essential electron optics, in J. T. L. Thong (ed.), *Electron Beam Testing Technology* (Plenum Press, London), pp. 142–145.

Dinnis, A. R. (1996). A high-resolution time-dispersive electron spectrometer for fast voltage-contrast measurements. *Microelectron. Eng.* **31**, pp. 101–107.

Dinnis, A. R. (1987). Detectors for quantitative voltage contrast on submicron devices, *Microelectron. Eng.* **7**, pp. 139–146.

Dinnis, A. R., Khursheed, A. and Smart, P. D. (1992). The time-of-flight spectrometer: First results, *Microelectron. Eng.* **16**, pp. 35–42.

Dubbledam, L. (1993). Electron spectrometers and voltage measurements, in J. T. L. Thong (ed.), *Electron Beam Testing Technology* (Plenum Press, New York), pp. 211–239.

Dubbledam, L. and Kruit, P. (1987a). First experimental results of an e-beam tester with dispersive secondary electron energy analyzer, *Microelectron. Eng.* **7**(2–4), pp. 231–234.

Dubbledam, L. and Kruit, P. (1987b). Signal-to-Noise ratio improvement in electron beam testing by using a dispersive analyser, *Scanning Microscopy* **1**(4), pp. 1647–1650.

El-Gomati, M., Zaggout, F., Jayacody, H., Tear, S. and Wilson, K. (2005). Why is it possible to detect doped regions of semiconductors in low voltage SEM: A review and update, *Surf. Interface Anal.* **37**, pp. 901–991.

Everhart, T. E. and Thornley, R. F. M. (1960). Wide-band detector for micro-ampere low-energy electron currents, *J. Sci. Instrum.* **37**, p. 246.

Fentem, P. J. and Gopinath, A. (1974). Voltage contrast linearization with a hemispherical retarding analyzer, *J. Phys. E: Sci. Instrum.* **1**, pp. 930–933.

Feuerbaum, H. P. (1979). VLSI testing using the electron probe, *Scanning Electron Microscopy* I, pp. 285–296.

Flemming, J. P. and Ward, E. W. (1970). A technique for accurate measurement and display of applied potential distributions using the scanning electron microscope, *Scanning Electron Microscopy* I, pp. 465–470.

Frank, L. and Müllerová, I. (1999). Strategies for low - and very-low-energy SEM, *J. Electron Microsc. (JEM)* **48** (3), pp. 205–219.

Fransen, M. J., Van Rooy, Th. L., Tiemeijer P. C., Overwijk, M. H. F., Faber, J. S. and Kruit, P. (1999). On the electron-optical properties of the ZrO/W Schottky electron emitter, in P. W. Hawkes (ed.), *Advances in Imaging and Electron Physics*, Vol. 111 (Academic Press), pp. 92–166.

Frosien, J., Lanio, S. and Feuerbaum, H. P. (1995). High precision electron optical system for absolute and CD-measurement on large substrates, *Nucl. Instrum. and Meth. in Phys. Res. A* **363**, pp. 25–30.

Frosien, J. and Plies, E. (1987). High performance electron optical column for testing ICs with submicrometer design rules, *Microelectron. Eng.* **7**(2–4), pp. 163–172.

Frosien, J., Plies, E. and Anger, K. (1989), Compound magnetic and electrostatic lenses for low-voltage applications, *J. Vac. Sci. Technol. B* **7**, pp. 1874–1877.

Garth, S. C. J. (1987). In-the-lens electron spectrometers for e-beam testing: A review, *Microelectron. Eng.* **6**, pp. 667–672.

Garth, S. C. J., Nixon, W. C., and Spicer, D. F. (1986). Magnetic field extraction of secondary electrons for accurate integrated circuit voltage measurement, *J. Vac. Sci. Technol. B* **4**(1), pp. 217–220.

Goldstein, J., Newbury, D. E., Joy, D. C., Lyman, C. E., Echlin, P., Lifshin, E., Sawyer, L. and Michael, J. R. (2003). Scanning electron microscopy and x-ray microanalysis, 3rd edn (Springer).

Gopinath, A. (1977). Estimate of minimum measurable voltage in the SEM, *J. Phys. E: Sci. Instrum.* **10**, pp. 911–913.

Gopinath, A. and Sanger, C. C. (1971). A technique for the linearization of voltage contrast in the scanning electron microscope, *J. Phys. E: Sci. Instrum.* **4**, pp. 334–336.

Goto, Y., Ito, A., Furukawa, Y. and Inagaki, T. (1981). Hemispherical retarding type energy analyser for I. C. testing by electron beam, *J. Vac. Sci. Technol.* **19**(4), pp. 1030–1032.

Goto, E. and Soma, T. (1977). MOL (moving objective lens) - Formation of deflective aberration free system, *Optik* **48**, pp. 255–270.

Granneman, E. H. A. and Van Der Wiel, M. J. (1983). Transport, dispersion and detection of electrons, ions and neutrals. In E. E. Koch (ed.), *Handbook on Synchrotron Radiation*, Vol. 1 (North Holland Publishing Company).

Green, T. S. and Proca, G. A. (1970). A parallel plate electrostatic spectrograph, *Rev. Sci. Instrum.* **41**(10), pp. 1409–1414.

Hafner, H., Simpson, J. A. and Kuyatt, C. E. (1968), Comparison of the spherical deflector and the cylindrical mirror analyzers, *Rev. Sci Instrum.* **39**(1), pp. 33–35.

Hannah, J. M. (1974). SEM applications to Integrated Circuit Testing. PhD thesis, University of Edinburgh, Scotland.

Heritage, M. B. (1975), Electron projection micro-fabrication system, *J. Vac. Sci. Technol.* **12**, pp. 1135–1140.

Henstra, A. and Chmelik, J. (2006). Particle source with selectable beam current and energy spread, US Patent No. 7,034,315 B2.

Hawkes, P. W. and Kasper, E. (1989a), *Principles of Electron Optics*, Vol. 1 (Academic Press, London), p. 363.

Hawkes, P. W. and Kasper, E. (1989b), *Principles of Electron Optics*, Vol. 1 (Academic Press, London), p. 416.

Hoang, H. Q. (2010). Energy spectrometers for the SEM, PhD thesis, National University of Singapore.

Hoang, H. Q. and Khursheed, A. (2008). The energy dispersion characteristics of a magnetic beam separator, *Physics Procedia* **1**, pp. 161–166.

Hoang, H. Q. and Khursheed, A. (2009). Toroidal spectrometer for signal detection in scanning ion/electron microscopes, *J. Vac. Sci. Technol. B* **27**(6), pp. 3226–3231.

Huchital, D. A. and Rigden, J. D. (1972). Resolution and sensitivity of the spherical-grid retarding potential analyzer, *J. Appl. and Phys.* **43**(5), pp. 2291–2302.

Integrated Engineering Software (2009), Lorentz3D-EM, 220-1821 Wellington Avenue, Winnipeg, Manitoba, R3H 0G4 Canada.

Jacka, M. (2001). Scanning Auger microscopy: Recent progress in data analysis and instrumentation, *J. Electron. Spectrosc. Relat. Phenom.* **114–116**, pp. 277–282.

Jacka, M., Kirk, M., El Gomati, M. M. and Prutton, M. (1999). A fast aparallel acquisition, electron energy analyzer: The hyperbolic field analyzer, *Rev. Sci. Instrum.* **70**(2), pp. 2282–2287.

Joy, D. C. (1995). *Monte Carlo Modeling for Microscopy and Microanalysis* (Oxford University Press).

Joy, D. C. (2008). The aberration corrected SEM. *Biological Low-Voltage Scanning Electron Microscopy*, in H. Schatten and J. B. Pawley (eds.) (Springer, New York), pp. 107–127.

Kanaya, K. and Okayama, S. (1972). Penetration and energy-loss theory of electrons in solid targets, *J. Phys. D.* **5**, pp. 43–58.

Kazemian, P., Mentink, S. A., Rodenburg, C. and Humphreys, C. J. (2006). High resolution quantitative two-dimensional dopant mapping using energy-filtered secondary electron imaging, *J. Appl. Phys.* **100**, p. 054901.

Kazemian, P., Mentink, S. A., Rodenburg, C. and Humphreys, C. J. (2007). Quantitative secondary electron energy filtering in a scanning electron microscope and its applications, *Ultramicroscopy* **107**, pp. 140–150.

Kelly, M. A. (1999). A new electron energy analyzer for electron spectroscopy, *J. Electron. Spectrosco. Relat. Phenom.* **98–99**, pp. 55–66.

Khursheed, A. (1983). The Computer Aided Design of Electron Detectors for the Scanning Electron Microscope, PhD Thesis, University of Edinburgh, Scotland.

Khursheed, A. and Dinnis, A. R. (1984). A comparison of voltage contrast detectors, *Scanning* **6**, pp. 85–95.

Khursheed, A. and Dinnis, A. R. (1990). A time-of-flight voltage contrast detector for measurements on VLSI circuits, *Meas. Sci. Technol.* **1**, pp. 581–591.

Khursheed, A. and Dinnis, A. R. (1992). Experimental results from a time-of-flight spectrometer for electron beam testing, *Microelectron. Eng.* **17**, pp. 451–454.

Khursheed, A. (1992). Multi-channel vs conventional retarding field spectrometers, *Microelectron. Eng.* **16**(1–4), pp. 43–50.

Khursheed, A., Dinnis, A. R. and Smart, P. D. (1991). Micro-extraction fields to improve electron beam test measurements, *Microelectron. Eng.* **13**, pp. 519–522.

Khursheed, A. and Karuppiah, N. (2001). An add-on secondary electron energy spectrometer for scanning electron microscopes, *Rev. Sci. Instrum.* **72**(3), pp. 1708–1714.

Khursheed, A., Karuppiah, N. and Koh, S. H. (2001). A high-resolution add-on lens attachment for scanning electron microscopes, *Scanning* **23**, pp. 204–210.

Khursheed, A. and Karuppiah, N. (2002). A high resolution mixed field immersion lens attachment for conventional scanning electron microscopes, *Rev. Sci. Instrum.* **73**(8), pp. 2906–2909.

Khursheed, A. (1996). Scanning electron microscope design for quantitative multicontrast, *Scanning* **18**, pp. 81–91.

Khursheed, A. (1999). The finite element method in charged particle optics (Kluwer Academic Publishers).

Khursheed, A. (2002a). Add-on lens attachments for the scanning electron microscope, in P. W. Hawkes (ed.), *Advances in Imaging and Electron Physics*, Vol. 22 (San Diego: Aacademic Press), pp. 87–171.

Khursheed, A. (2002b). Aberration characteristics of immersion lenses for LVSEM, *Ultramicroscopy* **93**(3–4), pp. 331–338.

Khursheed, A. and Hoang, H. Q. (2008a). A second-order focusing electrostatic toroidal electron spectrometer with 2π radian collection, *Ultramicroscopy* **109**, pp. 104–110.

Khursheed, A. and Hoang, H. Q. (2008b). Redesign of the scanning electron microscope for parallel energy spectral acquisition, *Ultramicroscopy* **108**, pp. 151–157.

Khursheed, A. and Osterberg, M. (2006). Developments in the design of a spectroscopic scanning electron microscope, *Nucl. Instrum. and Meth. in Phys. Res. A* **556**, p. 437.

Kienle, M. and Plies, E. (2004). An off-axis multi-channel analyzer for secondary electrons, *Nucl. Instrum. and Meth. in Phys. Res.* A **519**, pp. 325–330.

Knell, G. and Plies, E. (1998). Determination of the collection efficiency for combined magnetic-electrostatic SEM objective lenses, *Optik* **108**(1), pp. 37–42.

Knell, G. and Plies, E. (1999). Simulation of an improved magnetic-electrostatic detector objective lens for LVSEM, *Nucl. Instrum. and Meth. in Phys. Res. A* **427**, pp. 99–103.

Knell, G. and Plies, E. (2000). Initial resolution measurements of an improved magnetic-electrostatic detector objective lens for LVSEM, *Ultramicroscopy* **81**, pp. 123–127.

Kolarik, V., Mankos, M. and Veneklasen L. H. (1991). Close packed prism array for electron microscopy, *Optik* **87**, p. 1.

Konvalina, I. and Mullerova, I. (2006). The trajectories of secondary electrons in the scanning electron microscope, *Scanning* **28**, pp. 245–256.

Kruit, P. and Read, F. H. (1983). Magnetic field paralleliser for 2π electron-spectrometer and image magnifier, *J. Phys. E: Sci. Instrum.* **16**, pp. 313–324.

Liu, H., Munro, E., Rouse, J. and Zhu, X. (2002). Simulation methods for multipole imaging systems and aberration correctors, *Ultramicroscopy* **93**, pp. 271–291.

Luo, T. and Khursheed, A. (2007). Imaging with surface sensitive backscattered electrons, *J. Vac. Sci. Technol. B* **25**(6), pp. 2017–2019.

Luo, T. and Khursheed, A. (2008). Elemental identification using transmitted and backscattered electrons in a SEM, *Physics Procedia* **1**, pp. 155–160.

Matsukawa, T., Shimuzu, R. and Hashimoto, H. (1974). Measurements of the energy distribution of backscattered kilovolt electrons with a spherical retarding field-field energy analyser, *J. Phys. D: Appl. Phys.* **7**, pp. 695–702.

Meisburger, W. D., Brodie, A. D. and Desai, A. A. (1992). Low-voltage electronoptical system for the high-speed inspection of integrated circuits, *J. Vac. Sci. Technol. B* **10**(6), pp. 2804–2808.

Menzel, E. and Kubalek, E. (1983). SE detection systems for quantitative voltage measurements, *Scanning* **5**, pp. 103–122.

Menzel, E. and Buchanan, R. (1984). In-the-lens secondary electron analyser for IC internal voltage measurements with electron beams, *Electron. Lett.* **20**(10), pp. 408–409.

Mizuno, F., Yamada, S., Miura, A., Ohtaka, T. and Tsumaki, N. (1994). Ultra-high resolution in-lens type wafer inspection system. *SPIE* **2196**, pp. 109–115.

Mullerova, I. and Lenc, M. (1992). Some approaches to low-voltage scanning electron microscopy, *Ultramicroscopy* **41**, pp. 399–410.

Mullerova, I. and Konvalina, I. (2009). Collection of secondary electrons in scanning electron microscopes, *J. Microsc.* **236**(3), pp. 203–210.

Mulvey, T. (1982). Unconventional Lens Designs, in P. W. Hawkes (ed.), *Magnetic Electron Lenses* (Springer-Verlag, New York).

Munro, E. (1974). Calculation of the optical properties of combined magnetic lenses and deflection systems with superimposed fields, *Optik* **39**, pp. 450–466.

Nakagawa, H., Nomura, N., Koizumi, T., Anazawa, N. and Harafuji, K. (1991). A novel high-resolution scanning electron microscope for the surface analysis of high-aspect-ratio three-dimensional structures, *Jpn. J. Appl. Phys.* **30**(9A), pp. 2112–2117.

Narum, D. H. (1993). Schottky field emission based Auger microprobe, *J. Vac. Sci. Technol. B* **11**(6), pp. 2487–2492.

Oatley, C. W. and Everhart, T. E. (1957). The examination of p-n junctions in the scanning electron microscope, *J Electron* **2**, pp. 568–570.

Ohiwa, H., Goto, E. and Ono, A. (1971). Elimination of third order aberrations in electron beam scanning systems. *Electron. Commun. Jpn* **54B**, pp. 44–51.

Ose, Y., Taya, S., Todokoro, H., Otaka, T., Sato, M. and Ezumi, M. (2008). Monochromator and scanning electron microscope using the same, US Patent, No. 7,315,024 B2.

Osterberg, M. and Khursheed, A. (2005). Simulation of magnetic sector deflector aberration properties for low-energy electron microscopy, *Nucl. Instrum. and Meth. in Phys. Res. A* **555**, p. 20.

Osterberg, M. and Khursheed, A. (2008). Initial results on the spectroscopic SEM concept, *Physics Procedia* **1**, pp. 529–535.

Osterberg, M., Hoang, H. Q. and Khursheed, A. (2009). Initial experimental results on a magnetic beam separator spectrometer for the SEM, *Ultramicroscopy* **109**, pp. 1310–1314.

Pawley, J. B. (1990). Practical aspects of high-resolution LVSEM, *Scanning* **12**, pp. 247–252.

Pfeiffer, H. C. and Langner, G. O. (1981). Advanced deflection concept for large area, high-resolution e-beam lithography, *J. Vac. Sci. Technol.* **19**, pp. 1058–1063.

Plies, E. (1984). Opposing field spectrometer for electron beam mensuration technology, US Patent No. 4,464,571.

Plies, E. (1990). Secondary electron analyzers for electron-beam testing, *Nucl. Instrum. and Meth. in Phys. Res. A* **298**, pp. 142–155.

Plies, E. and Elstner, R. (1989). Computation of first-order and third-order properties of compound electron-optical systems, *Optik* **83**, pp. 11–18.

Plows, G. S. (1969). Stroboscopic scanning electron microscopy and the observation of microcircuit surface voltage, PhD thesis, Cambridge University, UK.

Plows, G. S. (1981). Specifications of Stroboscopic SEM system, Lintech Instruments Ltd, Cambridge Science Park, Milton Road, Cambridge CB4 4BN, UK.

Postek, M. T. (1997). The Scanning Electron Microscope, in J. Orloff (1st edn.), *Handbook of Charged Particle Optics* (CRC Press), pp. 363–399.

Rau, E. I. and Robinson, V. N. E. (1996). An annular toroidal backscattered electron energy analyser for use in scanning electron microscopy *Scanning*, **18**, pp. 556–561.

Rau, E. I., Khursheed, A., Gostev, A. V. and Osterberg, M. (2002). Improvements to the design of an electrostatic toroidal backscattered electron spectrometer for the scanning electron microscope, *Rev. Sci. Instrum.* **73**(1), pp. 227–229.

Read, F. H. (2002). The parallel cylindrical mirror electron energy analyzer, *Rev. Sci. Instrum.* **73**(3), pp. 1129–1139.

Read, F. H., Cubric. D., Kumashiro, S. and Walker, A. (2004). The parallel cylindrical mirror analyzer: Axis-to-axis configuration, *Nucl. Instrum. Meth. Phys. Res. A* **519**, pp. 338–344.

Reimer, L. (1998a). *Scanning Electron Microscope, Physics of Image Formation and Microanalysis*, 2nd edn. (Springer), chapters 6–8, pp. 207–328.

Reimer, L. (1998b). *Scanning Electron Microscope, Physics of Image Formation and Microanalysis*, 2nd edn. (Springer), p. 32.

Reimer, L. (1998c). *Scanning Electron Microscope, Physics of Image Formation and Microanalysis*, 2nd edn. (Springer), chapter 4.2, pp. 152–169.

Reimer, L. (1998d). *Scanning Electron Microscope, Physics of Image Formation and Microanalysis*, 2nd edn. (Springer), chapter 4.1.4, pp. 146–148.

Reimer, L. (1998e). *Scanning Electron Microscope, Physics of Image Formation and Microanalysis*, 2nd edn. (Springer), chapter 4.1.5, pp. 148–152.

Reimer, L. (1998e). *Scanning Electron Microscope, Physics of Image Formation and Microanalysis*, 2nd edn. (Springer), chapter 10.2, pp. 396–406.

Richardson, N. and Muray, A. (1988). An improved magnetic-collimating secondary electron energy filter for very large scale integrated diagnostics. *J. Vac. Sci. Technol.* **B6**, pp. 417–421.

Rose, H. (2009). *Geometrical Charged-particle Optics* (Springer-Verlag, Berlin, Germany).

Sar-el, E. Z. (1967). Cylindrical capacitor as an analyzer I. Nonrelativistic part, R*ev. Sci. Instrum.* **38**(6), pp. 1210–1216.

Shao, Z. and Lin, P. S. D. (1989). High resolution low-voltage electron optical system for very large specimens, *Rev. Sci. Instrum.* **60**(11), pp. 3434–3441.

Scherzer, O. (1936). Uber einige Fehler von Elektronenlinsen, *Z. Physik* **101**, pp. 593–603.

Scherzer, O. (1947). Spharische und chromatische Korrektur von Elektronen-Linsen, *Optik* **2**, pp. 114–132.

Schimtz, W. and Mehlhorn, W. (1972). Parallel plate analyser with second-order focusing property, *Rev. Sci. Instrum.* **5**, pp. 64–68.

Septier, A. (1966). The struggle to overcome spherical aberration in electron optics, *Adv. Opt. Electron Microsc.* **1**, pp. 204–274.

Staib, P. (1972). An improved retarding-field analyser, *J. Phys. E: Sci. Instrum.* **5**, p. 484.

Staib, P. and Dinklage, U. (1977). Recent developments on an improved retarding-field analyser, *J. Phys. E: Sci. Instrum.* **10**, p. 914.

Swanson, L. W. and Schwind, G. A. (1997). A review of the ZrO/W Schottky cathode, in J. Orloff (1st edn.), *Handbook of Charged Particle Optics*, (CRC Press), p. 81.

Taylor, N. J. (1969). Resolution and sensitivity considerations of an Auger electron spectrometer based on display LEED optics, *Rev. Sci. Instrum.* **40**(6), pp. 792–804.

Tee, W. J. and Gopinath, A. (1976). A voltage measurement scheme for the SEM using a hemispherical retarding field analyser, *Scanning Electron Microscopy* I, pp. 595–602.

Thong, J. T. L. (1993). *Electron Beam Testing Technology* (Plenum Press, NY).

Tromp, R. M., Mankos, M., Reuter, M. C., Ellis, A. W. and Copel, M. (1998). A new low energy electron microscope, *Surf. Rev. Lett.* **5**, p. 1189.

Tsai, F. C. and Crewe, A. V. (1998). A gapless magnetic objective lens for low voltage SEM, *Optik* **109**(1), pp. 5–11.

Tsuno, K. (1999). Optical design of electron microscope lenses and energy filters, *J. Electron. Microsc.* **48**(6), pp. 801–820.

Tsuno, K., Handa, N., Matsumoto, S. and Mogami, A. (1996). Optical properties of immersion objective lenses and collimation of secondary electrons in low voltage SEM, *Charged Particle Optics II, SPIE* **2858**, pp. 46–57.

Tokodoro, H., Yoneda, S., Seito, S. and Hosoki, S. (1986). Electron beam tester with 10 ps time resolution, *Proc. IEEE International Test Conference*, Washinghton D. C., USA, pp. 600–606.

Uno, S., Honda, K., Nakamura, N. and Matsuya, M. (2005). Aberration correction and its automatic control in scanning electron microscopes, *Optik* **116**, pp. 438–448.

Ward, B. W., Notte, J. A. and Economou, N. P. (2006). Helium ion microscope: A new tool for nanoscale microscopy, *J. Vac. Sci. Technol. B* **24**(6), pp. 2871–2874.

Wells, O. C. (1971). Low-loss image for the scanning electron microscope, *Appl. Phys. Lett.* **19**, pp. 232–235.

Wells, O. C. and Bremer, C. G. (1968). Voltage measurement in the scanning electron microscope, *J. Phys. E: Sci. Instrum.* **1**, pp. 902–906.

Wells, O. C. and Bremer C. G. (1969). An improved energy analyser for the scanning electron microscope, *J. Phys. E: Sci. Instrum.* **2**, pp. 1120–1121.

Yavor, M. I. (2010). Optics of Charged Particle Analyzers, in P. W. Hawkes (ed.), *Advances in Imaging and Electron Physics*, Vol. 157 (San Diego: Aacademic Press), pp. 169–281.

Zach, J. (1989). Design of a high resolution low voltage SEM, *Optik* **81**, pp. 30–40.

Zach, J. (1990). Resolution limits in low voltage scanning electron microscopes, *Nucl. Instrum. and Meth. in Phys. Res. A* **298**, pp. 255–259.

Zach, J. (2000). Aspects of aberration correction in LVSEM, *Proc. 12th Eur. Congress on Electron Microscopy*, III, Brno, The Czech Republic, pp.169–172.

Zach, J. and Haider, M. (1995). Aberration correction in a low voltage SEM by a multipole corrector, *Nucl. Instrum. and Meth. in Phys. Res. A* **363**, pp. 316–325.

Zhu, X. Q., Liu, H. N., Munro, E. and Rouse, J. (1997). Analysis of off-axis shaped beam systems for high throughput electron beam lithography, *Proc. SPIE* **3155**, pp. 47–61.

Zouros, T. J. M. and Benis, E. P. (2002). The hemispherical deflector analyser revisited. I. Motion in the ideal 1/r potential, generalized entry conditions, Kepler orbits and spectrometer basic equation, *J. Electron. Spectrosc. Relat. Phenom.* **125**, pp. 221–248.

Index

www.ingramcontent.com/pod-product-compliance
Ingram Content Group UK Ltd.
Pitfield, Milton Keynes, MK11 3LW, UK
UKHW020402150926
8624IPUK00004B/290

9 789812 83667